DISCARDED

de Gruyter Studies in Organization 82
Getting New Technologies Together

de Gruyter Studies in Organization
Innovation, Technology, and Organization

This international and interdisciplinary book series from de Gruyter presents comprehensive research on the inter-relationship between organization and innovations, both technical and social.
It covers applied topics such as the organization of:
- R & D
- new product and process innovations
- social innovations, such as the development of new forms of work organization and corporate governance structure

and addresses topics of more general interest such as:
- the impact of technical change and other innovations on forms of organization of micro and macro levels
- the development of technologies and other innovations under different organizational conditions at the levels both of the firm and the economy.

The series is designed to stimulate and encourage the exchange of ideas between academic researchers, practitioners, and policy makers, though not all volumes address policy- or practitioner-oriented issues.
The volumes present conceptual schema as well as empirical studies and are of interest to students of business policy and organizational behaviour, to economists and sociologists, and to managers and administrators at firm and national level.

Editor:
Arthur Francis, Glasgow University Business School, Glasgow, GB

Advisory Board:
Prof. Claudio Ciborra, University of Trento, Italy
Dr. Mark Dodgson, The Australian National University, Canberra, Australia
Dr. Peter Grootings, CEDEFOP, Berlin, Germany
Prof. Laurie Larwood, Dean, College of Business Administration, University of Nevada, Reno, Nevada

Getting New Technologies Together

Studies in Making Sociotechnical Order

Edited by
Cornelis Disco and Barend van der Meulen

Walter de Gruyter · Berlin · New York 1998

Editors:

Dr. *Cornelis Disco*, Assistant Prof., Department of Philosophy of Science and Technology, University of Twente, Enschede, The Netherlands

Dr. *Barend van der Meulen*, Senior Researcher, Department of Philosophy of Science and Technology, University of Twente, Enschede, The Netherlands

With 40 figures

∞ Printed on acid-free paper which falls within the guidelines of the ANSI to ensure permanence and durability.

Library of Congress Cataloging-in-Publication Data

> Getting new technologies together : studies in making sociotechnical order / edited by Cornelis Disco, Barend van der Meulen.
> p. cm. – (De Gruyter studies in organization : innovation, technology, and organization : 82)
> Includes bibliographical references and index.
> ISBN 3-11-015630-X (alk. paper)
> 1. Technological innovations–Social aspects.   I. Disco, Cornelis.   II. Meulen, Barend van der.   III. Series: De Gruyter studies in organization : 82.
> HM221.G47   1998
> 303.48′3–DC21   98-28235
>   CIP

Die Deutsche Bibliothek – Cataloging-in-Publication Data

> **Getting new technologies together** : studies in sociotechnical order / ed. by Cornelis Disco ; Barend van der Meulen. – Berlin ; New York : de Gruyter, 1998
> (De Gruyter studies in organization ; 82 : Innovation, technology, and organization)
> ISBN 3-11-015630-X

© Copyright 1998 by Walter de Gruyter GmbH & Co., D-10785 Berlin
All rights reserved, including those of translation into foreign languages. No part of this book may be reproduced or transmitted in any form or by any means, electronic or mechanical, including photocopy, recording, or any information storage and retrieval system, without permission in writing from the publisher.
Desktop Publishing: Regina Mundel, Berlin – Printing: Arthur Collignon GmbH, Berlin – Binding: Heinz Stein Buchbinderei, Berlin – Cover-Design: Johannes Rother, Berlin

Foreword

Getting a new book together. The parallels between the subject of this book and editing it are too obvious not to mention them. Just like new technologies, which emerge at many locations and levels at once, books like this would never be gotten together without a great deal of coordination. As becomes scrupulous editors, we tried to structure ourselves and the authors in frameworks which defined such things as the basic thrust of the book, the specific angle of each contributory case study, and time schedules. Some of these frameworks had their effects, others proved too weak to exert much coordinative force. Coordinating all the agents in time turned out to be a particularly difficult challenge. The ins and outs of this kind of work are of course familiar to any reader who has contributed to preparing an edited volume and if this book were not about coordination and if this text were not a foreword it wouldn't even be worth mentioning.

This book is a spin-off of the research program of the Philosophy of Science and Technology Department at Twente University. Some years ago, at the conclusion of a two day departmental seminar on the Dutch Heath, one of us (BvdM) suggested by way of a summary evaluation that results of the individual projects fitted more or less naturally into a book. The concept of a collective book became a forceful promise (and something of a global order) after a visit to Twente by Knut Sørensen — honor to whom honor is due. In reflecting on our research program he characterized it as "pragmatic constructivism." The book was not just a concept anymore but appeared to have acquired a definite theme and trajectory.

Since then the editors have been captives of a dilemma so well described by several contributors to this book. For a long period we felt we were on a see-saw with the book, trying to use our mass and effort to push the idea of the book up into a real book. Had we stuck with the original notion of merely collecting existing results, our weights would easily have been sufficient to the task. We would merely have had to collect the chapters and sandwich them between an introduction and a conclusion. But we proved more ambitious. Editors and authors alike began to pursue a book which would not only mark the progress we had made, but would also make new progress. This idea of the book proved to be much heavier and in fact succeeded several times in lifting the editors clear off the ground. Between the very first notes on coordination intended as a global framework to coordinate the local efforts of the authors and the present introduction and conclusion, our insights into the coordination of socio-technical order have become clearer and more profound.

This book, then, is not just a collection of earlier results. It is a step in the further development of the technology-dynamics thread in the research program at Twente and we hope it will be a step in the development of technology studies as such. The success and status of the various conferences at Twente and the volumes they have produced are a challenging benchmark.

In drawing the parallels between getting this book together and getting new technologies together, we should not forget the more quotidian elements that have made the task possible: the day-to-day practices, routines, relations and events in our department. We want to thank our colleagues at Twente for their efforts in making this book possible, for giving us the time and resources to accomplish the labor of editing, and for their trust that we would eventually find our feet, even when the weight of the book-idea had pushed us so far aloft that we seemed to have lost contact with *terra firma* altogether. Our special thanks go to Arie Rip, who at crucial moments not only kept the faith, but also added his weight to the see-saw to keep us on the ground and the book in the air. Annemarie Zwijnenberg deserves thanks for meticulous preparation of much of the manuscript and bibliography as does Jasper Deuten for struggling through the many drawings and illustrations. The editors at De Gruyter are to be commended for their stoic patience as we tried to master the see-saw over a period of many months. The result is not only a mark of the lively diversity of science and technology studies at Twente, it is also a collective contribution to the understanding of socio-technical order.

Cornelis Disco
Barend van der Meulen
Enschede, December 10, 1997.

Table of Contents

Introduction 1
Cornelis Disco and *Barend van der Meulen*

Chapter 1
Coordination of Ship Propeller Design: Technical Models and
the Relation of *T.Y. Draco* with *Queen Elizabeth II* 15
Barend van der Meulen

1	Introduction	15
2	Technical Model	19
3	Coordination of Design	24
4	Designers or ... Technical Models?	25
5	Representing GVW as Actor-networks	28
6	*T.Y. Draco* and *Queen Elizabeth II*	30
7	Dynamics of Coordination	33
8	In Conclusion: The Force of What Has Been Done	36

Chapter 2
Means of Coordination in Making Biological Science:
On the Mapping of Plants, Animals, and Genes 39
Dirk Stemerding and *Stephen Hilgartner*

1	Introduction		39
2	Coordination in Early Taxonomy		41
	A	Collecting	43
	B	Representing	46
	C	Standardizing	47
	D	In Search of a Natural System	50
	E	The Animal Economy	51
	F	The Cabinet, the Field, and the Scalpel	52

	G A Method of Correlation	53
3	Coordination in Modern Genomics	54
	A Collecting, Representing, and Standardizing	56
	B New Problems of Coordination	58
	C Sequence-tagged Sites as a Means of Coordination	60
	D A Politics of Coordination	62
4	Discussion	65

Chapter 3
Coordination in Military Socio-technical Networks:
Military Needs, Requirements and Guiding Principles 71
Wim A. Smit, Boelie Elzen, and Bert Enserink

1	Introduction	71
2	Innovation Networks in Military Technology	73
	A Needs, Requirements, and Guiding Principles as Coordination Mechanisms	78
3	Case Study: The European Fighter Aircraft	82
	A Episode 1: The Birth of the European Fighter Aircraft	83
	1 Military Requirements for EFA	83
	2 Industrial Requirements — Heading for Trouble	84
	3 Breaking the Deadlock	86
	4 Evaluation: The Birth of a New Network	87
	B Episode 2: The Battle over the EFA-radar	89
	1 The Continuing Story of EFA	89
	2 The Radar — EFA's Eyes and Brains	89
	3 Two Radar Consortia: ECR-90 Contra MSD-2000	90
	4 Fierce Competition	94
	5 Compromise Found: Increased Resilience Against New Threats	97
	C Evaluation: Surviving Adolescence	98
4	Conclusions	100
	A The Limits of Needs and Requirements as Coordination Mechanisms	101
	B Afterthought on Influencing Technological Development	102

Chapter 4
Getting an Experiment Together in High Energy Physics:
Big Plans, Big Machines, and Bricolage 107
Cornelis Disco

1	Introduction	107

Table of Contents IX

2	Building Experiments: Ethnomethods and Frames	110
3	Building Constructs and Unfolding Frames in the Muon g-2 Experiment	116
	A Getting a Collaboration Together: From Notional to Approved Experiment	116
	B Getting the Detector Together: From Proposal to Working Machine	123
4	Conclusion	136

Chapter 5
Why Are Chickens Housed in Battery Cages? 143
Ibo van de Poel

1	A Brief History of the Battery Cage	144
2	The Primacy of Efficiency	149
	A The Economic Argument	150
	B The Cultural Argument	151
3	Attempts to Change the Design of Battery Cages	152
	A Changes in the Market	153
	B A (New) Role for Ethologists	154
	C The Development of Governmental Regulations	156
	D The Development of Alternative Systems	157
	E The Spread of Alternative Systems	160
4	Discussion	161
	A The Economic Approach, New Markets for Eggs?	161
	B The Cultural Approach: Towards a New Guiding Principle?	163
	C Other Relevant Modes of Coordination	166
	D Towards a More Comprehensive Approach: Design Regimes	167
5	Conclusions	172

Chapter 6
Aligning Gender and New Technology:
The Case of Early Administrative Automation 179
Ellen van Oost

1	Introduction	179
2	Gender and Technology in the Making	180
3	The Gendering of Computers in the Process of Their Socio-cultural Embedding	181
	A Gendered Frames of Meaning	182
	B Aligning Computers with Male-Dominated Actor Groups	184
4	The Automation of the Dutch Postcheque- and Clearing Service	188

		A Punch Cards and the Construction of the Data Typist............	190
		B Shaping a Computer System and Computer Jobs	194
5	Conclusions		199

Chapter 7
Expectations in Technological Developments:
An Example of Prospective Structures to be Filled in by Agency......... 203
Harro van Lente and *Arie Rip*

1	Introduction	203
2	Technology and the Problem of Structure-Agency in Sociology....	204
3	Some Examples of Emerging Patterns in Technology	206
	A Example 1: Moore's Law as a Self-Fulfilling Prophecy	206
	B Example 2: The Emerging World of Membrane Technology in the Netherlands........................	208
	C Example 3: HDTV, a Self-Justifying Technology	212
4	From Promise to Requirement..	215
5	Mechanisms: Mutual Positioning and Agenda-Building	218
6	In Conclusion	222

Chapter 8
Boundary Maintenance and
Radioactive Waste Disposal Technology in the U.S., 1945-1970 231
Adri A. de la Bruhèze

1	Introduction	231
2	The Atomic Energy Commission and the Construction of a Radwaste World	233
	A Reconsidering Existing and Future Waste Management: The Sanitary Engineering Branch....................................	238
	B Geological Waste Disposal: The Hess Committee	239
	C The 1959 Hearings: Towards Stabilization........................	242
	D Radioactive Waste R&D in the 1960s.............................	243
3	Boundary Maintenance of a Radwaste World	246
	A Inside Maintenance: The U.S. National Academy of Sciences .	247
	B Inside Maintenance: The U.S. General Accounting Office......	248
	C Outside Maintenance: Plutonium Wastes in Idaho and Bedrock Disposal in South Carolina...............................	250
	D Outside Maintenance: The Implementation of AEC's Commercial Radioactive Waste Disposal Technology at Lyons, Kansas...	253
4	Discussion and Conclusion	256

Table of Contents XI

Chapter 9
Meaningful Boundaries: Symbolic Representations in
Heterogeneous Research and Development Projects.......................... 265
Elke Duncker and *Cornelis Disco*

1	Introduction	265
2	The Use of Generic Semi-specific Repertoires	270
3	Coupling and Transformation via Passive Dictionary, Agency of the Listener	273
4	Coupling and Transformation via Active Dictionary, Agency of the Speaker	280
5	Hybrid Repertoires...	284
6	Conclusions	290

Chapter 10
Antagonistic Patterns and New Technologies 299
Arie Rip and *Siebe Talma*

1	Introduction	299
2	Agonistic and Antagonistic Patterns Around Novel Technologies: Their Shape and their Long-term Evolution	302
3	Promises and Risks in the Cases of Nuclear Technology and Biotechnology..	307
4	Networks of Actors Linking up with Biotechnology (and Other Labels)..	312
5	Compartmentalized Cultures	316
6	In Conclusion	318

Chapter 11
Getting Case Studies Together:
Conclusions on the Coordination of Sociotechnical Order................. 323
Cornelis Disco and *Barend van der Meulen*

1	Introduction		323
2	Local Orders and Global Orders...		324
	A	Genesis of Global Orders	325
	B	Reversal	327
		1 Reversal of Global Constructs..................................	329
		2 Reversal of Global Agency	331
	C	Local and Global as Metaphors: Coordinating Heterogeneous Time and Discourses	333

	1 Time	334
	2 Discourse	335
	D Local-Global is More than Agency-Structure or Micro-Macro	336
3	Coordination and Sociotechnical Orders	338
	A Socio-technical Coordination?	341
4	Steering Technology is Steering Sociotechnical Orders	348

Bibliography	353
Notes on the Contributors	367
Index	371

Introduction

Cornelis Disco and *Barend van der Meulen*

In 1970 the United States Atomic Energy Commission announced a plan to bury long-term radioactive waste in the salt mines of Lyons, Kansas. Ten years of careful investigation had demonstrated the safety of salt mine disposal and the suitability of Lyons as a trial site. Public acceptance seemed a matter of course. However, efforts to implement the plan aroused a major public outcry. A well-researched and presumptively routine technology became a practical impossibility. What had gone wrong? Since its founding shortly after World War II, the Atomic Energy Commission had done its best to monopolize decision-making on nuclear waste disposal. Accordingly it had worked out a highly "bureaucratic" and "technological" solution for radwaste disposal which excluded other organizations and groups. Against the background of this legacy, Lyons was a bridge too far. The Atomic Energy Commission's actions there produced a vocal opposition composed of both old and new opponents. Lacking allies, the Atomic Energy Commission was forced to recall its carefully constructed plans (De la Bruhèze).

In 1985 the famous cruise ship *Queen Elizabeth II* was renovated. Among other things the owners were interested in upgrading the efficiency of the propeller system. An entirely new type of propeller system was proudly presented at conferences as an example of the manufacturer's capabilities. However, at the sea trials of the renovated *Queen Elizabeth II* the propeller blades broke off. Two years later, the manufacturer was able to market a similar propeller system only after assuring the public that the entire design process had been overhauled. The new design was publicly touted, stressing the R&D investments, the thorough tests and the systematic nature of the design process. Apparently, not only the blades of the *Queen Elizabeth II* had been damaged in that first voyage, but also the standards, design rules, testing procedures, and the reputation of the manufacturer — and all had to be repaired. The "misbehavior" of a couple of propeller blades had not merely destroyed a propulsion system, but an entire, carefully planned, socio-technical order and had forced the manufacturer to create a new one (Van der Meulen).

In 1964, Gordon E. Moore, the research manager of Fairchild Semiconductor, noticed a pattern in the development of memory chips. He observed that about every year and a half the number of logical gates in the largest chips had roughly doubled. He predicted that this trend would continue. Events have borne out Moore's prediction; so much so, that "Moore's Law" has become accepted as a kind of natural law describing the growth of memory chips. But in fact this growth can hardly be explained by the technology of chip-making itself. Instead, "Moore's Law" becomes true because the producers of memory chips themselves try to achieve what "Moore's Law" predicts — because this is what they expect the competition to be doing and this is what they expect customers (those who incorporate memory chips in their products) are planning for. So instead of the "law" being a description of an autonomous technological process, it frames expectations which shape actors' efforts and so it becomes a self-fulfilling prophecy (Van Lente and Rip).

What do these vignettes tell us about getting new technologies together? First, they all, in different ways, demonstrate that technological artifacts are inevitably embedded in social order. They show that it makes no sense to speak of, to design, or to implement technological objects without considering their societal context. In fact, they suggest that getting new technologies together requires shaping social orders no less than designing technological objects. These are simply different aspects of a single process, the making of "socio-technical order." Second, the vignettes show that getting new technologies together successfully requires coordination among many different actors at different locations and through time. This coordination can be achieved in many ways: by markets, through organization, by knowledge, by standards, and even by expectations about what will likely be the case. The stories suggest that the ways in which this coordination is achieved and maintained will have consequences for the future development of the technology.

The first story emphasizes that getting new technologies together also requires getting societal actors together and that poor strategy on the second count can come home to roost. The Atomic Energy Commission's failure at Lyons was not simply due to the ad hoc emergence of a powerful opposition to a potentially risky plan but was a result of the very way the Atomic Energy Commission had been developing radwaste technology over the years — especially its exclusion of meddlesome but important actors. The Atomic Energy Commission's pursuit of a bureaucratic and technological radwaste technology simultaneously antagonized important actors in the radwaste domain. In this context, Lyons was a bridge too far. By linking issues and actors throughout the years, opponents of the plan succeeded in mobilizing the legacy of antagonisms and the Atomic Energy Commission found itself facing a solid wall of nay-sayers — and a debacle for its carefully prepared salt mine disposal technology.

The second vignette radicalizes this point in showing that artifacts are never isolated objects, but always elements of complex networks of hybrid elements. Artifacts emerge from those networks, reflect back on them and in so doing shape their futures. This story shows that the new propeller system is an exponent of its design network; the artifact embodies the network's particular standards, modeling practices, and organizational arrangements. When the artifact is submitted to practical trial, the entire design network is too. And when, as in this case, the artifact fails, the entire network fails as well. In order to recover, to re-establish credibility, it is not enough simply to redesign the propeller system itself. The entire design network must be overhauled as well, for there is no reason to trust knowledge, procedures and norms that have produced a dramatic failure in the past.

The third story shows that actors involved in competitive games (like technological competition) may evolve norms to regulate the rate and nature of their R&D investments. The more objective such norms appear to be, the more predictable will be the behavior of other actors and the easier it is to adjust one's own behavior. With a naturalistic "law" describing technological progress, chip manufacturers tacitly agreed to cooperate in making it come true — hence enabling them to act rationally in a dynamic and potentially open-ended process. In this case, although the market is supposed to regulate competition among firms, it is clear that this does not structure the situation sufficiently. Simply trying to outpace competitors, with no idea of where they are going, can lead to self-destructive efforts like overinvestment, hasty marketing, poor R&D strategy. Instead, norms for expected (and desirable) outcomes coordinate the players. Their mutual expectations now constrain and structure their behavior. They have immersed their economic game in a well-defined structure of expectations which allows them to proceed with less risk.

Technological Determinism and Social Constructivism

These three vignettes illustrate a perspective on technology development which has emerged over the past decade, among other places at the Department of Philosophy of Science and Technology at the University of Twente, where most of the contributors to this volume are located. The stories — and the chapters of this book — exemplify the broadly shared conviction at Twente (and indeed within the field of Science and Technology Studies in general) that technology is not an exogenous and autonomous force but a contingently constructed phenomenon resulting from what actors do and think and with whom and how they interact. At the same time the contributions to this volume emphasize that this construction is also itself a structured process — because in the process of construction actors are always constrained and enabled — by history, by norms, by their own expectations, by the logic of competitive games. Of course, emergent and embedded technologies are themselves also an important structuring agent in society and history, but that is another story for another book.

The Department of Philosophy of Science and Technology at Twente is widely considered one of the founding centers of the new social studies of technology. This approach builds on a number of traditions which achieved an uneasy rapprochement at two international conferences held at Twente in the second half of the 1980s. The contributions to these conferences (Bijker et al. 1987; Bijker and Law 1992) exhibit a shared horror of technological determinism and a strong commitment to several varieties of social constructivism. Technological determinism avers that technologies develop according to an inner logic (or in more sophisticated versions as a reflex of scientific discovery) and are therefore more or less impervious to human influence. On this view you can't hurry technology, but neither can you restrain it once its time has come. Social constructivism argues, contrariwise, that technologies are shaped by the actions, strategies and interpretations of human actors. Now at Twente, almost ten years down the road, horror of technological determinism is as vigorous as ever, but so are concerns about how to avoid getting bogged down in utter contingency — in approaches which represent technologies as endlessly malleable and freely interpretable by groups of actors. Some strains of social constructivism seem to be taking this course.

Other and earlier social science explanations of technological development also attacked technological determinism. Long ago, Adam Smith, and after him Marx, Engels, and Schumpeter explained the development of production technologies as a reflex of capitalist competition. Contemporary adherents of labor process theory like Braverman (1974), Noble (1984), Edwards (1979), economists like Dosi (1982), Nelson and Winter (1982) and other students of the innovative behavior of firms like Abernathy and Clark (1985) have pursued different aspects of this tradition. At another level, social critics like Mumford (1963) and Ellul (1964) have made grand efforts to explain technologies as the outcomes of the political and cultural forces of an epoch. Studies like these are laudable because they view technology and its development as part and parcel of broader societal processes. In this sense they clearly adumbrate the program of social construction of technology. Nonetheless, these efforts are ultimately disappointing because — unlike social constructivist studies — they tend to black-box technological artifacts and their development and see technologies as direct reflexes of social structures and strategies — thus committing the inverse of the sin usually ascribed to technological determinism, i.e. economic, political, cultural, or social determinism. These are a small improvement over technological determinism. History of technology, in other words, cannot simply be reduced to economic, cultural or political history, but has its own specific dynamics.

Paradox or not, we would like to have our cake and eat it too. We want to retain the focus on technological detail and actor strategies characteristic of the social constructivist approach — that is, proximity to an actor-centered and materially detailed history of technology. On the other hand, we want to embed, so to speak, voluntarist accounts of technology development in structural accounts of sociotechnical order, we want to see technology development as constrained and enabled by structures of prior technology and history — which is the particular virtue of more traditionally sociological or economic approaches to technology. This book can be seen as a set of explorations into specific technologies, a sequence of dioramas portraying different kinds of technology in different settings, in which — for better or worse — these two approaches are mobilized simultaneously.

The three stories already reveal something of what progress along this track might involve. In a like manner, the contributions to this book place the constructive efforts of technology's actors within the constraints and possibilities of existing or emergent social structures. In describing how actors and networks construct technologies, in unpacking all these black-boxes, we pay particular attention to the ways in which such construction is limited and defined — we might say *pre-dicted* — by the particular ways actors are *already* connected to one another, by the cultural norms that prevail, by the technological heuristics that are available, by expectations they have about one another's behavior, by the "objective" constructions that emerge out of technological action itself — all of which make some outcomes more likely than others. Diverse features of the material and socio-historical context — few of them subject to the will of any individual actor — may determine what actors can see or do as they go about constructing new technologies. This framework, understood as a set of pragmatic possibilities by the actors involved, must also be understood by those seeking to understand technology development from an observer's or policy-oriented perspective. The rule now is: Follow the actor as the actor follows up on the opportunities it can perceive in the structure of its situation. But also, certainly if the aim is to exert *influence* on the process whereby actors produce new technologies, seek to know more about the structure of the situation than the actors themselves. Following this injunction carries investigations beyond mere documentation and begins to organize knowledge for reflexive interventions in the societal process of getting new technologies together.

Societal Constructivism

We can now make a distinction between *social* constructivism and *societal* constructivism (Rip 1990). *Social* constructivism emphasizes the interactive construction of meaning, the basic organization of perception and action. As a mode of comprehending technology it investigates how actors and networks of actors organize their perceptions of existing technologies and how they mobilize resources to negotiate new criteria and develop appropriate technologies. *Societal* constructivism emphasizes how perception and action are constrained and enabled by the situations in which actors find themselves or which they have produced as the result of earlier actions. It investigates how actors and networks of actors, embedded in a particular historical situation, possessing particular cognitive and other resources, create and exploit opportunities to negotiate criteria for, and to realize, suitable technologies. Our collective ambition in this book is to demonstrate the fruitfulness of a *societal* constructivism — and, by implication, to show how mere *social* constructivism falls short.

A key term in this endeavor is coordination. This concept has the virtue of capturing both the active dimension of "construction" and the passive dimension of "being structured." Subjects can actively coordinate other actors and processes by doling out rewards or punishments, but subjects can also be coordinated by constraints and opportunities that are already in place or which manifest themselves: other actors, the rules of the game, testing procedures, legal regulations, artifacts and technological systems etc. Hence, coordination has both an active and a passive voice. Moreover, coordination is ecumenical. It does not have to proceed from the intentions of a human actor but can be the outcome of structures of interaction (e.g., game structures like the prisoner's dilemma, markets or organizations) or the effect of an assortment of objects and object-systems like languages, symbol systems, design norms, technological systems, and a wide variety of artifacts. This, in combination with its active-passive symmetry, enables it to capture the fact that actors can simultaneously be engaged in arranging the elements of a successful science or technology while themselves being arranged by what has gone before or is emerging around them — partly as the outcomes of their own activities.

It is clear that coordination expresses the central notion of societal constructivism, that is, the idea that actors who are getting new technologies together are being influenced (by circumstances, by other actors, by rules and laws, by their perceptions of nature) even as they are influencing others. Coordination means that actors are somehow attuning their behaviors toward one another, that they are choosing to act in ways that are relevant to some goal shared with other actors or that are strategic moves in some game being played with other actors. But to what do actors attune their behavior? How do they know what moves contribute to collective technological goals — or to individual goals given the possible behavior of other actors?

In principle, actors can influence other actors in an immediate, direct way. Players of the board game "Monopoly," for example, are ongoingly coordinated in a face-to-face setting as they make moves and decisions which provide information enabling other players to make relevant moves (i.e. improve their individual strategic position and collectively carry on with the game). Now the question is: What actually provides the coordinative information to the other players? In the first place, of course, the rules of the game which determine what moves are allowed, what transactions must or may occur, and what meaning can be attributed to the various squares and markers. But in our admiration for rules we must not forget how important the board, the markers, the money, the banker is in playing a game of Monopoly. In principle one could imagine a game of Monopoly without all the physical paraphernalia supplied by the manufacturer. In this case players would have to constantly be informing one another on which square they were, which square they felt they were entitled to be going on their next turn, how many houses and hotels they had and where, who owed what to whom, etc. Clearly, even if the players could manage such a Lewis Carroll version of Monopoly, the stupendous effort required to coordinate the game would make it senseless as a pastime. Instead, the usual procedure is for players to delegate coordination both to physical objects and to human functionaries. The objects function as signs of the players' strategic position and as intimations of possible future moves. Other players now only have to read the distribution of signs on the board to know what the situation is and to plot their own strategy. The functionaries (in Monopoly, the banker) facilitate the progress of the game by administrating markers and in some cases adjudicating moves.

We can analyze a Monopoly game as consisting of a set of locations (the individual players) which are coordinated by an emergent global level of object-signs — and of course by the invariant rules of the game as thought up by its inventors. At the locations decisions are made to invest or not —

given an assessment of the state of the game, reflexive analysis of the position of the location, and the opportunities presented by the dice. At bottom this is all the game consists of — except the work necessary to communicate local action to the other locations and hence to coordinate the game. This requires some extra effort on the part of each location — making payments, placing houses and hotels, moving the marker. And one player must invest part of his time and attention in producing the collective good of a banking system. The global level, which in this case serves as a collective semiotic representation of the game, is brought into being and maintained by a "surplus effort" on the part of the locations.

This layered morphology, we contend, underlies social coordination in general. Stock exchanges are good examples of such layering at intermediate levels of aggregation. Here there is still face to face interaction, but there are also many tendrils connecting actors to far-away buyers and sellers. Hence the stock-exchange is itself a global actor serving to coordinate widely dispersed locations. Here too there are numerous physical constructs and markers which coordinate the buyers and sellers, but also specialized functionaries that process and recontextualize information to produce new global-level signs. Markets and politics, to take the most grand examples, are modes of coordination in which locations contribute information which is aggregated into coordinative constraints at a global level. Consumers buy specific qualities at specific prices and may have to expend effort to make the most rational purchase. Their behavior is transformed into market demand, prices, interest rates etc. by an elaborate global economic machinery. Citizens vote and pay taxes in order to maintain the representational and coordinative apparatus of democratic politics. In these sorts of intermediate and large-scale orders, like those in which new technologies are gotten together, the global level consists of far more than the physical objects and rules which hold a game of Monopoly together. Here, it can make sense for locations not only to contribute directly to the maintenance of a global level (e.g., attending a town meeting on the new sewer system) but also to delegate such concerns to global level actors, in this case, democratically responsible politicians.

Actors engaged in getting new technologies together are *nolens volens* situated at locations in large-scale global political and economic orders. These global orders exert their own coordinating force and thereby enable and constrain particular technological choices. In addition, however, actors involved in the design and implementation of particular technologies are also locations in very specific global technological orders — thus constituting what Van de Poel and De la Bruhèze (this volume) call *technical regimes*.[1] Analyzing how such global technological orders emerge and are modified by the global-level efforts of local actors, how local actors delegate agency to emergent global-level actors and how the global level in turn enables and constrains action at a local level (in part by re-delegating to locations) is the basic program of a societal constructivist approach to technology development. It is an effort to understand the mechanisms of coordination among the heterogeneous and dispersed actors engaged in getting new technologies together.

The accounts in this book may seem very diverse. Nonetheless they all contribute to defining a societal constructivist approach to technology development because they all explore the relationship between locations and global orders and hence investigate how the development of new technologies is coordinated. We consider this an improvement over the treatment of levels of action in most social constructivist approaches, where the doctrines of symmetry and seamlessness have reduced a dialectic of levels to simple hierarchies of aggregation. As will be elaborated in the conclusion, moreover, several of the contributions stretch the local-global notion in ways that emancipate it from its usual spatial metaphors. Locations may be defined not only with reference to spatial (or organizational) heterogeneity, but also with reference to temporal or discursive heterogeneity. That is, there are not only locations in space, but also in time and in systems of discourse that need to be coordinated via emergent global levels if technologies are to be gotten together successfully.

Introducing "coordination" as a key term opens a second polemic front toward the extensive literature on that subject. Fortunately we have been able to fight fire with fire, by confronting the opposing strengths and weaknesses of social constructivism with those of traditional theories of coordination. Social constructivism's strength is its emphasis on symmetry, its consequently almost naive empiricism and, in the case of technology, its special attention to the role of knowledge and artifacts. Its weakness is its inability to deal with supra-local structures and agency and its consequent suppression of local-global dialectics. The strength of theories of coordination focussing on, for example, markets, hierarchies, and networks, is precisely the centrality of such dialectics. Their weakness, from the point of view of technology dynamics, is that up to now they have said next to nothing about the special role of technology — particularly knowledge and artifacts — in effecting coordination, nor about the coordination of technology development itself — except as a contingent output of market, organizational, or network coordination. They have spoken of the coordination of social orders, but not of the coordination of socio-*technical* orders. In this context our arguments for societal constructivism amount to:

1. The fortification of social constructivism by the "structuralist" insights of theories of coordination and;

2. the fortification of theories of coordination by the heterogeneous empiricism and the sensitivity to the role of cognitions and artifacts we find in the program of social construction of technology.

The result is the enriched form of social constructivism we are calling *societal* constructivism.

The interest in coordination is directly related to an interest in *steering* sociotechnical processes. This is reflected in the literature on coordination as a sociological, political and economic phenomenon. Understanding how a distributed and possibly long-term process is coordinated may indeed provide insight into exerting effective influence — particularly where the available means depend on mobilizing cultural resources, political power or market alignments. It is however unlikely that traditional approaches to coordination can provide sufficient guidance for steering technology development. The specificity of technology as a societal phenomenon demands attention to current work in the field of science and technology studies — particularly in the area of technology dynamics. Societal construction of technology, as it is being defined here, can provide one essential bridge between scholarship and policy in this area.

The Organization of this Book

The book opens with "Coordination of Ship Propeller Design: Technical Models and the Relation of *T.Y. Draco* with Queen *Elizabeth II*" by *Barend van der Meulen*. This is an analysis of how the design process of ship propellers is coordinated in space and through time. Van der Meulen argues that the actor-networks which constitute specific designs (involving heterogeneous elements like artifacts, theories, insurance companies, journal articles) never emerge from a *tabula rasa* but are in fact discrete time-slices of a continually evolving actor-network in which all elements are continually subject to transformations. He illustrates this by analyzing how existing technical models constrain the construction of actor-networks in ship propeller design and are in turn modified in the course of subsequent design. Next, in "Means of Coordination in Making Biological Science: On the Mapping of Plants, Animals, and Genes," *Dirk Stemerding* and *Stephen Hilgartner* examine the role of "centers of calculation" and standardized material means in the coordination of two major international mapping projects in biology: the emergence of a taxonomic system in the late 18th and early 19th century and the current Human Genome project. They conclude that the success of such projects, which depend on the aggregation of dispersed results, require the prior establishment of means of coordination which enforce the standardization of work at numerous peripheral sites. However, they also insist that the construction of these coordinating means and their development cannot be explained simply as domination: the process depends on ongoing reconfigurations of center and periphery. Chapter 3, by *Wim Smit, Boelie Elzen,* and *Bert Enserink,* is called "Coordination in Military Socio-technical Networks. Military Needs, Requirements, and Guiding Principles." In this chapter the tribulations of achieving consensus on the design criteria for a European Fighter Aircraft and its radar system are analyzed as the construction of a supranational socio-technical network. This network necessarily emerges out of pre-existing networks at the national level. The persistence of these national networks constrains the freedom of actors at the supranational level and so coordinates the formation of the new network and the concomitant design criteria. Chapter 4, "Getting an Experiment Together in High Energy Physics. Big Plans, Big Machines, and Bricolage," by *Cornelis Disco,* argues that coordinating the activities of many physicists and engineers over many years is only possible if means of coordination are constructed in the process of experimentation itself. Such "frames" are both summations of past agency and constraints on future agency. Two frames are identified in the experiment under study: the text of the formal proposal and the completed big machine, the detector. Both, in different ways, coordinate subsequent episodes of bricolage so that experimental goals are conserved. In the next chapter *Ibo van de Poel* asks "Why are Chickens Housed in Battery Cages?" In answering this question, Van de Poel examines two popular types of explanation, an economic and a cultural one and concludes that while both are relevant, neither is sufficient. In order to explain the continued dominance of the battery cage and the limited success of efforts to abolish it, multiple modes of coordination must be charted. He invokes the concept of design regime to express how the hegemony of the battery cage is the outcome of several distinct and simultaneous modes of coordination impinging on designers and users. Chapter 6, by *Ellen van Oost,* "Aligning Gender and New Technology: Early Administrative Automation," shows how the management of the Dutch Postcheque and Clearing Service was able to introduce computers into their organization without causing disruption by capitalizing on the masculine gender typing of the new machines and associated jobs. Because the new computers would eliminate many jobs in the manual processing of transfer orders, management had promised to recruit applicants for programmers and operators from within the organization. In allocating the limited number of jobs, management was able to bypass female employees because computer jobs were universally defined as upwardly mobile male career tracks. Dutch women did not expect careers and so they could be passed up for the interesting jobs. Chapter 7,

"Expectations in Technological Developments: An Example of Prospective Structures to be Filled in by Agency," by *Harro van Lente* and *Arie Rip* show how technology actors are coordinated by storylines in which the promise and performance of new technologies are touted. Expectations are seen as virtual actor-worlds which are yet to be realized, but which, because of their promise, forcefully coordinate actors in making them come true. In this way, expectations become "prospective structures" which nonetheless forcefully constrain and enable specific kinds of future agency. In Chapter 8 exclusions, rather than expectations, coordinate the future. *Adri Albert de la Bruhèze*, in a chapter entitled "Boundary Maintenance and Radioactive Waste Disposal Technology in the US, 1945-1970," investigates how the Atomic Energy Commission's exclusionary approach to designing radwaste disposal technology gradually produced a socio-technical order which forcefully coordinated opposition to implementation of Atomic Energy Commission policies at a later point in time. Tracing the history of Atomic Energy Commission initiatives in managing radwaste, de la Bruhèze shows that the Atomic Energy Commission's bureaucratic and technological policy demanded the successive exclusion of more and more dissident actors from the process. This resulted in an excluded "radwaste world" most of whose participants were easily mobilizable against the Lyons attempt to implement the Atomic Energy Commission's ultimate solution of storage in salt-layers. In Chapter 9, "Meaningful Boundaries: Symbolic Representations in Heterogeneous Research and Development Projects," *Elke Duncker* and *Cornelis Disco* question how communication is coordinated among researchers with different symbolic repertoires. They investigate two examples of collaborative research involving different disciplines and professions and conclude that communication across boundaries is coordinated in the first instance by mutual recourse to more basic and generic symbolic repertoires. In the course of time, coordination is enhanced via the construction of passive and active "dictionaries" which allow researchers to translate among their different repertoires. Ultimately this can lead to the emergence of hybrid repertoires and possible new proto-disciplines. The penultimate chapter by *Arie Rip* and *Siebe Talma* on "Antagonistic patterns and new technologies" generalizes some of the lessons of previous chapters by considering how patterns of antagonistic coordination emerge around particular controversial technologies and how such patterns subsequently coordinate the conflictual introduction of successive risky technologies according to well-established scripts. The authors argue that antagonistic coordination may be productive as an occasion for social learning, inasmuch as both proponents and opponents will be interested in fortifying their points of view with robust arguments. The book closes with "Getting Case Studies Together" by *Cornelis Disco* and *Barend van der Meulen*. Here the editors return to the themes broached in this introduction and assess what the case studies have in fact contributed to our understanding of how technologies are gotten together.

Note Concerning the References

Throughout this book we distinguish between source references and scholarly references. The latter are listed in the bibliography and referred to in the usual way by the name of the author and the year of publication. Source references referring to empirical sources are alphabetically listed at the end of each chapter and referred by (A), (B), etc. Footnotes are listed in the usual way at the end of each chapter.

Note

[1] See also Disco et al. (1992).

Chapter 1
Coordination of Ship Propeller Design: Technical Models and the Relation of *T.Y. Draco* with *Queen Elizabeth II**

Barend van der Meulen

1 Introduction

In social studies of technology, design processes are conceived as social processes entailing heterogeneous elements; i.e. social processes that are therefore complex, varied, flexible and diverse. Within the design process all elements of the ultimate artifact and of the larger network that makes the artifact and its operations possible, have to be aligned (Bijker et al. 1987; Bijker and Law 1992). In this chapter we will consider the design of Grim Vane Wheels: a second propeller put behind the main propeller of a ship. The Grim Vane Wheel aims to use some of the energy lost by the first propeller and hence saves energy and decreases operating costs — at least if energy prices are high enough to justify the additional costs of a propulsion system with a Grim Vane Wheel.

In order to have a working Grim Vane Wheel, the two propellers have to be adjusted to each other and must have an optimal ratio of number of blades. Both propellers have to be linked to the ship by nuts, bolts and, shafts. The design has to be tested at the test station. The ship builder has to inform the propeller designer about the design of the whole ship in order to have an optimal ship-propeller interaction. Grim Vane Wheels can only be designed using hydrodynamic calculation programs. We can push this further: if there are nuts and bolts, calculation programs and testing stations, somewhere in the past these had to have been realized: some actors must have anticipated becoming an element that had to be *translated* in the Grim Vane Wheel-network. Maybe they did not anticipate this specific design, but at least designs like this. We can extend the design process as far as we want and make it as complex as can be encompassed in texts and figures (Callon and Latour 1981; Callon et al. 1986).

Within technical communities design processes are readily represented as a linear process, leading from idea, via requirements, calculations, and tests to the actual design. Figure 1 shows such a schedule, taken from a report of a design process of fixed propeller with a Grim Vane Wheel. Engineers simplify heterogeneous design processes in one simple linear scheme; a scheme that makes sense to their colleagues and seems to be a good representation of what has happened. Apparently, most of the complexity can be neglected and is self-evident for both author and audience. That there are calculation programs, that there is a test station doing its job, that the components of the artifact bought from outside fit into each other, that the ship builder cooperates, all this is self-evident and needs no discussion.

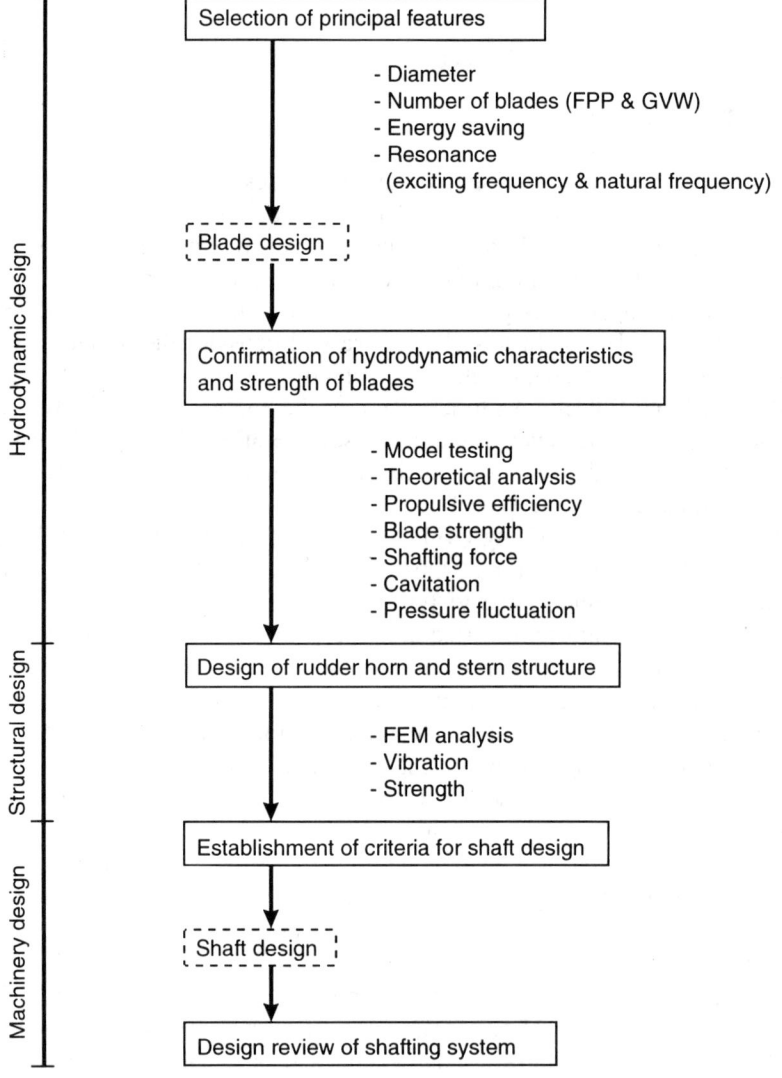

Figure 1: Diagram of a Design Process of a Ship Propeller with GVW (C)

There is a striking difference between the reconstructions of design processes in social studies of technology and "native" accounts by designers. SST reconstructions open a black box that at first sight remains closed in the descriptions by the designers themselves. Within the black box one can see the design process as it *really* was. It is certainly true that while designers reduce a lot of the complexity, they do not do so haphazardly; they have to think about what to reduce and what to open up all the time. But the schemes are a good device to tell about design experiences, decisions and results in a structured way.[1] Apparently how the propeller actually was designed is not of interest to the engineers. What matters is the qualities of the design and the design process with regard to the design traditions.

In technical fields like that of shipbuilding, in which design traditions last for decades, the success of a design process never depends just on the attempts of a single entrepreneurial designer to combine and align all elements in one design. In fact, most of the elements were aligned long before the design of this specific artifact took place. The design process is not just an alignment of heterogeneous elements at that specific time and specific place. Rather it is a joint project of people and things distributed all over the world, doing their work at different times but within the same technical field. The designer did not have to align all these elements himself to accomplish this specific design process. Most of it was aligned already and the results of the work, the products and experiences of men at the shop floor, of consultants, ship owners and builders, of test stations and calculation programs, and of engineering scientists in fluid dynamics, shipbuilding and applied mathematics was available for use without further ado. Of course, the simple need to align all these elements does not align them. In principle, divisions of labor, standardization and other patterns within the heterogeneous network are a result of historical, contingent, processes in which coordination problems have emerged and have been solved. But the more durable the patterns turn out to be, the less contingent the subsequent processes are (Dosi 1982; Sahal 1985). And, as a consequence, in project plans and in summary reports most of the complexity of design processes can be neglected without doing injustice to the actual design practices.

Vincenti (1990b, 1991) reflecting on his book *What Engineers Know and How They Know It* (Vincenti 1990a), has suggested that the fascination with novelty prevailing in technology studies might be due to the background of most of its scholars in philosophy of science and science studies. He, being an engineer, stresses the importance of normal technology:

[M]y career as a research engineer and teacher has been spent producing and organizing the kind of knowledge [normal] technology requires. [...] The kind of knowledge I dealt with was conditioned accordingly. The fact that an experienced engineer would focus, automatically and at first unconsciously, on normal technology for study of engineering knowledge may itself say something about the nature and importance of such knowledge (Vincenti 1990b: 3).

For the sociological analysis of technology development, the challenge is to analyze the effects of these stabilized patterns without reverting to straightforward technological determinism. After all, design is an innovative process which, to some degree at least, entails the breakdown of existing sociotechnical patterns and the creation of new ones. The aim of this chapter is to show that a sociological analysis of technology development is possible — and necessary — which merges the apparent innovativeness of design processes with the potency of stabilized design traditions.

In the first section I introduce the idea of technical models. Within local design practices, designers use models to elaborate design ideas, verify possible solutions and find optimal combinations of design parameters. These models may be scale models, computer models, drawings or even formulae in which certain parameters are linked to each other. These models function on a local level, but their development involves global division of labor and organization. The models incorporate a range of design experiences and thus embody stabilized technical patterns. Taking technical models as a focus, we can accomplish a first assessment of the sociological and technological consequences of design traditions.

It is possible to go one step further. Recent sociology of technology has convincingly argued that technological development is *socio*technical development. One of its important strains, actor network theory, has incorporated within its conceptual framework the radical idea that *non-human* elements can be social actors (Callon 1986a, 1986b; Latour 1992). Consequently, an analysis of technological developments should be sensitive to the roles of technical elements, as non-human actors, within design processes (Latour 1992). I suggest that by looking at the development and potency of technical models, we can study these roles. In the second part I will unravel some of the dynamics behind the development of technical models and how they enable technical elements to become actors.

The specific design history which I will use to ground my argument, can be recounted in two paragraphs. In 1983, the owner of the *Queen Elizabeth II*, world's largest passenger ship, faced with the decision whether or not to keep the ship in operation for an additional 20 years, decided that an extensive feasibility study was necessary. The feasibility study incorporated an assessment of the economic, technical and operational advantages and disadvantages of several engine and propeller systems. On the basis of this study it was decided to refit the *Queen Elizabeth II* with new engines and a new propeller system, including two Grim Vane Wheels. The Grim Vane Wheels were designed by Lips, one of the world's major ship propeller manufac-

1 Coordination of Ship Propeller Design 19

turers. Lips used existing design tools and test methods that had been successful in the design of many other propeller systems. As before, the simulations and scale-model tests predicted unproblematic cost-effective Grim Vane Wheels. However, during the first voyage of the renovated *Queen Elizabeth II*, the blades of the Grim Vane Wheels broke off.

Just after this accident, Lips convinced the builders of a new generation of VLCCs (very large crude oil carriers) that it would be worth their while to study the possibility of fitting them with Grim Vane Wheels. In order to overcome previous problems with GVWs it was decided to incorporate several in-depth studies and new design methodologies within the usual design process. In 1989, at its triennial conference on ship propellers, Lips was able to announce that the first of these new VLCCs, the *T.Y. Draco*, had successfully passed sea trials and was now in operation. The GVW, in particular, performed extremely well.

This little design history contains both technical success and technical failure, both technical continuity and technical innovation. Below we will use it to develop the idea of technical models and assess their sociological relevance for understanding design processes. In the second part I subject the history to an actor-network analysis of the dynamics of technical models and their functioning.

2 Technical Model

Stabilization of technical experiences and achievements implies that in one way or another they acquire significance at the global design level. From anecdotal occurrences with local and temporal significance they are translated into general knowledge that can be used at other places and in other times. Structuring at a more abstract level is a crucial step. Disco et al. (1992) have introduced the concept of technical model to conceptualize structured design processes and meta-modelling for the development of technical models as technological work in its own right.

Technical models are representations of technical artifacts. They define functional dependencies between parts of the artifact and between critical parameters, dependencies between various performances of the artifact and dependencies between the parts, the parameters and performances. These technical models are communicated at conferences and in journal articles, they are embedded in design methods like CAD tools and model tests, and articulated in educational activities. Components of technical models, moreover, can become standardized and obligatory as a result of formalized technical evaluation (Disco et al. 1992).

As technical models become better articulated and well-founded, designers can design more easily. The models cumulate good past practice and successful designs. Technical models can be made at any level of abstraction. Designers use simple drawings of the basic configuration and working of an artifact to structure their creativity and to communicate design ideas (Henderson 1991; Duncker and Disco, this volume). With more abstract theoretical models like CAD programs based on hydrodynamic theoretical and experimental knowledge, they can simulate performances. Thus technical models allow designers to engage in manipulation of design parameters with an eye to optimization. Actors involved in the design process can make assessments along relevant evaluative dimensions of unconventional requirements, innovative ideas, proto-designs and the like before putting the artifact into use. Several solutions can be tried at low cost until the best solution is found. The proof of the pudding is no longer in the eating, but is already available in the recipe.

Figure 2 can be considered as a technical model of the Grim Vane Wheel. It relates the basic principles of the performances of the Grim Vane Wheel (reclaim some of the lost energy in the slipstream of the first propeller as additional thrust) to the shape of the blades and to its basic position within the propulsion system. The model structures the thinking of the designer: at the same time it defines the minimal size and the basic forms of the propeller and in essence it delineates some of the design freedom like the distance between the two propellers and the numbers of blades. In the accompanying text this technical model is further articulated:

> The system propeller plus vane wheel has a larger diameter than the propeller alone. From this follows an entrainment of a larger amount of water, and hence in accordance with jet theory an additional efficiency improvement can be expected. A similar effect can be achieved by using a larger propeller diameter, but this requires a lower shaft speed. Furthermore, a larger propeller soon meets a limit owing to the tip clearance which is needed to avoid vibration problems. On the other hand a vane wheel can be located much closer to the hull because it is a lightly loaded device and hance no vibration problem will arise from it. The main parameter which indicates the amount of efficiency improvement is the thrust load coefficient c_{TH}. With the increasing of c_{TH} the gain will increase. c_{TH} should be at least 0.5 (B: 5).

The quote reveals some of the considerations and choices designers have to make in their quest for an optimal propeller system. If we look in more detail at the design of the Grim Vane Wheels for the *Queen Elizabeth II* and the *T.Y. Draco* respectively, we find that other technical models were used as well. The design of the new propulsion system for the *Queen Elizabeth II* started in 1984, when it was concluded from feasibility studies that, despite the high costs, a complete re-engining of the ship was to be preferred. A new propulsion system, including an energy-saving propeller system, would

1 Coordination of Ship Propeller Design

result in a 50% reduction of fuel consumption. Because the Grim Vane Wheels for the *Queen Elizabeth II* had to fit behind the existing ship and were part of a larger propulsion system, some parameters were constrained. It was decided to design a Grim Vane Wheel of the maximum diameter that the shape of the *Queen Elizabeth II* allowed. Subsequently, the diameter of the main propeller could be calculated, taking shaft speed, ship speed and engine power into account. Once given their diameter, the blade geometry could be calculated from hydrodynamic propeller theory. Although a GVW has an atypical geometry, conventional propeller theory was initially used to calculate the *Queen Elizabeth II* wheels. Rather than elaborating hydrodynamic theory for the specifics of GVW — which requires all kind of estimations and approximations because of the non-linearity of hydrodynamic equations — the design team trusted in the adequacy of the existing design tools. This choice was reaffirmed by the nautical classification society responsible for quality control of the design process on behalf of the ship owner.

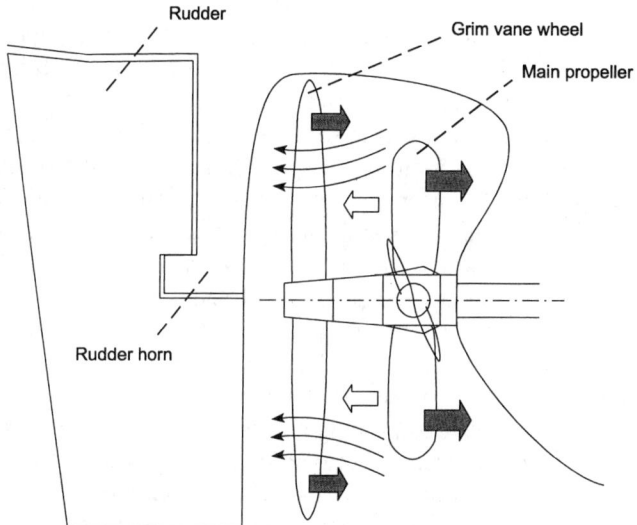

The second propeller converts energy losses of the first propeller into extra thrust. The inner part of the Grim Vane Wheel functions as a turbine that is driven by the slip stream of the first propeller. The outer parts of the blades act as an additional propeller.

Figure 2: Principle of the Grim Vane Wheel

The optimal shape of the Grim Vane Wheels as calculated by hydrodynamic theory was tested extensively for its expected performance, cavitation characteristics[2] and noise and vibration behavior. To conduct these tests, several scale models were built for measurements within well controlled circumstances. Performances were measured in a towing tank — a large basin through which the scale model is towed. Cavitation was observed in a cavitation tunnel and calculated by the design team and the classification society. In order to predict noise and vibration behavior, a scale model in a vacuum tank was used to measure pressure amplitudes.

Each of these scale models can be considered as a technical model of the designed propulsion system, inasmuch as it represents a relationship between the geometry of the GVW and certain aspects of its technical performances. The scale models are not just small versions of the ultimate artifact with similar behavior. The capacity to predict future behavior at full scale, depends on the possibility of performing reduced-scale tests in closely controlled circumstances. For towing tanks, knowledge of conditions under which reduced-scale results can be extrapolated to full-scale goes back to William Froude and Osborne Reynolds, who found that relations between a number of key parameters have to be similar at both reduced and full scale. Both the Froude number[3] and the Reynolds number[4] are universally considered significant for validating model tests. Nevertheless, each towing tank, which are often subdivisions of national technical institutes and well-integrated into national maritime communities, have developed additional (local) similitude rules for interpreting measurements on scale models.

Mutatis mutandis, this holds true for other technical models as well. Each technical model serves as a check on specific aspects of the operation of a GVW. From the initial concept to the ultimate artifact, several models are employed to ensure optimal working. Indeed, one can conceptualize a process of the design of an artifact as series of *reality tests*. In each reality test the feasibility of the initial concept is tested and after each reality test the concept has become more real. Technical models frame the reality tests. They embody explicit and tacit knowledge whether and how certain combinations of performances, geometries and operations are possible.

In each stage of the design process an appropriate representation is used that is robust enough to detect failures in earlier stages, that allows for sufficient flexibility for improvements and that provides enough certainty to move the design to a next stage. Design processes might start with simple drawings and sketches in which only a few actors are linked provisionally or with a list of design criteria that links the eventual design to the user situation. In each phase new actors are linked into the actor network and the force of established linkages is tested. Eventually when the design is able to withstand all tests, it ends up as a description of the artifact itself.

If technical models are well developed, design becomes a routinized practice and can even be computer-based.[5] It is tempting to suggest that technical models assure good design. However, the design of the GVWs of the *Queen Elizabeth II* is an example in which the technical models did not suffice. The usual practices to test for noise and vibration behavior were in fact *a priori* unsatisfactory insofar as the large diameter of the "real" proposed GVW compromised the results of the now relatively too small scale model tests. As a result the design team had to perform experiments and reference tests to improve its knowledge of the relationship between noise and vibration behavior in the scale model and at full scale. Thus some meta-modelling, the purposive development of a technical model, was incorporated in the design process. Still the design team had to conclude that: "More consistent results can probably be obtained when a refined theory is used." A conclusion that is followed by a reassuring: "In general, the design tools have proved to be reliable enough to be used even for this type of vessel" (A). With hindsight we know that the ostensibly optimal GVWs could not resist the final test. During their first voyage the blades broke off.

Subsequently the design process of the GVWs for the *T.Y. Draco* included extensive study of the causes of earlier failures of GVWs. Most failures were due to mechanical problems in the shaft system. Therefore, it was decided to mount the GVW on the rudder horn instead of on the shaft of the main propeller. The problems with the *Queen Elizabeth II* were due to unexpected stresses in the GVW that ultimately broke the blades. Contrary to former assumptions, GVWs turned out to be significantly loaded in open water conditions. The design process included an adaptation of the propeller analysis programs to the peculiarities of Grim Vane Wheels and a comparative analysis of excitation frequencies of several combinations of the number of blades of the main propeller and those of the GVW. A four bladed propeller with a nine bladed GVW turned out to be the combination that was most unlikely to induce resonance. This solution was tested in two different model basins and by several stress calculations.

During the design of the GVWs for the *T.Y. Draco*, technical models of GVWs were considerably improved. In order to arrive at a new propeller, experiments were done to improve the technical model, improvements which could subsequently be used to design a better GVW. The sequence is significant for understanding the role of technical models in design processes: before a reliable GVW could be designed for specific purposes, the relation between GVWs and propellers in general had to be modeled. Also significant is the way the failures are used. Apparently these experiences were translated from technical failures into relevant design knowledge. In general then, technical models allow local designers to take advantage of previous experiences, experimental results and theories.

3 Coordination of Design

Obviously, engineers have developed technical models of Grim Vane Wheels with some general validity, in the sense that they can be used in various local design practices and for the design of distinct Vane Wheels. The several models allow for a design process that takes place at different localities and which makes use of as much design and test capacity as possible. The capacities of hydrodynamic and mechanical designers, of several model basins and tanks and of the owner, wharf and classification society are incorporated in the design process. Each of these actors work on specific aspects of the design. Thus, because of the different representations, the design work can be divided into several separate design tasks that can be worked on at different places by different people (Law 1992).

But division of labor goes beyond the level of a specific design process. It is part of the structure of the naval architecture community. The different representations are embedded in the ongoing practices of the various actors. These actors contribute to the several pieces of knowledge, developing different parts of the technical model, and they use different parts of the technical model in their own design — or, more generally, technical — practice. For some, the use, production and reproduction of certain representations is implicit in their activity, for others it is the *raison d'être* of their activities.

For example, to a ship owner the propeller is only a part of his ship. He does not bother about hydrodynamic representations. Ship owners just want guarantees that the performances are in accordance with the requirements; a guarantee the test station can give them. Hydrodynamic designers and scientists have some interest in the performance of the artifact in use, but as these experiences are seldom restricted and quantified, they will hardly be able to use these in their design work, research and meta-modelling. Rather, they will use the results of the scale models, which were generated under restricted conditions. Consultants are deeply interested in all kinds of knowledge and experiences, inasmuch as they make their living off the division of labor in the design process. Their work can be considered as structuring and articulating technical models and transforming these into recommendations for ship owners and ship builders.

The technical models and the division of labor in the design of propellers structure relations between actors within technical fields. An example is the central position of the maritime research institutes within design processes, if only because each design is separately tested in their towing tanks. On the basis of the test results the propeller manufacturer and the ship builder can prove that the design meets the requirements. The ship owner uses the test results to convince the classification society that his ship can be insured. On a meta-level, model tests contribute to the development of technical models as they make controlled comparisons between designs possible. The crucial function of the test results give test stations a central position in the field of ship design.

Note, however, that this position is under fire because of enormous improvements in the possibilities of computer aided design. Computer aided design tools have now been developed to the point that they can compete with scale model tests in regard not only to reliability and validity, but also as regards costs and time. Obviously, positions in the design process will change in favor of the designers when computer simulations come to be considered as valid as model tests. There is in fact an ongoing struggle about the validity of the CAD-programs and the possibility of designing without model tests. The test stations themselves play a curious role in this discussion. On the one hand they actively pursue the development of these CAD-programs. On the other hand they try to underplay the value of these programs in order to maintain their central position in the design network.[6]

Because representations are of the same artifact, results generated at one place can easily be moved to and understood at other places. At a micro level this implies the possibility of dispersing design processes among different places and of burdening actors with specific tasks. At a macro level, different representations come to coincide with long standing practices and there are actors like journals and engineering schools, focused in whole or in part on the production and articulation of specific representations (see Stemerding and Hilgartner on STS, this volume).

4 Designers or ... Technical Models?

Thus far, the analysis of the design process of the ship propellers implies two conclusions. First, for the design of ship propellers several technical models are used. Each of the models reflects a specific layer of cognitive patterns: From user experiences to hydrodynamic theory. They reflect the degrees of freedom for a new design and thus to some extent determine future designs. Second, technical models are flexible and can be and are adapted to design processes. In both design processes (i.e. for the *Queen Elizabeth II* and *T.Y. Draco*) existing technical models were improved and checked against others. Within technical models earlier experiences with designs and design tools are structured and offered as resources with which designers can improve their designs.

In the description of the design processes I have assumed the existence of technical models and asserted that designers make use of technical models in order to design. In principle, it is possible to give a fully sociologically rational explanation of the use of technical models. Conformation to existing technical models increases the likelihood of acceptance of results, increases the possibility for cooperation and increases the trust of other actors in the design network in the outcomes (Smit et al., this volume). This would be an explanation that takes the designers as central actor and corresponds to the formulation of the second conclusion. Consequently the coordination that takes place within the design process is due to the actions of designers, not to the technical model.

The first conclusion, however, emphasizes the potency of the technical model. Together, the two conclusions depict a subtle balance between technical models determining the design outcome and technical models providing the designer with the means for optimal design. This balance is the quintessence of the role of technical models within design processes and not a mere difference in formulation. The balance reflects the crucial relation between the designer and the technical model, in which both can be seen as actors acting upon each other and others. The challenge then is to analyze design processes so that technical models appear as actors or at least so that the potency of technical models to coordinate can be explained.

In order to accomplish this we need to make what in social studies of technology is called a *symmetrical* analysis (Bijker et al. 1987); i.e. one in which society and technology are treated equally. Although authors in social studies of technology now widely claim that in sociotechnical networks, the social is technical and the technical social and that human and non-human actors are indistinguishably connected, most studies still focus on engineers, firms, consumers, in short, on human actors as acting agents. The human actors remain the subjects, those who act, and the technical, non human actors are objects, those upon whom is acted. However, a truly symmetrical analysis would allow for a non-human actor to be a subject, acting upon human beings. But can we think of ship propellers making rational choices, flexibly interpreting socio-technical networks, deciding about strategies, using people to realize their objectives? As long as we do not have an acceptable vocabulary that can simultaneously be applied to both human and non-human actors, we are forced to analyze technological development asymmetrically.

In the field of social studies of technology it is only within actor network theory that methods and concepts have been developed for a truly symmetrical analysis. Authors like Callon (1986a, 1986b), Latour (1991, 1992) and Akrich (1992) have drawn the radical conclusion that sociological theory has to be reconsidered. In three related chapters Akrich and Latour, for instance, have developed a semiotic vocabulary to describe sociotechnical development. Focusing on such things as photoelectric lighting kits, electricity networks, seat belts and doors, they analyze how "the composition of a

1 Coordination of Ship Propeller Design

technical object constrains actants in the way they relate both to the object and to one another" and "[...] the extent to which [these actants and their links] are able to reshape the object" (Akrich 1992: 206).[7] Basically, their method of analyzing is observing the juxtapositions of actants within networks and the consequences of the juxtaposition for the definition of these actants and their possibilities for acting.

Akrich analyzes these juxtapositions for photoelectric lighting kits designed in France, and intended for Ivory Coast. In France the kit is linked to a network including a government agency promoting new energy sources and a photoelectric cell industry in which the optimal configuration of technical elements is more important than a relation to a market. In Ivory Coast the lighting kit got introduced in a user context which appeared to lack the appropriate actants to establish functional relations. Spare batteries were not available, wires were too short, users and local electricians were forbidden to repair in case of breakdown. The configuration of technical elements within the lighting kit prescribed the existence of certain actants and relations in the user context in order to create a socio-technical network for a working lighting kit.

Akrich (as does Latour) concentrates on designed artifacts: black boxes inscribed with *scripts* that are imposed upon the users and user contexts. As far as the analysis focusses on design, the designer, the human actor, is again the only subject that acts. "Thus, many of the choices made by designers can been seen as decisions about what should be delegated to a machine and what should be left to the initiative of human actors" (Akrich 1992: 216). It is striking that although actor network theory allows for a symmetrical analysis, ultimately Akrich again places a human actor, the designer, in the center and reduces the artifact to a mere tool by which the intentions and interest of the designer can be forcefully imposed upon users, i.e. other human actors.

My focus is on design contexts, which implies that at the very least the question of symmetry shifts to an earlier phase in the development of artifacts. But that would only be a spurious bit of comfort, if in the end technical models still proved to be the creatures of other human actors and to embody their intentions and interests. The crucial question is whether the juxtaposition of certain technical elements influences the development of social action, regardless of the inscribed intentions of designers. To address this I will pursue an analysis that focuses on how technical relations and the performances of technical elements influence designers and design contexts — rather than the other way around.

5 Representing GVW as Actor-networks

Within actor network theory, design processes should be conceptualized as processes in which heterogeneous actors are linked to each other in texts, drawings, and other representations or intermediaries. In effect, these intermediaries impose meanings or roles upon actors by linking them to other actors and hence translating them to specific positions in a network of relations (Callon et al. 1986; Callon 1986a). The translation of actors into actor-networks can be exemplified by a conference paper describing the design of the *T.Y. Draco's* improved Grim Vane Wheels. It is in fact sufficient to deconstruct only the introduction to the text. In the introduction (Figure 3) a spectrum of important and heterogeneous actors (Law 1986a, 1986b) are linked to the Grim Vane Wheel: new generation VLCC, IHI, *T.Y. Draco*, Tonen Energy & Marine Pte Ltd, energy-saving technology, rudder horn and Lips. The Grim Vane Wheel is defined in this text by its relations with these actors: With IHI and Lips a producer-product relation is made, with the VLCC, *T.Y. Draco*, the technical hierarchical relation component-artifact is made, with energy savings technology the relation category-instance. In addition relations are stipulated among the heterogeneous actors themselves, for instance a producer-consumer relation between IHI and Tonen Energy & Marine Pte Ltd and an owner-property relation between Tonen Energy & Marine Pte Ltd and the *T.Y. Draco*. The text constitutes a network of heterogeneous actors with heterogeneous relations defining the Grim Vane Wheel as an actor-network (Figure 4).

The text exemplifies how actor network theory can extend beyond usual technology studies into the technologies themselves. Indeed technical elements occur as actors or *actants* within the network. We can see how the GVW is defined not only by the designers, the ship builder and the ship owners, but also by the VLCC, other energy saving technologies, the rudder horn and several performances.

Albeit on a schematic and aggregated level (because this is based only on an abstract of a fuller text), we can see how patterns between technical elements are established. Figure 4 suggests that it is possible to delineate these relations specifically and to concentrate on them, ignoring relations between human actors. Both kinds of actants seem to be loosely connected in the GVW network itself. But still it is a matter of empirically grounded analysis whether a focus on technical relations only is valid. Note also the definition of the GVW as being reliable, which implies it is endowed with a character which it will carry from context to context and which allows other actors to anticipate its behavior.

1 Coordination of Ship Propeller Design

> **IMPROVED GRIM VANE WHEEL SYSTEM APPLIED TO A NEW GENERATION VLCC**
>
> M. Tanaka, R. Fujino and Y. Imashimizu
> Ishikawajima-Harima Heavy Industries Co., Ltd.
>
> [abstract:]
> In september 1988, Ishikawajima-Harima Heavy Industries Co., Ltd. (IHI) delivered the first of 6 of a new generation of VLCCs, the 'T.Y. Draco', to Tonen Energy & Marine PTe Ltd. (Singapore). The 236.000 dwt tanker features the integration of a number of energy-saving technologies, of which the Grim Vane Wheel (diameter 11.64 m) is especially remarkable for being the largest one ever made, mounted on a rudder horn. This paper discusses the improved Grim Vane Wheel system. The system was developed jointly by IHI and Lips B.V., with full attention given to reliability.

Figure 3: Title, Authors, Introduction of Paper on Improved GVW

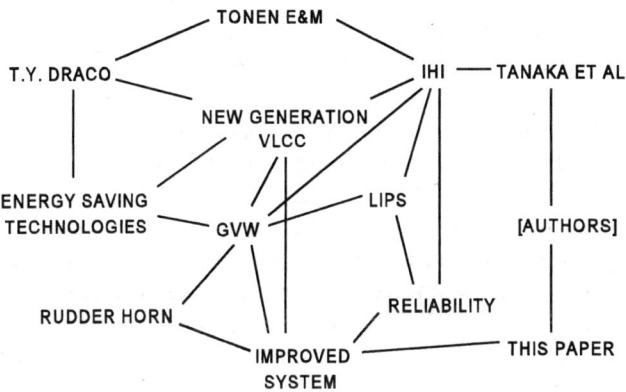

Figure 4: Grim Vane Wheel Network

Translations made in texts and text fragments like the one on *T.Y. Draco's* Grim Vane Wheel, are just a few translations in a whole range of previous, simultaneous and subsequent translations.[8] Through these translations similar actor-networks are produced. The texts and other representations in which these translations are made cite each other, make use of, reproduce or contradict each other and create more or less robust relations between the actors. All these translations have created a network of actors that is larger and stronger than one that can be represented in a single text. The actor-network of Figure 4 is (synchronically) part of that larger network of relations and actors and (diachronically) one in a series of translations. Hence, a full analysis of the design of GVWs reveals not only the several stages by which the design goes from first idea to artifact, but also how the translations made are embedded in the larger network of relations.

But pragmatic reductions are necessary, even were I to concentrate simply on the design and mounting of a GVW propeller system as a process of translating actors into a network. Even in this relatively simple case I would have to analyze the negotiations and collaborations between Lips and IHI, the design work of the Lips designers, their use of calculation programs, and the work of the test station. Indeed, such events are important elements in stories of technology development. Our selection will aim at elucidating the construction of a technical model of the GVW and at understanding the solidifying of relations of sufficient strength to structure design processes of similar artifacts. In principle, this could result in an archaeology of the GVW in which the old technologies on which the GVW is based are dug up from the contemporaneous GVW network. We limit ourselves to a history of Lips-designed GVW networks and look back only a couple of years.[9]

6   *T.Y. Draco* and *Queen Elizabeth II*

The decision to mount a GVW on the *Queen Elizabeth II* was based on technical models of propulsion systems with GVW and alternative propulsive systems. The technical models enabled a comparison of aspects like compatibility with the rest of the power system, costs, speed and efficiency. After the decision for a propulsion system with GVW was made, further design studies were done on the relative dimensions of propeller and GVW, the exact size of the GVW, the number of blades and their optimal shape. These were considered to be the critical parameters for achieving optimal vibration and cavitation performances.

This is the point from which I will follow the technical model, and thus especially the models that structure the hydrodynamic design of a GVW. The text reporting the design process is illuminating.[10] It starts with Prof. Grim as the main actor in the design process. "The design has been made by Prof. Grim who is the inventor of this device." Then within a few sentences he disappears. At first, his presence is still implicit. "In the process of arriving at the optimum configuration the diameter is varied for given thrust, number of revolutions and ship speed." From this, Prof. Grim concludes that the main propeller should have a diameter of 5.6 m. Then technical elements and technical models take over. During several paragraphs they are the subjects, defining the propeller and the GVW. The first thing they do is to redefine the diameter at 5.8 m. "Subsequent design calculations and the results of model tests indicated that to improve upon the cavitation properties a diameter of 5.8 was necessary." The text proceeds in a similar vein. "The high ship speed causes the frictional losses to be higher than for most

1 Coordination of Ship Propeller Design

other Grim Wheel applications. Frictional losses are reduced with a lower number of blades and a lower number of revolutions. Therefore the *Queen Elizabeth II* Grim Vane Wheels have 7 blades and their design number of rotations per minute is about 31%." It is a relation between the ship and the propeller, modelled in terms of speed, frictional losses and number of blades and revolutions, that defines the latter for the GVW. The designer reappears only when conventional hydrodynamic theories have to be used to calculate the details of the blades. And later he is joined by the manufacturer Lips and the classification society on behalf of the owner in order to carry out extensive calculations and model tests, apparently to verify the decisions of the calculations and model tests.

Could the designer have remained the subject within the texts? Could he have kept his position of main actor within the first phases of the design process? Certainly, but it makes a difference when a designer decides that a diameter of 5.8 m is necessary or when such a figure results from calculations and model tests. In the latter case the design gets linked to technical models that have proven their value again and again. Just as the relation between the GVW and the ship is not merely an ad hoc local decision, but comes from a generalized established relation between ships and Grim Vane Wheels, so is the 5.8 m a result of established technical models.

As part of the extensive evaluations, the pressure amplitude and cavitation performances were tested. Indeed, for this part of the design process, technical models were ill-developed. After "general experience" has defined the model size normally used for propulsion tests as too small, the designer and especially the authors have an important role to play. The next paragraphs show a remarkable interplay between the author of the text, the designer and all kinds of technical elements expressing themselves within experiences, test results, calculations and formulas. But it is especially the author who implicitly and explicitly decides on the outcomes. "Taking into account the different model scales and test setups it may be concluded that the resulting hull pressures agree very well with each other." And just when a formula on maximum allowable hull pressures has established that the value for the *Queen Elizabeth II* can be 5.6 kPa, the authors warn that "the formula should be regarded with some caution." Nevertheless, the aggregation of these separate pieces of engineering knowledge and results seem to be forceful enough to conclude that "the design tools (used for the hydrodynamic design, BM) have proven to be reliable." The change in perspective is again striking.

Established models that could define the pressure amplitudes and cavitation performances were clearly lacking. Instead, local experiences, research results, test results and the like are brought together in order to verify the hydrodynamic design collectively. As in the earlier design phase, the GVW design is linked to an aggregated level, but one that is more ambiguous and in which the relations are mediated by the designer and the author. If we think of technical models as developing from locally established relations between elements that are reproduced and that develop reliability and resistance, this piece of text illuminates some of the dynamics of the way these locally established relations become forceful. What if the GVWs of the *Queen Elizabeth II* had resisted the first and subsequent trials? Obviously, the design tools would then indeed have been reliable, and the collectivity of technical elements and experiences would have been confirmed in their capacity to verify. The collectivity would then not only have been kept together by interpreting actors, but also by the GVWs of the *Queen Elizabeth II*.

But as we know, the GVWs did not resist the trials at sea. The carefully built relations within the analyzed text became obsolete. What remained was a negative experience which demonstrated that the technical models used were less reliable than they seemed. In the case of the *T.Y. Draco* the designers faced, again, a challenge for which an established technical model was lacking. The conference paper reporting the design of *T.Y. Draco's* propulsion system states: "Troubles frequently occurring with GVW systems in use required the return to a basic starting point of design before a safe and reliable new system could be produced." Again, I will focus on the hydrodynamic design, and the establishment of some principal particulars.[11]

Again, the text starts with designers — implicitly — being the subject. The models have moved to the second plan. "The design study was carried out using theoretical calculations and the results of model and full-scale tests." The size and the number of blades of the Grim Vane Wheel are not defined by technical models, but chosen and selected by the designers. However, the selection is constrained by relations between GVW, propeller and rudder horn in terms of excitation frequencies. "Figure 6 represents the relation between each exciting frequency and the natural frequencies of the structural part calculated during initial state." These relations are tested at an IHI model basin from which: The tests confirmed the exciting forces and blade stress variations of the GVW." With reference to this figure and the models tests the designer, "after full evaluation," selected a combination of a four-bladed propeller with a diameter of 9.5 m and a nine-bladed GVW, a selection that was verified and confirmed by several model tests and sea trials.

The study on exciting frequencies and the model tests at the IHI model basin define some general relations between GVWs, rudder horns and propellers. This general relation guided designers in selecting a configuration for the *T.Y. Draco* GVWs. The situation closely resembles the verification of the design for the *Queen Elizabeth II* GVWs. An established technical model is lacking and "thus" the designers assemble a local technical model from several bits of technical knowledge. But there is a difference as well. The calculations and tests provide the same outcomes, and mediation by the designer is hardly necessary to link these two. Thus it is the designer who starts to mobilize other actors to define the GVW and in the end decides on the definition by selecting a certain configuration. But the introduction of outcomes of calculations and tests within the network immediately changes the relation between the designer and the GVW. The GVW has been defined to some extent by these new actants and thus the possibilities of the designer are limited.

What we see here is the development of a new technical model. Indeed, the same design process has been used for a text titled: "Designing Vane Wheel Systems." The technical model developed for the GVWs gets translated from the local design context to a global designing context. It is a translation which again is made by Lips and its designers:

[...]As a propulsion efficiency improver the concept of the Vane Wheel is beyond doubt. The promising economic application of the wheel however, was troubled by some failures. Contradictory requirements have to be fulfilled. On the one hand the system has to show maximum efficiency, whereas on the other hand sufficient strength and margin to resonance frequencies must be guaranteed. Lips Propeller Works have adapted their proven propeller analysis programme in order to be capable of designing these wheels. Reliable prediction of hydrodynamic and consequently mechanical loads has resulted (A: 1).

Thus, relations that were demolished because of the failure of the blades of *Queen Elizabeth II's* GVW seem to be restored again. A design tool, embodying generalized relations between technical elements of the GVW, has been developed and can be used to design reliable GVWs. Still, it is a tool to be used by a designer. But one may wonder what happens when it has proven to be reliable again and again. It is likely it will become forceful enough to establish configurations of GVWs on its own.

7 Dynamics of Coordination

In order to understand the coordinating potential of technical models, I performed the actor network analysis of the design processes with the intention of focussing on the technical elements and how they influence design processes. Obviously, it was not possible to limit the analysis to the technical elements alone. Human actors like the designers, the ship owner, the classification society took part in the development of technical models and appeared instrumental in several design phases as mediators for the influence of technical elements on the design process. This conclusion confirms results of other technology studies: The social and technical are interwoven and cannot be separated except by means of *a priori* classifications.

But contrary to what is often concluded from this finding (e.g., Hughes 1986; Bijker et al. 1987), we can still both make distinctions and follow certain processes adequately. The web is not seamless, at least not if we are willing to observe the seams made within the processes. The final description of the design processes reveal not only that heterogeneous actors define the ultimate design, but also reveal some of the interactions among these heterogeneous actors. At first an existing technical model defined the main characteristics of the GVW for the *Queen Elizabeth II*, without mediation by the designer. Then, further on during the design, technical elements and their relations became ambiguous, and a design team was needed to confirm their reliability. Note that similarly decisive forces can also be exerted by model tests and sea trials, as was the case within the second design process (and in fact also, but negatively, for the *Queen Elizabeth II*). *T.Y. Draco's* design process shows how new technical relations are aggregated from several experiences and experiments and how these to a certain extent define the improved GVW. These are subsequently translated to generalized relations in order to influence other designs at other places and other, future, times.

From these interactions we can distill how technical elements and technical models coordinate design processes. Coordination results from the translation of actors and linkages. The cases show at least three generalized processes that account for coordination by technical models. The first two, *generalization* and *localization*, conceptualize, respectively, the transformation of locally established relations between technical elements into technical models and the use of technical models in a design context. The third process, *reversal*, conceptualizes the transformation of agency between generalized relations on the one hand and designers on the other.

In order to verify the design of the GVW for the *Queen Elizabeth II* the designers aggregated all kind of experiences, experiments, tests and calculations. Together they produced a model sufficient to buttress the design, but too uncertain to become a generalized model. For the design of the second GVW the design team again aggregated several technical elements and relations into a new GVW configuration. During the design, established models embodied within tests and calculations confirmed the resulting relations and, as a result, reinforced the established configuration. This time the configuration survived its final trials. The more local trials that were passed, the more general the new aggregation could become. Eventually Lips inscribed the generalized configuration into a design tool and thus linked it to such powerful actants as calculation programs and computers.

The generalization of relations results in a meta-network of stable relations. These are the relations that can easily be reproduced in subsequent designs. Within stable design traditions, most of the translations of heterogeneous elements into a design network are easy translations. Callon (1992) has introduced for this meta-network the notion of *translation regime*. The designers are translated into this larger meta-network as well, instead of having to build an entirely new actor-network (Latour 1987) — a process on which actor network studies tend to concentrate — they can limit their effort to creating only the necessary additional relations.

Aggregation and generalization are still an insufficient basis for coordination. Coordination occurs when the generalized configurations, the technical models, enter design localities and link these up with earlier and other ones. This might be done with design localities within the same design process, but as we argued before, also on the level of the design community. So there is a second process, which I call localization. The generalized configuration is (re)translated into a specific local configuration. The generalized relation between ship speed and number of blades is translated into a relation of the *Queen Elizabeth II* with a seven bladed GVW.

The notion of localization has consequences for our understanding of the operation of technical models as opposed to fixed technical relations. In social studies of technology such fixed technical relations are often regarded as black boxes that come into new contexts without revealing their constitution and thus appearing as unchangeable. Technical models do not necessarily operate this way. First, because of the localization they have to reveal their internal constitution. Thus, while they might be unchangeable, they are certainly not black but transparent. Second, they might have to compete with other models where localization might lead to another local configuration. In such cases designers, those upon which established technical models could act, are offered a decision space in which they can decide on the ultimate configuration. Indeed, the descriptions of the GVW design processes reveal how designers could control design processes in general, once confronted with indefinite technical relations.

Obviously, if designers can take over, we need a third process to understand coordination by technical models. The coordination is also a result of the relation between the designers and the technical models. One of the main dynamics in the development of technical models and a source of their potency is the possibility for reversal in the relation between designers and technical models (Staudenmaier 1989). Designers deliberately construct technical models to facilitate their design work. As long as these technical models have not matured, designers use them. The more a technical model increases its competencies by withstanding more and more trials, the better it can dominate the designer and correct his or her design solutions. It was by means of such reversal that the propeller for the *Queen Elizabeth II* got a diameter of 5.8 m instead of 5.6 m.

8   In Conclusion: The Force of What Has Been Done

The cases are not only relevant for the understanding of design processes and the coordinative role of technical models. The results have some consequences for the study of technology and especially actor-network theory as well. In fact, actor-network theory conceptualizes reality as a continuous flux of sociotechnical processes in which networks are created by successful translations, get solidified into a (macro-) actor or dissipate again into multiple actors. The processes are linked to each other: they make use of each other's results, try to translate the same actor or merge into a new process. A representation of the GVW network like the one in Figure 4 is like a thin cross-section orthogonal to the axis of the processes, allowing the analyst to have a closer look at some of the relations. The engineer's account of the design process as a structured process in a limited number of phases is a (backward) perspective parallel to the processes. I have looked back as well, but only to follow the dynamics of some of the design processes. Within each account a selection was made of the totality of the relations and events, all the while reducing complexity as much as possible. Any account of the construction of an actor-network has to select from the numerous processes and multiple events.

Most actor–network studies make a selection in which heterogeneity and complexity is maintained as much as possible and in which contingency is stressed. Although it is not a theoretical principle, the focus on technology in the making at least suggests that each artifact has to be constructed out of its rudimentary elements. Technology is created in a *tohu wa bohu*, an earth without form and void, in which in principle any sociotechnical network can be constructed. Classical distinctions between rationality and irrationality, between society, technology and science, between society and nature, between human beings and artifacts are denied. *A priori* no distinctions are allowed.

But as time within the case descriptions proceeds, the constructed network becomes a constraining and enabling matrix for subsequent actions. On an actor level one can see actors who elaborate successes, avoid previous failures and exploit existing relations. Such recurrent actor-strategies result in a reproduction of network relations, stabilization of action in rules, routines and standards, materialization of relations in technology, or generally in coordination of future actions. It takes only a little bit of reflexivity to realize that if socio-technical networks work like this halfway through a case description, there will probably have been networks that worked that way at the start. The recognition of the social force of constructed sociotechnical networks, of artifacts and standards, of rules, routines and relations, challenges the methodological rule of allowing no *a priori* distinctions. We do not need solid distinctions between, for example, social worlds and technical worlds, but we need to recognize the heterogeneity, asymmetries, transparent and black boxes, generalizations and localizations already made within the ongoingly ramified sociotechnical network.

Notes

* I am grateful to Lips Propeller Systems for their willingness to participate in the study and their readiness to provide me with information and documentation. Figures and quotes on ship propeller design are from the proceedings of the 6th Lips Propeller Symposium, Drunen, The Netherlands, 1986, and the 7th Lips Propeller Symposium, Noordwijk, The Netherlands, 1989.

1 But it should be noted that on a methodological level these kind of representations are not as useful and valid. Most general design methodologies (Pahl and Beitz 1977; Hubka 1982) are based on such linear representations but they have only limited impact on how designers structure their work (Hales 1987; Bucciarelli 1988). It has been noted that Computer Aided Design-methods based on such models can hardly be implemented in existing design practices. See Henderson (1991).

2 Cavitation: when the propeller rotates, an underpressure originates at one side of the blades and water in that area starts to boil at normal temperature. The resulting bubbles implode when they enter areas with normal pressures. The energy required for this process implies an efficiency reduction and the forces of the implosions may cause damage to the blades. Well designed propellers show minimal cavitation.

3 Froude number: $F_n = V/\sqrt{(g.L)}$, where V is the velocity of the stream, g the gravitational constant, and L the length of the body. In order to get extrapolatable results the Froude number in the model test has to be equal to the Froude number at full scale. William Froude (1810-1879) developed this key to interpreting scale-model tests.

4 Osborne Reynolds (late 19th century) argued that not only must the Froude number be equal for model and full scale, but also what we now call the Reynolds number, R_n. $R_n = V.L/v$, where V is the velocity of the stream, L the length of the body and v the viscosity of the medium.

5 Indeed, the reverse occurs as well. Design tasks may be so complicated that development of well-articulated technical models is a *sine qua non*. Examples are for instance hydrodynamic designs and, from another technical area, the design of integrated circuits.

6 On conflicts in design and development of technology, see Hård (1993).
7 Akrich and Latour both use "actants" in these chapters rather than 'actors. An actant is defined as "whatever acts or shifts actions, action itself being defined by a list of performances through trials; from these performances are deduced a set of competences with which the actant is endowed [...]; an actor is an actant endowed with a character (usually anthropomorphic)"(Akrich and Latour 1987: 259). The addition that actors are usually anthropomorphic suggest at first sight that actor remains a concept reserved for human actors only. However as Akrich and Latour show, technologies may be anthropomorphic as well. Below we will see that actants like design tools and GVWs are defined as being reliable.
8 Most actor-network studies on technological development share a focus on technology-in-the-making with other technology studies. See however for a comparable elaboration of ANT, Callon (1990, 1992).
9 Such necessary reductions demand some reflectivity, because they force the analyst to make pragmatic choices about whom to include in the network, whom to exclude, which processes to study and which not and, thus, to make implicit distinctions. We can follow the actors and intermediaries only within a limited time and space. In my account I leave aside, for instance, factors such as would be stressed by SCOT analysts: the changing interpretations within the texts and the design history of the GVW as an energy-saving device, a problem and danger for the ship, an innovation, an *ad-hoc* solution for poor energy-wasting propeller design. It would also be possible to relate decisions to mount GVWs to a level of energy prices that would justify the investments. The issue is not what is the right account, the issue is what kind of aspects of technology development can we highlight by excising certain relations from the flux of sociotechnical developments.
10 The following section is based on (A).
11 The subsequent section is based on (C).

Sources

A. Beek, T. van: "Hydrodynamic features of the new QE2 Propulsion System," Proceedings 6th Lips Propeller Symposium, 1986.
B. Blaurock, J: "An appraisal of unconventional aftbody configurations and propulsion devices," Proceedings 7th Lips Propeller Symposium, 1989.
C. Tanaka, M., R. Fujino and Y. Imashimizu: "Improved Grim Vane Wheel System applied to a new generation VLCC," Proceedings 7th Lips Propeller Symposium, 1989.

Chapter 2
Means of Coordination in Making Biological Science: On the Mapping of Plants, Animals, and Genes

Dirk Stemerding and *Stephen Hilgartner*

1 Introduction

During the late 1980s, the idea of mapping and sequencing the complete human genome evolved from a visionary proposal into a worldwide research effort, as genome programs of various sizes and organizational forms were established in the United States, Europe, and Japan (Cook-Degan 1994; Jordan 1993; Hilgartner 1995a). In the United States, the official timetable calls for completing maps of all human chromosomes and determining the complete sequence of human DNA by the year 2005 at a cost of $3 billion, roughly $200 million per year. The genomes of several "model organisms" (e.g., E. coli, yeast, C. elegans, and the mouse) are also being analyzed. The information generated by the human genome project is expected to be "the source book for biomedical science in the 21st century" (J).

What is novel about the human genome project is not simply its aim of mapping and sequencing genes, but rather the conviction that this aim requires a coordinated project focused on mapping and sequencing entire genomes within a short time span. This vision of a comprehensive, coordinated effort distinguishes the human genome project from earlier programs of research in molecular genetics. In the words of one of the leaders of the project, the effort is "similar to the 1961 decision made by President John F. Kennedy to send a man to the moon inasmuch as here also the United States has committed itself to a highly visible and important goal" (Watson 1990: 44-49). To achieve this goal, a number of organizational measures have been taken, including the establishment of national and international advisory committees to provide overall planning and advice, the creation of research centers and collaborative networks, the organization of regular meetings and workshops to stimulate the coordinated collection and analysis of data, and comprehensive research and spending plans "to move the human genome program forward at an optimal rate" (J).

In this chapter, we compare the present human genome mapping and sequencing project with an earlier large-scale mapping effort in biological science: the search for a natural system of classification in eighteenth and early nineteenth century natural history. Although centuries apart and involving different objects and means of investigation, both these efforts are attempts to create a new, generalized knowledge object — a comprehensive source book — by developing systematic collections, and by describing and comparing a totality of things in the form of a comprehensive inventory of biological entities. Both these efforts have been inspired by grand visions of the true aim of doing science, requiring coordination on a scale that had not been encountered in previous research programs.

However, whereas the complete sequence of human DNA is the anticipated result of a carefully planned and strategic research effort to be completed in a scientist's immediate lifetime, systems of classification in natural history were the subject of long-lasting battles which involved many generations of naturalists (Daudin 1926a, 1926b). In other words, whereas the human genome project represents coordination that is being deliberately organized with the support of the modern state, coordination in natural history only gradually emerged in a long history of attempts to find a truly natural system of classification. What can we learn about coordination in science, when comparing the example of highly institutionalized and strategic genome science at the end of the twentieth century, to the emergent endeavors of naturalists, shaping and reshaping systems of classification, during the eighteenth and early nineteenth century?

With a range of organizational measures in place, the human genome project strikes the observer as a most salient example of coordination in science. Coordination, however, should not be seen as a purely social phenomenon of managing interactions; it is achieved through heterogeneous means, including all kinds of technical and procedural practices that build coordination into the fabric of scientific work (Latour 1987; Star and Griesemer 1989; Fujimura 1992; Kohler 1994). In our comparison of the recent history of genomics with the early history of taxonomy, we take up this point by focussing the analysis on the various practical means or tools through which coordination has been achieved. First of all, our comparison shows that, in achieving coordination, both individual actors striving for order in eighteenth century natural history and macro-actors in the form of agencies and organizations taking the lead in a well-directed human genome effort, depended on a variety of tools and means that were (made) part of normal scientific practice. Moreover, in comparing these cases, we find striking similarities between the means of coordination that shaped the search for systems of classification in early taxonomy, and the means that have proved important in the early years of the human genome project.

On the theoretical level, however, coordination in science can be viewed from several perspectives, each of which emphasizes different pictures of how heterogeneous means contribute to coordination. These divergent perspectives are reflected in the debate about means of coordination among Latour (1987), Star and Griesemer (1989), and Fujimura (1992). An important issue in this debate is whether the coordinating role of heterogeneous means can be understood through a conflict model that focusses on domination, or through a cooperative model of collective work. In other words, should coordination be understood as occurring through the "imperialist imposition of representations, coercion, silencing and fragmentation," or is it the result of flexible, mutual accommodation (Star and Griesemer 1989: 413). In this chapter, we view these different images of coordination not as alternatives to decide among, but as sensitizing concepts for understanding the complex dynamics at work in achieving coordination. As we will show below, in both early natural history and contemporary genomics, we find coordinating activities that arguably lean toward each of these extremes — or, even more tellingly, have an ambiguous status with respect to these positions. We examine coordination in these very different contexts as a way of exploring in concrete situations the role of means in achieving coordination. In the concluding section we will discuss the results of our comparative exploration more explicitly in the light of the different theoretical perspectives on coordination.

2 Coordination in Early Taxonomy

In the history of Western civilization we find a tradition of descriptions of plants and animals going back to antiquity. From the sixteenth century onward, however, this tradition underwent a significant change. The first signs of this change can be found in the works of the so-called herbalists (Larson 1971). Traditionally, herbals were chiefly compilations of descriptions or "histories" of individual plants. But, since the sixteenth century, the authors of these books increasingly sought to arrange plants in a systematic way on the basis of an analysis of external resemblances and differences. A variety of circumstances promoted these early attempts at systematization. Voyages of exploration, the introduction of printing, and technical advances in illustration offered means to compile specimens of plants and to disseminate botanical descriptions on an unprecedented scale. The need to manage a growing number of plant descriptions, of which many had evidently not been known to the Ancients, as well as the need of correct identification of plants in view of their medicinal virtues, led to a gradually emerging search for methods of classification according to which any given plant could be readily identified. Moreover, during the seventeenth and eighteenth centuries, collections of specimens and descriptions of plants and animals — gathered together in gardens, cabinets, and libraries — increased in number and in scale. As these collections grew, problems of order acquired new dimensions and culminated in attempts to construct comprehensive and rigorous systems of classification.

The growing interest in systematic descriptions and classifications also reflected the ambition of seventeenth and eighteenth century naturalists to create a veritable science of natural history. Among these naturalists, the aim of a natural system of classification — a system that would represent real relationships between the objects of nature — became an important and controversial issue that preoccupied many of the most prominent authors in the field (Sloan 1972, 1976). In their attempts to arrange a growing number of known species, naturalists working in different countries and with different collections often used different names and different systems; indeed "hardly any two persons observed the same system of classification" (Allen 1976: 39). This situation not only raised practical problems of communication, but it also motivated naturalists to search for a truly comprehensive and natural system of classification, a system that would align both a variety of natural objects, as well as a variety of naturalists.

The French naturalist Mathurin-Jacques Brisson, portrayed by Farber in his book on the emergence of ornithology as a scientific discipline, was on this count one of the more successful eighteenth century naturalists (Farber 1982). In 1760, Brisson published an *Ornithologie* in which he described fifteen hundred species and varieties of birds, three times as many as in Linnaeus' tenth edition of the *Systema Naturae*, published two years earlier. For the arrangement of his descriptions, Brisson introduced a new system of classification. Although his system was constructed for convenience and made no claims to representing natural relationships, it was used by a number of naturalists, and was considered to be the most important system in ornithology up to the early nineteenth century. As Farber points out, Brisson's work "set a new standard for the classification of birds" (Farber 1982: 26). Brisson's work was thus more than an expanded catalogue of birds; it also embodied an attempt at coordination, both of an increasing amount of data and of the endeavors of a small lot of naturalists. How to understand the impact of his achievement? As Brisson's career and work clearly demonstrate, three kind of practices were especially instrumental in coordinating mapping efforts in early taxonomy: the practices of collecting, representing and standardizing.

A Collecting

A first key to understanding the basis of Brisson's coordinating achievement may be found in the article on Natural History in the French *Encyclopédie* (Diderot and d'Alembert 1966). If it is the task of the naturalist to describe and compare every animal, plant and mineral, so the article wonders, how can this be achieved? The number of natural entities is immense and no one is able to go and search everywhere. Progress in natural history clearly is not and cannot be the work of a single genius. Even all the inhabitants of a nation will not suffice. What is needed is the collective effort of many nations over a period of centuries. The author of the article thus rejoices at the popularity that natural history has gained in the eighteenth century, a popularity which has expressed itself not only in a zeal for observation, but also for collecting. The result is a growing number of natural history cabinets and the article emphasizes the great value of this development for the progress of natural history. Now, more than ever, the naturalist can study the dispersed productions of nature within the walls of a cabinet:

[...] one has found the means to abbreviate and flatten the surface of the earth on behalf of naturalists; of every species of animals and plant, and specimens of mineral, individuals have been gathered in natural history cabinets. There one can find productions of every country in the world, and so to say an abridgement of complete nature [Diderot and d'Alembert 1966, VIII: 229, translation by the authors].

Thus, as the article indicates, natural history was not only an endeavor of lonely naturalists wandering in the field at home or abroad, but was shaped by collections of various sorts. Exploration and colonial expansion had become an important source of "exotic" specimens that were collected, grown in gardens, kept in herbaria, and exhibited in cabinets. Indeed, as David Allen observes, toward the end of the seventeenth century, "a cabinet of natural (and artificial) curiosities had come to be regarded as one of the essential furnishings of every member of the leisured classes with claims to be considered cultivated" (Allen 1976: 30).

One of the finest collections of natural history objects in mid-eighteenth century Europe was the natural history cabinet of the French Prince of Naturalists, Rene-Antoine Ferchault de Réaumur (Farber 1982). Réaumur succeeded in solving some of the problems of preserving bird specimens, problems which were responsible for the generally small number of quadrupeds and birds represented in the cabinets of the day. Thanks to his method of preservation and with help of an extensive network of correspondents throughout Europe and the colonies, Réaumur assembled the largest bird collection in Europe. Since 1749, Mathurin-Jacques Brisson had been *"garde et demonstrateur"* of this cabinet and thus he had, as Farber observes, the "unusual opportunity of being placed in a private museum with an unparalleled bird collection that contained recently mounted birds in excellent condition and that included large numbers of exotic species, many unknown to science" (Farber 1982: 10). Thus, Brisson's expanded knowledge of birds was a direct result of his connection with Réaumur's cabinet, a connection which enjoined and enabled him to solve difficult problems of order. Indeed, any naturalist who was in charge of a garden, herbarium or cabinet could hardly escape the difficult problem of organizing the collection. But the ambitions of those naturalists who faced such problems often extended beyond the walls of their particular cabinet. They tended to aim at more than just a convenient arrangement of this or that collection. It had to be an arrangement according to general principles applicable to the world of plants or animals as a whole. Brisson's work is a conspicuous example of this search for system inspired by what Farber calls a "museum curator's perspective." In an earlier, more general, work on classification, *Le Règne Animal divisé en IX classes* (1756), Brisson explained his aims in the following words:

> The position that I have had the good fortune to occupy for several years, which has put me in daily contact with the richest collection of nature's objects that has ever been made, has allowed me to make a great number of observations on the animal kingdom; to compare them and to examine the closest and most distant relations. I have been led to think of arranging the animal kingdom into an order different from those used up to the present time. My intention in this labor was solely to instruct myself and to place myself in the position of being able to judge the most convenient place to put a specimen of a new animal which would arrive to be placed in a cabinet (cited and translated by Farber 1982: 10).

As Brisson's observations clearly illustrate, his position in the midst of a large natural history collection not only prompted him to search for a system that might contain all animals in forms that were easy to grasp, but it also gave him a unique opportunity to do so. In other words, Réaumur's network of correspondents and his preservation method had allowed his cabinet to become a "center of calculation" where someone could compare a great variety of animal specimens at ease by taking a few steps, going from one case to another (see Figure 1).[1] Thus visually dominating a rapidly accumulating variety of animal forms, Brisson was able to coordinate a great number of data in a previously unparalleled fashion.

2 Means of Coordination in Making Biological Science   45

INTÉRIEUR DES GALERIES D'HISTOIRE NATURELLE
(Jardin des Plantes)

Exhibition of a collection of mounted birds. Réaumur's famous collection of birds became part of this natural history collection.

Figure 1: A gallery of the Paris Muséum National d'Histoire Naturelle in the Early Nineteenth Century

B   Representing

The coordinating impact of Brisson's achievement cannot be understood, however, only on the basis of the physical resources that were at his disposal. Through his published work, careful descriptions and illustrations of these resources became available to every naturalist. Thus, as Farber points out, knowledge of Réaumur's collection was preserved well after most of the original specimens had disappeared. Indeed, by examining Brisson's *Ornithologie*, every naturalist could compare the specimens from Réaumur's bird collection with specimens from collections elsewhere, or again with descriptions based on these collections, without it being necessary to visit Réaumur's cabinet, or any other. Now, an interested observer only had to turn a few pages. This may seem a trivial observation. But the coordinating force of the ability to translate things into inscriptions — words and images on paper — that may be endlessly reproduced and compared, becomes very clear when we look at how the history of taxonomy has been affected by what Eisenstein has called the shift from script to print (Eisenstein 1979; Latour 1986).

It is very interesting to take another look — but now from Eisenstein's point of view — at how, from the sixteenth century on, attempts at classification in botany began to take shape. That story begins with the sixteenth century herbals in which the historian can trace early attempts to establish groupings of plants. The history of those herbals in fact goes back to antiquity. But as Eisenstein points out, it was a history that could only be preserved in the form of written documents, locked up in a few libraries and copied from time to time through the work of scribes until the advent of printing at the end of the fifteenth century. As such, it was a very fragile history. Libraries could be destroyed, manuscripts could be lost or fragmented, scribes could err when copying a text. In this scribal culture, the history of compilations or catalogues like herbals was one of "increasing confusion and disorganization" (Eisenstein 1979: 108). Moreover, to consult a substantial number of such texts a scholar had to be able and prepared to wander from one library to the next.

As Eisenstein shows, this situation began to change rapidly in the sixteenth century. Now many herbals, including classical texts derived from Greek authors, appeared in print and thus became widely available to all scholars who wished to study and compare them, or to confront them with their own observations. Although at first printed editions were often faithful copies of original manuscripts, such texts did not remain unaltered for long. With so many different readers checking books against each other and against their own observations, and with printers eager to produce revised and enlarged editions, sixteenth century herbals marked a turning point in the gradual deterioration which had tainted those works in the age of scribes. Subsequent editions of printed herbals began to show a rapid accumulation of data, not merely as a result of information and material received from readers or correspondents, but also from expeditions mounted by printers, publishers or editors themselves in order to get first-hand information on the plants that were so often confusingly described in ancient books. Printing also made it possible to rely on pictures which, now that they could be reproduced without corruption, considerably relieved difficulties of identification. Indeed, as subsequent editions "became bigger, more crammed with data, and more profusely illustrated," it became evident that many plants had not been known to the Ancients, and it also became necessary to provide each work with "more tables, charts, indexes which made it possible for readers to retrieve the growing body of information that was stored" (Eisenstein 1979: 109).

Moreover, printers' workshops not only produced more and more books, but also catalogues and indexes which satisfied the needs of readers no less than the commercial interests of printers. With the help of such catalogues, the Swiss scholar Conrad Gesner even made an attempt, in the mid-sixteenth century, to produce a universal bibliography and a comprehensive reference guide covering all Latin, Greek and Hebrew works published in print. In doing so, he also collected an "immense mass of information relating to the animal kingdom," leading to the publication of his famous seven-volume *Historia Animalium*. Thus, as Eisenstein puts it, "a long-lived desire to comprehend the divine scheme for creation and to classify and order all of God's creatures was given a new impetus" (Eisenstein 1979: 99). The practice of printing enormously enhanced the power of naturalists to mobilize and manipulate data, and it greatly augmented the volume and ease of communication between different observers. In other words, thanks to new and powerful means of inscription, naturalists were confronted both with a rapidly increasing amount of data and with a growing variety of systems used by observers working in different times and places. Again, as in the case of collections of physical objects, this circumstance created the need as well as the possibility for solving difficult problems of coordination.

C Standardizing

Nevertheless, in order to understand the coordinating impact of Brisson's published works, we need to add to our analysis another aspect that made his work especially effective. The point here is that Brisson was not just describing the variety of forms in Réaumur's cabinet. He also standardized his many hundreds of descriptions in such a way that naturalists could compare them almost at a glance. Thus, in his *Ornithologie*, Brisson explained how he had described every specimen in the same format: "First the size and proportions of the bird; next its colors, starting with the head and finishing with the tail" (Farber 1982: 12). In addition, Brisson arranged his descriptions according to a new method of classification on the basis of which he defined 26 orders, which all together accommodated 115 genera, and 1,500 species and varieties of birds.

Indeed, as the great classifier of the century, Linnaeus, observed, even with books full of descriptions and illustrations, it may still be utterly impossible to gain a comprehensive view. Without system, natural history would be chaos (Stafleu 1971). Nothing was more important in this respect than the work of Linnaeus himself, who owed his success largely to the precise way in which he standardized his botanical descriptions (Figure 2). Linnaeus was clearly annoyed at what he described as "the wide-spread wild confusion" prevalent in early taxonomy (Sloan 1972: 50). It was, as Eriksson observes, "the dream of young Linnaeus that, with his method, it would be possible for anyone who had learned the system to place any plant anywhere in the world in its right class and order, if not in its right genus, whether the plant was previously known to science or not" (Eriksson 1980: 58). But Linnaeus did more than simply propose a system. From the very beginning of his career, he set out to work on "a broad front at a program for the development of standards, not only of classification in the narrower sense, but also of nomenclature and description with its definitions and its terminological procedure" (Eriksson 1980: 76-77). Thus he invented a whole array of terms designating in a very detailed and concise way the many different parts of plants. Linnaeus' attempt at standardization was directed to what Eriksson describes as "the curse of never knowing for certain whether two authors are talking about the same plant or different plants when they use plant names of their own and describe their genera and species in different words and with regard to different details and organs" (Eriksson 1980: 59).

TRIANDRIA MONOGYNIA.

53. CROCUS.* *Tournef.* 183. 184.

CAL. *Spatha* monophylla.
COR. *Tubus* fimplex. *Limbus* fexpartitus, erectus: laciniis ovato-oblongis, æqualibus.
STAM. *Filamenta* tria, fubulata, corolla breviora. *Antheræ* fagittatæ.
PIST. *Germen* fubrotundum. *Stylus* filiformis, longitudine ftaminum. *Stigmata* tria, convoluta, ferrata.
PER. *Capfula* fubrotunda, triloba, trilocularis, trivalvis.
SEM. plura, rotunda.
OBS. Crocus *T. Corolla monopetala, germine oblongo infra receptaculum floris, tubo corollæ longiffimo filiformi.*
  Bulbocodium *T. Corolla fere hexapetala & germine fupra receptaculum floris, tubo floris vix ullo.*

54. IXIA.*

CAL. *Spathæ* vagæ, oblongæ, perfiftentes, germina diftinguentes.
COR. *Petala* fex, oblonga, æqualia, lanceolata.
STAM. *Filamenta* tria, fubulata, corolla breviora, fitu æqualia. *Antheræ* fimplices.
PIST. *Germen* ovatum, triquetrum, fub receptaculo floris. *Stylus* fimplex, erectus, longitudine ftaminum. *Stigma* trifidum, craffiufculum.
PER. *Capfula* fubovata, triquetra, trilocularis, loculis compreſſis, trivalvis.
SEM. fubrotunda.

55. GLADIOLUS.* *Tournef.* 190.

CAL. *Spathæ* vagæ
COR. fexpartita. *Petala* oblonga, obtufa, quorum tria proxima fuperiora conniventia, inferiora autem patentiora, omnia unguibus in tubum brevem incurvum connata.
STAM. *Filamenta* tria, fubulata, divifuris alternis petalorum inferta, omnia fub petalis conniventibus, adfcendentia. *Antheræ* oblongæ.
PIST. *Germen* infra receptaculum. *Stylus* fimplex, longitudine ftaminum. *Stigma* trifidum, concavum.
PER. *Capfula* oblonga, ventricofa, rude trigona, obtufa, trilocularis, trivalvis.
SEM. plura, fubrotunda, calyptra involuta.

In this work, which appeared in print in 1737, Linnaeus described 935 genera of plants — all the then known genera of the world. The descriptions were presented in a strictly methodical and standardized way.

Figure 2: A page from Linnaeus' Genera Plantarum

Like Brisson's work in ornithology, Linnaeus' botanical work offers a prime example of a successful attempt to coordinate both an unprecedented mass of (new) data and the efforts of a growing number of naturalists. Even more than Brisson's career, Linnaeus' rise as a naturalist illustrates the importance of the practices of collecting, representing and standardizing in achieving this coordination. First, thanks to the very tools of standardization he had himself made, no one could match Linnaeus' productivity in creating descriptions of genera and species which, to use Eriksson's apt metaphor, "were fabricated according to the principles of serial manufacture, thus allowing mass production" (Eriksson 1980: 61). Moreover, Linnaeus himself moved about in order to promulgate his standards and disseminate his methods. Thus, his decision to go to Holland was of enormous importance for the broader acceptance of his standards — for two reasons, as Stafleu points out. First, thanks to the large botanical gardens in Amsterdam and Leiden which, at the beginning of the eighteenth century, were unsurpassed for the extent and quality of their collections, Dutch botanists were leading taxonomy in the discovery and description of the non-European plant world. And, secondly, Holland was a major center of printing and publishing in those days (Stafleu 1971).

D In Search of a Natural System

In the foregoing we have seen how coordination in early taxonomy gradually emerged from the initiatives of individual naturalists, profiting from a variety of means which allowed them to collect and disseminate a rising amount of data. These initiatives not only urged naturalists to create comprehensive and rigorous systems of classification, but also convinced them that a truly natural system of classification was a desirable and attainable goal. Many eighteenth century naturalists, for example, eagerly accepted Linnaeus' work because of his uniform and concise definitions of genera and species. At the same time, however, the foundations of his system were often criticized, notably in France (Daudin 1926b; Stafleu 1971). According to these critics, the relations among the multiplicity of beings were too numerous and subtle to be caught in a system based on only a few arbitrarily chosen parts such as stamens and pistils. In their view, one might hope to achieve a true, natural system of classification only by taking into account all parts of a plant or animal.

However, the aim of a natural system entailed new problems of coordination. How to represent in a system the totality of complex relationships of a complete inventory of beings, that is, the relationships that can be found when observing all parts of every being instead of only a few, easily accessible, external parts? In the early nineteenth century, the French naturalist George Cuvier managed to create a classification of the entire animal kingdom that was generally accepted as a decisive step in the search for a natural system in zoology (Daudin 1926b; Coleman 1964; Outram 1984, 1986; Appel 1987). Cuvier's *Règne animal*, in which this classification was articulated, quickly became, as Coleman put it, the "standard zoological manual for most of Europe" (Coleman 1964: 13). In the following, we will see how Cuvier's method of "correlation" (D) enabled him to solve difficult problems of coordination, involving not only different kinds of data, but also different approaches used by different groups of practitioners in natural history.

E   The Animal Economy

In one of his earliest publications, a memoir on the Linnaean class of Mammals, Cuvier (and his co-author Geoffroy Saint-Hilaire) had already criticized the Linnaean approach to classification for not respecting the multiple affinities between different species of animals (F). That is, naturalists had sought to represent animal forms in such a way that they could be easily compared and, to that end, had been prone to base systems of classification on characteristics taken only from one and the same part. Cuvier and Geoffroy conceded that, in principle, this was a valuable approach. In practice, however, naturalists, in classifying animals, had allowed themselves to seize upon this or that part of an animal in a completely arbitrary way, thus creating a great variety of systems. Rather than classifying animals on the basis of a few of the most conspicuous, external characteristics, as naturalists so often had chosen to do, one should, according to the authors, rely on those parts of an animal that have a preponderant influence on the organization of the animal as a whole. It was this idea which Cuvier, from the very beginning of his career, made the cornerstone of attempts to create a natural system in zoology. In a functional system "so much connected as the animal economy" it was, in his view, the conformation of the physiologically most important parts, like the organs of circulation and sensation, which determined the constitution of most others. Animals that resembled each other in these primary parts, would necessarily resemble each other for the majority of the other parts and so would constitute a natural group.

Thus, by making the animal economy a relevant consideration in the classification of animals, Cuvier attempted to combine the advantages of a rigorous system of classification (in which taxonomic groups were indicated by a few distinctive characteristics) with the aim of a natural system (in which taxonomic groups represented multiple affinities between the variety of forms). That is, known functional relations between organs suggested to him a way to represent the various characteristics of the whole animal by a description of only its primary parts. The basic idea was that one had to find characteristics indicating the features of the animal organization as a whole, in order to work out a system of classification in which the majority of characteristics could be stipulated "in just one word" (B). In this way the creation of a system became a means to reduce, as Cuvier put it, a science to "the greatest possible economy of words," that is, "to elevate its propositions to the greatest possible generality" (C).

F   The Cabinet, the Field, and the Scalpel

Cuvier's search for a natural system in zoology was more than a mere attempt to coordinate different kinds of taxonomic maps and data. The latter presupposed coordination of practices embodying interests and activities standing largely apart in the world of natural history. In the eighteenth century, the practice of classification was dominated, especially in zoology, by naturalists working with collections in natural history cabinets. Naturalists, therefore, often based their classifications on a few salient, external characteristics, not only because such characteristics could be most readily identified, but also because it was often the external parts of an animal that were most easily preserved in a cabinet (Daudin 1926a). Brisson's previously noted ornithological work is an illustrative example. In classifying the birds in Réaumur's cabinet, Brisson had nothing to work with but taxidermic specimens and thus it is hardly surprising that his system of classification was based on well-preserved features such as beaks and claws (Farber 1982). To be sure, animals were also being studied in a living or fresh state, but as a rule there was a great gulf in eighteenth century zoology between the naturalist nomenclators, who ruled the practice of classification from their cabinets, and those naturalists who observed animals in the field or who undertook anatomical investigations. Those naturalists who observed and dissected animals in a living or fresh state, were often indifferent to exact classifications, whereas naturalists with an interest in classification were seldom inclined to undertake such dissections themselves. Indeed, as far as anatomical dissections were undertaken, it was not by naturalists aiming at classification, but mostly by physicians hoping to perfect the practice of medicine (Daudin 1926b).

Not everybody, however, accepted these divided worlds. For example, physician anatomists who had earned a name for themselves in natural history, like Louis Daubenton and Felix Vicq-d'Azyr in France, objected to the neglect of the study of anatomy, especially comparative anatomy, and criticized their fellow naturalists for being overly concerned with the surface of things (Daudin 1926a; Corsi 1982). Proclaiming the animal economy as the principal object of natural history, Daubenton upheld the view that naturalists first of all had to appreciate the integrity of the animal as a whole. Instead of describing and comparing animal forms on the basis of only a few external parts, one had to make complete descriptions, including all the important organs, internal as well as external (Farber 1975). At the end of the century, Vicq-d'Azyr expressed similar feelings when he noted the distance between a natural history that considered only external parts, and an anatomy that was restricted to the study of internal parts. It was a distance which, in his *Système anatomique* (1792), he proposed to eliminate, noting that:

[...] the method to be preferred when classifying animals after the manner of naturalists is probably not the one to be preferred when one proposes to arrange them in a way which is most in tune with anatomy. [...] In natural history, nothing else is being taken into account than external forms. In anatomy proper observations are restricted to internal structures. Neither of these classifications constitute the true natural method. I made them go together in the conviction that the study of the interior and exterior of an animal should belong to the same science [cited in Daudin 1926a: 224- 225, translation by the authors].

G   A Method of Correlation

When Cuvier, at the end of the eighteenth century, proposed a new approach to the classification of animals, seeking to justify classes and orders with reference to the mutual dependencies of the animal economy, he evidently moved in a direction that had already been hinted at by some distinguished naturalists. Like Daubenton and Vicq-d'Azyr, Cuvier took it for granted that in classifying animals one had to consider both external and internal parts. Thus, aiming at a natural system in zoology, Cuvier attempted to incorporate the points of view of both naturalists and anatomists in his project. As a naturalist, he wished to unite the multiplicity of beings into a rigid hierarchy of progressively more comprehensive taxonomic groups: as an anatomist he sought to define those groups in such a way that they would neatly indicate all the important features of animal organization (Daudin 1926b). We have seen how he succeeded: he based the most general groups of his system on those features of an animal which appeared to be highly correlated to the majority of other features and thus to the organization of the animal as a whole.

However, to fully appreciate Cuvier's success in solving difficult problems of coordination, it will not do to reduce his achievement to the force of one particular idea, that is, his conception of mutual dependencies as a characteristic feature of animal organization. Cuvier's innovation was not, in fact, a concern with the so-called animal economy, but rather the means or tools through which he managed to link this concern to the endeavors of eighteenth century naturalists. Seeking to align the observations of a naturalist with those of an anatomist, Cuvier enriched the already bountiful Paris Museum of Natural History with an impressive collection of anatomical preparations, which gave him access to all the parts of the animal body taken from a great variety of species. His collection, as Outram observes, was different from the older natural history collections, because it was full of objects "to be looked not at, but into" (Outram 1984: 176). It was a collection which allowed him not only to describe all the different parts of the animal body, but which also, and primarily, served him as a means to compare the various ways in which vital functions (like locomotion, sensation, respiration, etc.) were fulfilled in different groups of animals. In this way one could, as Cuvier put it in his famous *Leçons d'anatomie compareé*, exploit the experiments of nature, and see how in different groups of animals the vital functions are performed by diversely formed organs, and how in each group of animals the observer will indeed also find constant relationships between the form of one kind of organ and that of others.

Thus, by revealing particular correlations between parts, comparative anatomy became, for Cuvier, the pre-eminent method for clarifying the mutual dependencies of the animal economy. At the same time, with his method of correlation, Cuvier added a minor innovation to the tools of the naturalist, an innovation which allowed him to establish anatomical laws as well as a natural system. In other words, by means of his comparative anatomy collection and his method of correlation, Cuvier succeeded in rendering visible all the features of the animal body without betraying the rule of parsimony in the definitions of a taxonomic system, thus combining practices and conceptions from anatomy and natural history into one durable whole.

3 Coordination in Modern Genomics

The international human genome project may be seen as a self-conscious effort to transform both biology and medicine. One of the early advocates of the project, Walter Gilbert, envisions a dramatic "paradigm shift." With all the genes catalogued in biomolecular databases, biology will increasingly become a theoretical science, and scientists will cease to attack problems in a solely experimental manner (G). A new computational biology will emerge, as computers search for patterns in huge collections of map and sequence data, comparing the molecular anatomy of different species, and tracing the evolution of biological mechanisms. The genome project is also intended to contribute to changes in the practice of medicine that will reach well beyond advances in the management of disorders now classified as genetic. The genome project is intended to enable these transformations by providing an "infrastructure" of map and sequence data; by improving mapping and sequencing technology; and by creating the data, theory, and techniques required to build a new, molecular medicine (Caskey 1992; Hood 1992).

Mapping and sequencing the complete human genome is a formidable task by any standard. The human genome consists of some three billion nucleic acid base pairs. The genome can be compared to a text, written in a four-letter alphabet (corresponding to the four nucleic acid bases). In printed form, this text would occupy about a million pages (K). To sequence an entire genome means to determine all of the "letters" in the text and lay them out in order. To map a genome means to locate identifiable landmarks throughout the genome to use as points of reference. There are a number of kinds of maps based on different kinds of landmarks. Physical maps, which show the spatial relations among landmarks, are of several types, such as restriction maps, contig maps, and radiation hybrid maps.[2] Another kind of map — the genetic linkage map — is based not on spatial relations but on patterns of genetic inheritance; these maps measure the frequency with which two landmarks are co-inherited. As in traditional cartography, genome maps can be constructed at varying levels of resolution, with resolution defined as the number of landmarks per unit of mapped territory.

The creation of comprehensive genetic and physical maps was among the central goals of the first five years of the genome project (J). This task required building on extant practices in molecular genetics and confronting new problems of coordination that emerged from the goal of comprehensive mapping and sequencing. In this respect, it is interesting to compare the means used to achieve coordination in modern genomics with the means used in early taxonomy. As we will see, modern genome science — however different from natural history in terms of its objects, means of investigation, and organization — depends crucially, like natural history, on the practices of collecting, representing, and standardizing for its coordinated development.

A   Collecting, Representing, and Standardizing

Efforts to construct well-defined collections of specimens have been a salient feature of biological science since the early days of taxonomy. Thus, before the human genome project took shape, researchers in molecular genetics and related fields had established a variety of collections that later became relevant — and even essential — to the human genome project. Some of these collections were of modest scale. Many human geneticists, for example, gathered samples from patients with genetic diseases. In addition, however, resource centers emerged to serve broad groups of molecular biologists. These housed various types of collections: collections of DNA samples from large, multi-generational human families for use in genetic linkage studies (E); stock centers with well-characterized, inbred strains of model organisms from which to extract DNA; repositories of cloned DNA fragments; "libraries" of DNA from each human chromosome; and computerized databases containing biomolecular data, such as map and sequence information (Colwell 1989).

Whatever the form of these diverse collections, they invariably enable researchers in molecular genetics to work on shared problems (Fujimura 1992; see also Clarke and Fujimura 1992). For example, collections of well-defined, genetically identical strains of experimental organisms permit biologists to conduct research on the same experimental objects. Thus, networks of researchers studying a particular organism, such as the mouse, the fruit fly, or the worm, *C. elegans*, have created stock centers from which researchers throughout the world can obtain experimental materials. In the case of human genetics, ethical and practical considerations preclude creating inbred strains or maintaining collections of living organisms. Hence, human geneticists have established collections of DNA samples that may serve as shared research materials. During the past decade, DNA samples from large, multi-generational human families have been an important resource for creating genetic maps that describe patterns of inheritance. Researchers from around the world can study DNA from the same families, and the resulting data can be compared and aggregated because they are based on shared materials. Similarly, publicly-available collections of clones have facilitated research by producing results that can be easily compared among laboratories. Generally speaking, these collections expanded rapidly during the 1980s and the genome project was expected to accelerate this expansion.

During the 1970s and 1980s, molecular biologists also developed an increasingly impressive collection of laboratory methods and, ultimately, automated machines for representing and manipulating DNA molecules (Judson 1992). At the heart of molecular biology are recombinant DNA and cloning techniques, an ensemble of practices through which DNA molecules of interest can be cut, reassembled in different forms, inserted into bacteria or yeast, and replicated. DNA that has been tagged with radioactive or fluorescent labels can be used to detect the presence or absence of specific sequences in unanalyzed samples. The technique of DNA sequencing enables researchers to observe the nucleotide sequences of DNA molecules and represent them as written inscriptions — strings of the letters A, C, G, and T. Oligonucleotide synthesis, the inverse of sequencing, permits scientists to take an inscription representing a short DNA sequence (e.g., AAGTCCATGCCAATGGG) and synthesize the molecule (H).

The full repertoire of techniques, which can barely be hinted at here, provides a set of tools for building and taking apart molecules and for reading and writing the "language" of DNA. As the new molecular biology techniques have spread, the means to represent DNA molecules in terms of their sequences have become available to many laboratories, thus facilitating communication and accelerating research. Moreover, the power of these representational practices in coordinating research worldwide has been greatly enhanced by the development of electronic databases containing a rapidly increasing amount of map and sequence information. During the 1980s, the number of entries in sequence databases grew exponentially (Colwell 1989). These databases are not only a form of electronic publishing that can distribute large volumes of information rapidly (A; Hilgartner 1995b) they also permit a variety of mathematical analysis of sequence data and automated searches that can be used to compare new data from the laboratory bench with the database of known sequences (Bell and Marr 1990).[3]

The foregoing implies that, since the emergence of molecular genetics, the ability to create reproducible representations of DNA molecules has depended on an ensemble of standardized practices. Indeed, practices of standardization are deeply embedded in the history of molecular genetics and genomics.[4] Throughout the century, the use of standardized organism strains has been an important feature of genetics, as Kohler (1994) has shown nicely in his history of the early Drosophilists (see also Clarke and Fujimura 1992). As the methods of molecular biology transformed genetics, the use of standard protocols has been a means for making research "do-able" and for coordinating research programs among laboratories (Fujimura 1987). The emergence of repositories of publicly-available clones, computerized map and sequence databases, and a repertoire of powerful, reproducible, and standardized techniques provided a material infrastructure that enabled many laboratories to conduct and coordinate the mapping and sequencing of genes. Moreover, this infrastructure — in conjunction with shared expectations for its continued growth — provided support for the vision of a comprehensive genome project and a new kind of biological science.

B   New Problems of Coordination

In the second half of the 1980s, when the idea of mapping and sequencing the complete human genome was being actively discussed in the world of science policy (Cook-Degan 1994) the proponents of the project recognized that it would entail new problems of coordination. Many of these problems were foreseen and discussed in a report by the United States National Research Council (K) that was also influential in winning government support for the American genome project. The novel problems of coordination have a variety of interconnected sources: the scale of the task, the distributed nature of work, the problems of technological uncertainty and diversity, the recalcitrance of research materials, the problems of success, and the extant culture of molecular genetics.

We have already noted the scale of the task of mapping and sequencing the entire genomes of complex organisms, such as the human or mouse. Wholesale mapping and sequencing creates a vast array of mapping landmarks, sequence data, and other entities that have to be catalogued and managed. Indeed, the sheer volume of work in large-scale mapping and sequencing can only be accomplished through a well-coordinated effort. But even given the large scale of the task, the work is unlikely to be housed in one or two enormous facilities. For one thing, nationalistic considerations have discouraged the creation of monolithic centers. Moreover, genome mapping and sequencing is not (at present or in the foreseeable future) dependent on extremely expensive or unique instrumentation, but relies on bench top technology.[5] In the case of mapping, there is no technological or economic pressure to organize work around very large facilities that employ hundreds of scientists, as is typical of e.g., high-energy physics or astronomy. Sequencing may ultimately be performed in large facilities in order to take advantage of economies of scale. However, in the early 1990s, research has been conducted at many laboratories distributed throughout the world, most of which have at most a few dozen employees (a few are somewhat larger). As a result, many geographically distributed researchers must contribute to a sustained effort to reach a long-term goal. These researchers are funded by various nations or government agencies. In other words, there is no single central management presiding over a single facility, but instead a network of laboratories coordinated, sometimes loosely, by various funding agencies, advisory committees, and a leadership that is not integrated into a single, well-defined hierarchy.

With the technology of the late 1980s, sequencing the human genome would be prohibitively expensive, so the genome project was predicated on significant improvements in technology for mapping and sequencing. In addition, there was limited experience in creating maps of entire genomes, and it was unclear how large-scale mapping could be efficiently performed. In this context, laboratories experimented with diverse technological strategies, and techniques evolved rapidly during the first years of the project. From the point of view of getting the work done, the brisk pace of technological change was encouraging, because without major improvements the project could simply not be completed. At the same time, however, technological diversity and uncertainty contributed to coordination problems by complicating planning and making it necessary to link results produced by diverse means.

Moreover, although the techniques of molecular genetics are powerful, much mundane work is required to overcome the recalcitrance of the research materials used to produce maps and sequence data. Hybrid cell lines may spontaneously lose the DNA of interest; clones can be subject to rearrangements; experimental outcomes can be ambiguous; signals may be clouded by noise. In addition, the information provided by different techniques may not be readily combinable. In genome mapping, for example, landmarks produced by different means may not be directly comparable. Physical maps, based on spatial relations, cannot be mathematically related to genetic maps, based on patterns of inheritance. To complicate things further, the different kinds of physical maps — which use different landmarks based on different kinds of information — cannot be tied together by a simple formula. In short, the task of coordinating the genome project entails handling a variety of messy materials and incomplete or ambiguous information.

Ironically, novel problems of coordination were also expected to arise from future success in solving such problems. During the 1980s, the exponential growth of the rate of sequencing placed significant demands on the computerized databases that housed sequence information, and informatics specialists expected to encounter a continuing flood of new data as a result of the genome project (Hilgartner 1995b; A). In addition, mapping efforts were expected to identify many new genes and mapping landmarks, raising issues of nomenclature: how to assign these items unique and usable names? If this process was to be handled by nomenclature committees, for example, how could an exponentially increasing volume be processed without causing backlogs that would propagate throughout the research system?

An analogous problem of success was expected to surround collections of critical biological materials. Prior to the start of the genome project, all mapping landmarks depended on clones. Accordingly, if successful, the genome project was expected to require the construction of a repository, capable of archiving and distributing hundreds of thousands of clones (K). Scientists involved in planning the American genome project were concerned about the practical problems and financial burdens associated with building and running such facilities. To protect clones from contamination and to detect problems on that score, such a repository would need an extensive quality control program. A major mail-order facility would also have to be set up to ship materials on request (see Figure 3). One estimate of the cost of operating such a repository was $250 million over 30 years (K).

Finally, a number of cultural features of molecular genetics contribute to the complexity of coordinating the genome effort. For one thing, a long-term infrastructure project of the scale of the genome project is unprecedented in late twentieth century biology, which has typically been based on research conducted by small, independent groups. As a consequence, the genome effort was initiated in the absence of established organizational mechanisms for managing such a project. A second salient cultural feature of contemporary genetics is its extremely competitive nature, which at times conflicts with the need for collaboration and sharing of research materials and data in a long-term project of this type (Hilgartner and Brandt-Rauf 1994; Hilgartner forthcoming).

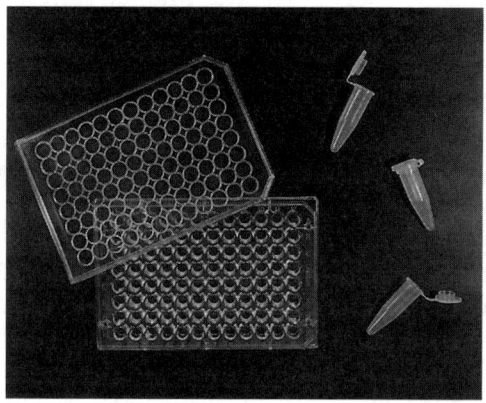

Individual clones are often shipped in small plastic tubes. Larger arrays of clones are frequently shipped in 96-well microtiter plates. In either case, clones must be carefully handled and packed for mailing.

Figure 3: Micro-centrifuge Tubes and Microtiter Plates Used to Ship Clones

C Sequence-tagged Sites as a Means of Coordination

As genome researchers have sought to cope with the new requirements for coordination that we have described above, have they identified opportunities for novel, and potentially more powerful, methods of coordination? Even at this very early stage in the history of genomics, this question can be answered in the affirmative. In doing so, one might point to a number of features of contemporary genomics, such as the computerization of laboratories or the integration of robotics and automation into laboratory practice. However, we see a particular socio-technical artifact — the STS, or sequence-tagged site as a particularly salient new means of alignment.[6] The idea of an STS was first fully articulated at a 1989 meeting to develop the first five-year plan for the American genome program. The concept was described publicly a few weeks later in a brief position paper in *Science*, now known informally as the STS proposal (I).

What is an STS? According to proponents, STSs are a new kind of landmark that can solve one of the critical problems facing genome mappers; namely, "the inability to compare the results of one mapping method directly with those of another and to combine maps constructed by two different techniques into a single map" (J). Prior to the advent of STSs, restriction maps, contig maps, and genetic maps could not be neatly integrated with one another or with sequence data. STSs, however, offered a way to join data from "any of a variety of physical mapping techniques," reporting it in a "common language" (J).

The STS proposal called for translating "all types of mapping landmarks into the common language of STSs" (I: 1434). Each mapping landmark would be examined to find a sequence-tagged site, or STS — a short DNA sequence that only occurs once in the genome and that, by virtue of its uniqueness, can be distinguished from all other sequences. A site on the genome that contained an STS could thus be tagged by a unique identifier. STSs thus transformed the ultimate goal of physical mapping, redefining the end product as an STS map that would display the locations of a set of sequence-tagged sites. Such a map would tie together maps of all other types, thus becoming the centerpiece of the mapping effort (I).

What, in the eyes of their proponents, accounts for the ability of STSs to become the centerpiece of an integrated genome map? The key reason: STSs offered means for representing mapping landmarks wholly as written inscriptions, (Latour and Woolgar 1979; Latour 1987) without depending on biological materials. Earlier mapping landmarks were all based on clones. For example, each landmark in a map of a human chromosome would consist of an identifiable piece of human DNA, spliced into bacteria or yeast. One could not determine whether a particular landmark existed in a DNA sample without using the clone containing the landmark to perform a test. Thus, to use a clone-based map, it was not sufficient to have a copy of the written map; one also needed copies of the clones used to create it. In contrast, STSs are based on the polymerase chain reaction (PCR), and they do not require cloning. In fact, STSs can be expressed entirely as sequence information, which can be readily distributed in printed or electronic form.

For example, an STS on human chromosome 4 is fully described with the sequences of two PCR primers, AGTTCGGGAGTAAAATCTTG and GCTCTATAGGAGGCCCTGAG. To test a DNA sample for this STS, all one needs is this information and some commercially available PCR equipment. As the STS proposal argued:

> The overwhelming advantage of STSs over mapping landmarks defined in other ways is that the means of testing for the presence of a particular STS can be completely described as information in a database. No access to the biological materials that led to the definition or mapping of an STS is required by a scientist wishing to assay a DNA sample for its presence (I: 1434).

The authors of the STS proposal saw STSs as much more than a means for manipulating data; they also saw them as a technology of coordination. STSs could simplify the problem of coordinating a research effort taking place in dispersed laboratories. STSs also could contribute to coordination across time — an important consideration given the pace of technical change in genomic technology. Thus, the common language of STSs was expected to provide a means for tying together not only *data* of diverse forms, but also to tie together diverse and dispersed *laboratories*. The STS proposal suggested that these landmarks could facilitate the smooth evolution of a physical map with inputs from many laboratories and methodologies, a process that would allow the dichotomy between big and small laboratories to disappear. The STS proposal also called for international discussions to "maximize the likelihood that the new common language would cross national boundaries" (I). STSs thus were self-consciously perceived, in Star and Griesemer's (1989) terminology, as a "boundary object:" a tool for tying together diverse laboratories — "small" labs and "big" ones, genetic mapping and physical mapping labs, restriction mapping labs and contig mapping labs, the labs of 1989 and the labs of the future, American labs and labs in Europe and Asia. STSs would allow a diversity of scientific strategies — including approaches based on techniques not yet available — to proceed together toward a common goal (Hilgartner 1995a: 308).

D A Politics of Coordination

The American five-year plan established program goals that, in effect, required mapping groups to construct STS maps, and (as one might expect given this directive), the new landmarks were rapidly incorporated into the sociotechnical order of genome mapping. The leadership of the American program saw STSs as a way to address a variety of problems. For one thing, because STSs are expressed as inscriptions, they provided a way to make

2   Means of Coordination in Making Biological Science   63

map data extremely mobile. Because STSs could be published in journal articles or biomolecular databases, they could prevent mapping labs from being inundated with requests for clones, eliminating the need for technicians to pack clones on ice and dispatch them by air courier. More significantly, STSs promised to obviate the $250 million clone repository that the 1988 National Resource Council report had envisioned. An STS database — accessible via the Internet — could replace the giant repository with its banks of freezers, its army of technicians and robots, its mail order house, and its quality control program (Figure 4).

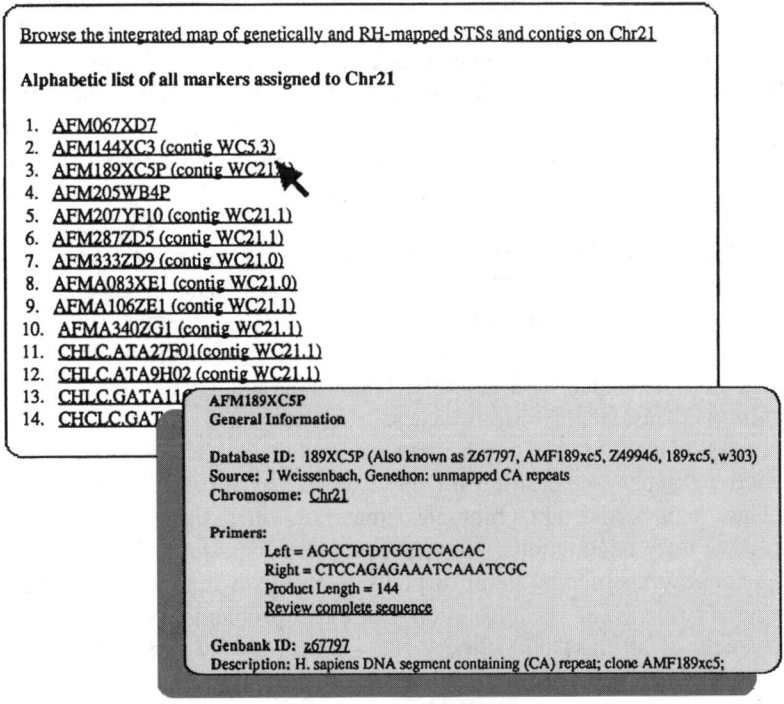

An STS database can be searched in various ways, including by the standardized names of markers, their genomic location, or their sequences. The example above shows the first few entries in a list (alphabetized by name) of several hundred STSs from human chromosome 21. Clicking on item 3, an STS known as AFM189XC5P, brings up the PCR primers and provides hypertext links to additional information about the marker and to other relevant databases, such as Genbank, a collection of nucleic acid sequences.

Source: Adapted from the database of the Whitehead Institute/MIT Center for Genome Research, 1996, http://www-genome.wi.mit.edu/.

Figure 4:   Sample Pages from an STS Database Accessible via the World Wide Web

Beyond this, STSs were seen as a tool for managing science. As the notion of a common language suggests, STSs were seen as a way to enhance coordination by facilitating cooperation among laboratories. But the new landmarks also offered a means for the American funding agencies to hold laboratories accountable. Thus, STSs not only facilitated communication among the nodes of the research network; they also tightened the link between the center and the periphery (Hilgartner 1995a). Moreover, the comparability of STS maps produced by different methods made it possible to identify discrepancies and detect the sources of errors, hence improving the accuracy of data. STSs also enhanced accountability by providing a common metric for measuring a laboratory's progress. No matter what technology was used, labs could be compared by counting the number of STSs they produced. The common language of STSs, as one mapper explained in a 1991 interview, provided funding agencies with the means "to hold people's feet to the fire."

STSs also provided a means of addressing the politics of access to genome data. In the genome project, funding agencies have sought to encourage rapid publication and to prevent researchers from monopolizing valuable data and resources. Thus, questions of access to maps, sequence data, and biological materials have been recurrent and prominent concerns (K; Hilgartner forthcoming). Tensions about access to data have emerged repeatedly in the history of human genetics, and there has been considerable variation in the ways researchers manage the disposition of the evolving assemblages of data under their control (Hilgartner and Brandt-Rauf 1994, forthcoming). Consider the case of access to clones. As a condition of publication, most major journals require authors to make public any clones used to produce the paper, but waiting for publication may result in a significant delay. Thus, clones and other biological materials often figure prominently in inter-laboratory negotiations. Also, at times, the departure of lab personnel has been accompanied by bitter disputes about the ownership of clones.

Proponents saw STSs as a solution for these problems. Since an STS consists solely of an inscription, there is no conceivable way to restrict access to an STS once it has been published. By insisting that grantees use STSs, the funding agencies could tighten data-sharing requirements and move somewhat closer to a world of "free access" to map data.[7] STSs should thus be understood as a tool for enhancing coordination by shaping data access practices within the field. Consequently, the new landmarks demonstrate how social control in science can be achieved through means that build accountability, directly and materially, into the very fabric of scientific production.

In the brief account above, we have seen how STSs functioned to enhance coordination on multiple levels. STSs simultaneously linked data and laboratories. They tied together not only pieces of DNA, but also mapping techniques. They permitted scientists to perform tests on DNA samples and also to "test" the performance of research teams. STS made it easier to share data and easier to compel sharing (Hilgartner 1995a). In a number of interacting ways, this single sociotechnical artifact provided new means for increasing the mobility, stability, and the combinability of data (Latour 1987); for enhancing control over a dispersed network of laboratories; and for integrating the many entities that constitute the world of genomics into a more cohesive whole.

4 Discussion

We have examined efforts at coordination in two very different contexts: the early history of taxonomy and the recent development of genomics. What can be learned about the different roles of means of coordinating by comparing these cases — especially in the light of different theoretical perspectives on coordination? Before examining this question, we first need to say more about the positions of Latour, Star and Griesemer, and Fujimura in the debate about means of coordination mentioned in the introduction.

The Latourian perspective (Latour 1987; Callon et al. 1986) focuses on the question of how scientists build "black boxes" — that is, stable facts and artifacts — and how they extend them into the world. Here, coordination is understood as enrollment. Heterogeneous means enable successful network builders to create stable facts and dependable machines, thus permitting them to make themselves indispensable, to compel assent, to extend their networks, to position themselves at the center of centers of calculation — in short, to dominate others.

Star and Griesemer offer a different frame for understanding coordination. Advocating a social world perspective, these authors direct their analytic attention toward the problem of bringing diverse social worlds where scientific work is conducted into sufficient alignment to permit them to perform collective work. The role of heterogeneous means in this process is expressed in their concept of boundary objects — entities that inhabit several intersecting social worlds and that are "both plastic enough to adapt to local needs and the constraints of the several parties employing them, yet robust enough to maintain a common identity across sites" (Star and Griesemer 1989: 393). In their case study of the Berkeley Museum of Vertebrate Zoology, Star and Griesemer provide a number of examples of boundary objects, including repositories (i.e. organized collections), ideal typical concepts that are vague enough to be adapted to local contingencies, and standardized forms. The focus on collective work and its requirements for interfacing among multiple worlds produces a very different image of coordination than the image of enrollment — one that is built around flexibility and mutual interaction, rather than domination.

Fujimura (1992) offers yet another take on the problem of coordination. Her approach shares features of both perspectives just described. On the one hand, she shares Star and Griesemer's interest in explaining the processes that permit the performance of collective work across social worlds. She also notes that the Latourian focus on enrollment directs attention away from the question of how collective work involving multiple social worlds is possible. On the other hand, Fujimura does not want to lose the ability to account for the stabilization of facts, something she suggests a model of coordination based on boundary objects is liable to. Boundary objects, she argues, are too malleable to account for fact stabilization. As a solution to this problem, she introduces the concept of a standardized package which combines several boundary objects with standardized methods in ways which "narrow the range of possible actions and practices, but also do not entirely define them" (Fujimura 1992: 176).

Even the brief discussion above suggests something of the complexity of the debate among different perspectives on the coordinating role of various means. In part, the debate reflects the fundamental ontological differences between the actor-network theory of Latour and the social worlds theory that underlies the work of both Fujimura and Star and Griesemer. More significantly, the debate reflects a set of questions about how the politics of coordination should be understood. To what extent can the coordinating role of heterogeneous means be understood through a conflict model that focusses on domination versus a cooperative model of interfaces and collective work? Another interesting issue concerns the extent to which coordination emerges from interactions that expand outward from a center, versus mutual interactions among the many nodes of decentralized networks (or worlds). Clearly, these extreme images — of domination versus cooperation, of outward expansion versus node-to-node interaction — all offer incomplete views of a picture that is rather more complex.

Together, these different perspectives on coordination may help us to highlight some interesting similarities between the two mapping efforts in biological science and also allow us to explore the diverse and complex ways in which means of coordination have contributed to processes of coordination in both these cases. As we have seen, early taxonomy and contemporary genomics can both be understood as Latourian efforts to construct new centers of calculation that depend on — and indeed embody — new means of coordination. Taxonomists and genome scientists both created a variety of new tools that enabled coordination. In each case, these tools included new collections, new representational practices, and new techniques of standardization that together enhanced the mobility, stability and combinability of data and thus formed the basis for coordination on a new scale. Indeed, for all of their differences, both the natural history cabinet and the computerized biomolecular database can been seen as a novel type of collection that facilitates rapid comparisons of a wide array of natural and experimental objects. Similarly, in both taxonomy and genetics, the ability to produce standardized descriptions of objects was of critical importance to the coordination of work — whether we are talking about the standardized descriptions of naturalists like Brisson and Linneaus or the common language of STSs.

However, the cases of taxonomy and genomics not only demonstrate the importance for coordination of such means as collections, representational practices, and standardization in a way that is consistent with a Latourian actor–network perspective. The two cases also highlight the importance of ongoing center–periphery reconfigurations. Brisson's connection to Réaumur's natural history cabinet, for example, gave him immediate access to a large bird collection and put him at the center of a network of correspondents which contributed to the growth of the collection and thus reinforced the centrality of his position. In this context, Latour's concept of a center of calculation dominating both an accumulating volume of data and other actors in the field, does indeed provide a useful image of how coordination may occur. On the other hand, through his standardized descriptions and illustrations, Brisson introduced a means that — owing to printing technology — could de-localize the collection, distributing a representation of it to the many nodes of a taxonomy network and thus facilitating cross-nodal communication and mutual accomodation. Here, the social world perspective of authors like Star and Griesemer, Fujimura, and Clarke offers a more appropriate understanding of coordination than Latour's model of domination.

Dynamic interactions between local centers and distributed nodes are also evident in the story of STSs, as becomes clear from the different meanings that may be, and actually have been, assigned to this tool. Like the standardized descriptions of Brisson and Linneaus in natural history, the common language of STSs can be seen as a way to facilitate interaction among the nodes of a network (of laboratories). On the other hand, STSs were also seen as a means of heightening the accountability of geographically distributed mapping labs to a center responsible for managing coordination (Washington-based funding agencies). Moreover, they were seen as a means of eliminating the construction of a central collection of clones. The same artifact, therefore, can be seen as a way of facilitating node-to-node communication, of linking nodes to a center, and of obviating construction of a center. Together the two cases suggest that means can play an important role in shaping dynamic interactions among centers and peripheries that instantiate coordination.

There is another theme emerging from the two cases which is important for our understanding of coordination in science, that is, the temporal dynamics of coordination. In both taxonomy and genomics, researchers faced what one might call the problems of success. In natural history, voyages of exploration, the practice of printing, as well as technical advances in preservation and illustration, offered means to compile plants, animals, descriptions and illustrations on an unprecedented scale, which in turn made problems of classification more salient. A broadly parallel story can be told in the case of genomics, as expanding biomolecular databases and growing collections of clones posed new coordination problems that stemmed in large part from the scale of success. These cases suggest that coordination can be understood not as something that is achieved once and for all, but as a series of ongoing adjustments for coping with evolving problems of order.

These observations about the ways in which means of coordination can destabilize and reconfigure center and periphery and about the temporal dynamics of coordination, help to draw together two messages about the relationship between means and coordination that have appeared in the above discussion. On the one hand, we have seen how means allow for more extensive interaction among researchers and thus create the need for new, large-scale forms of coordination. On the other hand, we have seen that means themselves embody forms of coordination; that is, coordination can be secured by the use of particular means, such as standardized descriptions or standardized genome mapping markers, like STS's. Viewed from a static perspective, these images would appear to present a paradox: in the first case, novel means cause new problems of coordination; in the second, case, novel means solve problems of coordination. The paradox disappears, however, when one adopts a view that recognizes the temporal dynamics of coordination and the potential of means of coordination to destabilize the global and local.

Notes

[1] The notion of centres of calculation we take from Latour (1987). For a much more detailed Latourian analysis of the mapping effort in early taxonomy see Stemerding (1991).

[2] Restriction maps display sites on the genome where specific enzymes will cut the DNA. Contig maps link clones that partially overlap one another, into a contiguous array. Radiation hybrids are constructed by breaking chromosomes with ionizing radiation; analysis of breakage patterns permits construction of a map.

³ Automation of genomic technology, however, has not only transformed the powers of representation. One may argue that it has re-engineered the division of labor by reconfiguring tasks once performed by humans to fit the "skills" of machines, and by re-engineering tasks that were once performed by individuals so that they can now be performed by organizations.
⁴ On the limits of standardization, see Jordan and Lynch (1992).
⁵ For an early discussion of this issue, see Heilbron and Kevles (1988).
⁶ For a more detailed discussion of STS's, see Hilgartner (1995a), which we draw on here. See also Hilgartner (1995b).
⁷ STSs have not, in the event, permanently solved the problem of data access. On the contrary, data access will be an area of ongoing tension and negotiation as the genome project proceeds.

Sources

A. Cinkosky, M.J., J.W. Fickett, P. Gilna, and C. Burks (1991): "Electronic data publishing and GenBank," *Science*, 252: 1273-1277.
B. Cuvier, G. (1795): "Mémoire sur la structure interne et externe, et sur les affinités des animaux auxquels on a donné le nom de Vers," *Décade philosophique*, 5: 386.
C. Cuvier, G. (1800): *Leçons d'anatomie comparée*, C. Duméril (ed.). Paris: Baudouin: I, 62.
D. Cuvier, G. (1817): *Le règne animal distribué d'après son organisation pour servir de base à l'histoire naturelle des animaux et d'introduction à l'anatomie comparée.* Paris: Déterville, 4 volumes.
E. Dausset, J. (1986): "Le centre d'etude du polymorphisme humain," *Presse Med*, 15: 1801-2.
F. Geoffroy Saint-Hilaire, E., G. Cuvier (1795): "Mémoire sur une nouvelle division des mammifères et sur les principes qui doivent servir de base dans cette sorte de travail," *Magasin Encyclopédique*, 2: 164-190.
G. Gilbert, W. (1991): "Toward a paradigm shift in biology," *Nature*, 349: 99.
H. Horvatch, S.J., J.F. Firca, T. Hunkapiller, M.W. Hunkapiller, and L. Hood (1987): "An automated DNA synthesizer employing deoxynucleoside 3'-phosphoramidites," *Methods in Enzymology*, 154: 314-26.
I. Olson, M., L. Hood, C. Cantor, and D. Botstein (1989): "A common language for physical mapping of the human genome," *Science*, 245: 1434-5.
J. United States National Institutes of Health & DOE (1990): *Understanding our genetic inheritance. The U.S. human genome project: The first five years FY 1991-1995.* Springfield VA: National Technical Information Service. U.S. Department of Commerce.
K. United States National Research Council (1988): *Mapping and sequencing the human genome.* Washington: National Academy Press.

Chapter 3
Coordination in Military Socio-technical Networks: Military Needs, Requirements and Guiding Principles

Wim A. Smit, Boelie Elzen, and *Bert Enserink*

1 Introduction

After World War II, for the first time in history during a period of relative peace, weapon innovation and its accompanying military R&D became a large-scale institutionalized process. It did so, in fact, on a scale previously unseen, and was legitimized as well as fueled by the climate of the Cold War. In the decades following the war, weapons were replaced in a rapid process of planned obsolescence. The necessary R&D was carried out at national laboratories in the defense industry, at laboratories of the military services and (in the U.S. at least) at many universities. It has been estimated that in the late 1980s some $100 billion were being spent annually on military R&D worldwide, involving some 600,000 scientists and engineers (Sen and Deger 1990). In a 1981 United Nations study, it was estimated that at least 20% to 25% of all R&D was devoted to military purposes and that a similar percentage of all scientists and engineers were engaged in military R&D (quoted in Thee, 1988).

After the collapse of the Soviet Union, the legitimizing role of the Cold War vanished. But whereas the budgets for procurement of weapons have indeed diminished substantially (though not in all countries, for instance not in France) this has not been the case for military R&D. Here reductions have been modest at best. The reason is that the continuous development of advanced weaponry is still considered of vital importance. Now, however, the legitimation of the Cold War has been replaced by the notion of risks — risks that have to be faced due to an uncertain international situation.

One major legitimation, if not a driving force, for a vigorous military technological innovation process, is the general conviction that "technology is our strong suit and must be vigorously pursued both for commercial and for national security purposes" (Ruina 1988). Long and Reppy have also referred to this long-standing tradition in the U.S.:

The Department of Defense is committed to a United States position of overall technological superiority in military technology; technical leadership is regarded as being essential in itself as well as being necessary to preclude technological surprise. Moreover, advanced technology is considered an area of comparative advantage for the United States, making it possible to rely on smaller numbers of superior weapons (Long and Reppy 1980b:4).

Even if a particular military R&D program does not result in a weapon or military application, it often is still considered as a necessary safeguard against unanticipated technological innovations by the adversary.

Several former participants in the armaments innovation process have described its driving force as the push of ever new technologies emanating from military laboratories and the defense industry. For instance, Lord Zuckerman, recalling his experience as Chief Scientific Advisor to the British Minister of Defense, has expressed as his view:

[...] that the basic reason for the irrationality of the whole [arms race] process is the fact that ideas for a new weapon system derive in the first place, not from the military, but from different groups of scientists and technologists [...] At base the momentum of the arms race is undoubtedly fueled by the technicians in governmental laboratories and in the industries which produce armaments (Zuckerman 1982: 103).

Herbert York, former Director of U.S. Defense Research and Engineering, has stated that "the technological side of the arms race has a life of its own, almost independent of policy and politics" (York 1971: 180). Allison and Morris, in a classic article, have emphasized the influence of the domestic scientific-technical community and of industrial and bureaucratic forces in the weapon innovation process:

We observe that each nation pursues its weapons strategy through large organizations for research, development and use. What drives these institutions are not only estimates of the other side's activity, but also their own internal dynamics (Allison and Morris 1975: 118).

These observations suggest that designing new weapon systems is an autonomous process, in which the military laboratories do not depend on others and do not orient themselves to an outside world. But this picture is too simple. Actually all organizations involved in military technology are only autonomous to a certain degree: there is also a strong mutual dependency. President Eisenhower, in his farewell address, described this mutual dependency as a military-industrial complex (MIC) referring to the network of military laboratories, the military, politicians, scientists, engineers, bureaucrats, and defense industry. Military technology emerges from and is shaped by the interactions between these various actors; that is to say, military technological development is basically an inter-organizational process.

Many studies, in one way or another, have tried to cope with the rather amorphous image of the military–industrial complex. Only a few have tried to explore the connection between the shaping of military technology and social processes within the MIC. Exceptions are MacKenzie (1990) and Evangelista (1988). One of the social processes shown to be relevant is the formation of specific networks of organizations and institutions linked to the development of a particular weapon system.

In this chapter we will show that the development of military technology and weapon systems is actually a co-evolutionary process of both technology and networks of involved organizations which become tied together in what we call a socio–technical network. Particular attention will be paid to the role of military requirements as a vehicle for coordinating activities during the formation of such a socio-technical network.

We will discuss the formation of the socio-technical network and the accompanying coordination processes in the case of the international collaboration on the development of a European Fighter Aircraft (EFA) in the 1980s. The case is an illustration of how actors and technology become tied together in such socio-technical networks. We will also show how the relevant coordination processes failed to include all of the five countries originally committed to the development of the EFA. Moreover, it will become evident that one of the problems turned out to be linking, i.e. coordinating, (pre-existing) socio-technical networks, rather than individual organizations.

2 Innovation Networks in Military Technology

Military technology and weapons development is distinct from most civilian technologies simply by virtue of the fact that the national government (as represented, for instance, by the Ministry of Defense and by the military services) is by far the major, if not the only, customer of weapons produced by the domestic defense industry. Moreover, weapons exports are subject to governmental regulations. These features are a consequence of the government's legal monopoly of the use of force and of the role of weapons in preventing or waging a war, in coercive actions, and, more recently, in United Nations peace–keeping and peace–enforcing operations.

The fact that the defense industry is producing not for an anonymous market, but for a specific customer, implies that weapon innovation and military R&D has to be attuned to the wishes of this customer. On the other hand, the military have to adapt both to technological limitations and to technological opportunities: they are dependent on the laboratories and defense industry for information on current technological possibilities. These technological possibilities are not an abstract given but are tied to the existing capabilities of military laboratories and the defense industry.

Because of their dominant market position, Ministries of Defense and the military Services are strongly involved in the weapon innovation process itself and even in military R&D, if only because they provide the funding. Still, industry does carry out so-called independent research. It is prepared to do so because most of the incurred R&D expenditures will be reimbursed by the government if the effort results in weapon systems that are actually procured. If the Ministry of Defense and the military did not, in one way or another, support such dedicated military R&D, the necessary investments might appear too risky to industry and, as a consequence, innovation might grind to a halt.

Whereas Ministries of Defense and the defense industry are formally autonomous, the exchange of funding of military R&D for information on new technological possibilities, and of procurement money for the production of real weapon systems shows that these actors are at the same time strongly mutually dependent. This interdependency requires that their actions in the weapon innovation process become attuned, that is, become coordinated to a certain extent.

The network type of relationships prevailing among organizations involved in the weapon innovation process has recently become more common in other areas as well, like industrial production and policy-making.[1] The main characteristics are that networks deal with interactions between relatively autonomous entities (often organizations, human actors, or even non-human actors in the actor-network approach of Latour (1987) and Callon (1986b), or the Techno–Economic Networks approach (Callon 1992)). These entities are interdependent as to resources and information — the main *intermediaries* that flow between the entities. The interactions typically have the characteristics of negotiating or bargaining: often, there is no central decision-making or power center (Van Waarden 1992), but rather a multitude of such centers. These features distinguish network relations from other types of relationships, like markets or hierarchies (Powell 1990). An additional feature is that in networks the interactions (or exchange relations) are durable over time and certainly not incidental. A major advantage of the network approach is that it recognizes the importance of structure, while also leaving room for individual initiative and strategic actions by corporate actors (Smit 1995).

What are the important mechanisms in the emergence, functioning and possible dissolution of various socio-technical networks? The main features of a socio-technical network are, first, the existence of a relatively stable pattern of interactions among the actors (individuals or organizations) and, second, their common involvement in some specific technological development — in this case the European Fighter Aircraft (EFA). The interactions among actors are characterized by the exchange of what Callon has called *intermediaries*: An intermediary is anything that passes from one actor to another, and which constitutes the form and the substance of the relation set up between them — scientific articles, software, technological artifacts, instruments, contracts, money, etc. (Callon 1992).

An actor receives a variety of intermediaries while sending out a variety of others. Actors can therefore be seen as processors of intermediaries. Actors recombine incoming intermediaries and send out other ones. What characterizes the actors is the nature of this recombination process. Some examples relevant to our case are:

- Engineers recombine, for instance, (scientific) papers, money, instruments, raw materials, artifacts, etcetera, into other artifacts, patent applications, etcetera.
- The military recombine, for instance, information, money, soldiers and artifacts into military strategy, training programs, organized units of soldiers, artifacts (weapons), etc.
- Politicians recombine, for instance, tax-payers' money and a host of information into money allocations for R&D, social security programs, etc.

A stabilized pattern of interaction is an important characteristic of socio-technical networks.[2] How can this stability be explained in view of the fact that, given their (relative) autonomy, network actors are principally free to make choices on how to interact, e.g., free to decide which intermediaries will be picked up and how these will be processed. The key is that a (socio-technical) network, through its *function*, provides a structure that both enables and constrains actors in carrying out their own strategies.[3] Van Waarden (1992) has specified function as one of the major dimensions of networks, adding that these functions depend on the needs, intentions, resources and strategies of the actors involved. Whereas network structures do not have goals of their own, actors do and, depending on these, the network acquires various functions. The meaning of the function of a network will thus vary among the network actors. The overlap in these different meanings may then be denoted as the function at the level of the network as a whole.[4] Van Waarden concludes that the concept of "function" forms the bridge between structure and actor perspectives on networks.

A necessary condition for a viable network is thus the presence of sufficient overlap in the various meanings that actors attribute to the function of the network. Network actors tend to act according to what is acceptable within the common interpretation of the function of the network, thus creating and reinforcing a certain pattern of interaction. Where network actors nonetheless act in ways that tend to jeopardize existing patterns of interaction, other actors in the network will often react so as to counter the destabilizing threat and reestablish the original order. In that sense socio-technical networks are characterized by a certain *resilience*.[5]

More specifically, the functioning of networks, including their resilience, can be derived from three basic rules governing the behavior of the individual actors within the network:

- Actors can only act on the availability of appropriate resources as embodied in intermediaries. This dependency upon incoming intermediaries directly constrains their room for manoeuvre. At the same time, network relations provide a stable flux of appropriate resources, i.e. in accordance with the function attributed to the network by the actor. This enabling feature of the network stimulates actors to build or maintain these relations.
- Even when constrained by incoming intermediaries, actors still have a wide variety of options on how to process them, depending on their particular interest. To be able to do that in a potentially successful way actors have to evaluate which new intermediaries will be picked up by which actors. They have their expectations about what will or will not sell and on this they base their strategy for which intermediaries to send out. Because of experience in past interactions, actors are able and tend to adapt their actions to what is considered relevant by other actors. This can be seen as a form of self-constraint by the actors. Thus within the network, actors are not just exchanging intermediaries, they are also adapting to each other in the interaction processes. This implies mutual orientation, which helps to strengthen bonds between actors (Johanson and Mattson 1987).
- Still, actors can be wrong in what they think (or hope) others will pick up. If the intermediaries an actor broadcasts are not picked up, the actor can put in extra effort to make that happen; nonetheless it will not be able to simply sell anything it wants. In this way the network indirectly limits an actor's room for manoeuvre. Thus, actors are corrected by the other actors in the network with the effect that the pattern of ongoing interactions is sustained.

At the socio-technical network level these three rules add up to the emergence of resilience, i.e. to a tendency for existing patterns of interaction to be perpetuated. They reflect a process of *longer-term* (socio-cognitive) coordination of activities by various actors, in contrast to occasional interaction.

Whereas socio-technical networks show resilience, empirical evidence also shows that socio-technical change can be quite drastic and may even lead to the complete disintegration of existing networks. This can be accounted for by looking at the interaction between a socio technical network and its environment. The characterization of socio-technical networks given above does not imply that interactions that are not included in the network are considered irrelevant. On the contrary, exogenous interactions must be taken into account in order to fully explain inter-network dynamics.

Exogenous interactions are an important source of instability for the network. Destabilization may occur, for instance, because actors are simultaneously part of more than one socio technical network and may therefore carry issues from one network to another. Being part of several networks, they may have a *low inclusion* (see Bijker 1987) in one or more of these networks.[6] In conflicting situations they may then give priority to acting in the context of one of the networks at the expense of one or more of the others. Strategies of actors with low inclusions may thus endanger the continuity of the network, in particular if they occupy a position that is particularly crucial to the functioning of the network. We refer to such actors as *critical actors*. Parliaments in European countries (or Congress in the U.S.) are examples, as will be illustrated in the empirical case of the EFA. Furthermore, there is always the possibility of change triggered by the incidental actions of actors outside of the socio-technical network.

Interactions with the environment are distinguished from the stabilized interactions within the network. These interactions with the environment can, however, develop into patternlike behaviors, thus making them part of network interactions. Eventually, completely new networks can develop in this way. This bit of dynamics is essential for the development of many new technologies where innovation goes hand in hand with the development of new configurations of actors, i.e. new socio-technical networks. Thus, on the one hand, existing socio-technical networks show a tendency to perpetuate themselves (i.e. resilience) while, on the other hand, their interactions with the environment are essential for understanding important parts of technological change, especially radical technological change.

Networks can emerge in various ways. They may emerge gradually in an unplanned way. Networks of industrial production, for example may arise through new combinations of several companies (Johanson and Mattson 1987). By contrast, we will show that in the case of international collaboration on military technology, where one often has to fight the pervasive influence of national networks, *dedicated network builders* have a crucial role. In the sphere of military technology, it is particularly governments who play that role, as they link up a heterogeneous set of actors, including the military services, defense companies and Ministries of Defense of various countries. In contrast to the civilian sector, the initiatives for such collaborative networks thus proceed from the customers rather than the producers of the technology. The EFA case also shows that in such an effort of dedicated network building, negotiation about the function of the network plays a major role. In that sense perceived function fulfills a coordinating role during the emergence of such a network.

A Needs, Requirements, and Guiding Principles as Coordination Mechanisms

Networks reflect a process of *longer-term* (socio-cognitive) coordination of activities by various actors. Often, in the literature on coordination, trust is considered as the coordinating principle within networks, like price within markets and rules within hierarchies (Thompson et al. 1991). Indeed, if relations continue and intermediaries are exchanged that fit the functions attributed to the network by the different actors, trust will likely increase and will facilitate further network activities.

However, our interest is not just in networks as a way of organizing inter-organizational interactions. For technology studies, networks are interesting because of their ability to produce technologies that serve the needs of heterogeneous sets of actors. This ability hinges on the coordination of the content of interactions. Within military technology development, there are two basic instruments for such socio-cognitive coordination — a formal and an informal one. The first are military needs and requirements. The second are guiding principles as paradigms for interaction and hence as means for governing the direction of the innovation process, at least to a considerable extent.

For a national government, military technological development is fraught with two basic uncertainties. First, it is impossible to foresee all the new military technologies that might conceivably emerge from the numerous national and defense industrial laboratories, let alone to control them. Second, given the international anarchy of autonomous national states, governments may at any time be confronted with new weapon developments in hostile countries.

Largely because of these uncertainties, national governments want to be militarily prepared. This at the very least requires a mutual attuning of military doctrines, military needs and military technological innovation. On the latter count, a very heterogeneous set of actors are involved (e.g., a great variety of defense companies, the Ministry of Defense, Military Services, Parliament, think tanks, lobbyists, and so on). It is out of the question that the required coordination could be a one way traffic of directives, for instance from the Ministry of Defense to the defense industry. Military technological innovation is basically an interactive process. Still, and precisely because of, its interactive nature, it can be governed by what has been called guiding principles (Smit 1989, 1991).

Guiding principle is an analytic concept that observers or analysts may use to describe and analyze processes of military technological developments. Two characteristics are of particular importance:

- The guiding principle is taken into account (shared) by various actors, who may have very different positions in the weapon innovation process (for instance politicians and the military, or R&D engineers and politicians). It thus plays an important role in the interaction process between those actors, and in that sense it is a way of conceiving *inter-organizational* coordination.
- Guiding principles are an interface between military doctrines or strategies and the actual shaping of weapon systems. For the various actors, guiding principles function as a heuristic for making choices with respect to military technology and as a touchstone for their evaluations of it.

Taking into account does *not* necessarily imply that the actors agree on the content of these principles or on their importance; the only claim is that in order to stay in business, the actors involved feel compelled to comply with them. For example, when in the early sixties U.S. Secretary of Defense McNamara introduced (system) flexibility as a major criterion for arms procurement decisions, aerospace engineers and the military in the Air Force staff started proposing multi-purpose flexible systems, even though, in this case, the Air Force preferred its traditional high speed, high altitude systems. Flexibility *in effect* then started to work as a guiding principle (Enserink et al. 1990). Guiding principle is not a rigid, but rather a *dynamic* concept: it has to be made operational again and again in view of the ever-changing (technological and political) circumstances. But in that flexible sense it invariably plays a role, implicitly or explicitly, in the interaction between the various actors involved in weapon innovation processes.[7] As an informal means of coordination, guiding principles govern the pattern and direction of military R&D through particular technological-strategic "eras." Their influence is, however, less easy to trace in one single technological project like the European Fighter Aircraft.

This contrasts with formal military requirements which are more explicit and more specific than guiding principles. Military requirements define new weapon systems in detail, either in terms of operational capabilities or in terms of technical specifications (e.g., the milspecs). Analysts often make a distinction between defense (operational) *needs* and military (operational) *requirements*. A RAND study gave the following description:

> Expressions of operational needs are supposed to describe a user's need for a capability to perform military tasks that cannot be satisfied with his existing and planned capabilities. [...] Operational requirements statements provide more details about the characteristics of the solution to the user's needs. To do so, those statements may refine, extend, or expand expressions of operational need. Thus they provide more details about the means to the end result desired (Stanley and Birkler 1986: 4–5).

In other words, military needs are expressions of missions to be performed, whereas military requirements translate these needs into (technical) specifications of weapon systems that are to fulfill these missions.[8] These specifications, subsequently, are the base for contractual requirements *vis à vis* industry. Program managers and acquisition decision makers use needs and more detailed expressions of requirements in the management and evaluation of weapon system acquisition programs (Stanley and Birkler 1986: 8-10). Military requirements play an important role in the interactions between the various actors involved in the weapon innovation process, for instance, between the military services and the defense industry, or between the Ministry of Defense and Parliament.

The distinction between needs and requirements is not only an analytical one but also one that is used in practice. The milestones decision structure of the United States weapon acquisition process starts with a Mission Need Statement, which identifies a need for a new technology or system, or an upgrade of an existing system, to meet current or anticipated threats (Hatchett and Keuter 1992). The same is true for many of the European NATO countries. Subsequent steps in the milestone process are formulated in terms of military requirements. Needs and requirements are thus complementary elements, together bridging the gap between doctrine and strategy on the one hand and technological hardware on the other.[9] Aside from a difference in emphasis, the main difference between them is in the actors that make use of them: requirements figure particularly in interactions between, for instance, the military services and the defense industry, whereas needs are more important in interactions between Ministries of Defense and the services, and in exchanges with Congress or Parliament.

From the perspective of military security it seems a logical sequence that first Mission Need Statements are formulated, which then, via operational requirements, are translated into desired weapon systems or technology and their specifications, thus constituting a linear process. If the required technology is not at hand this can lead to a demand-pull in military technological development: the military requirement encourages greater investment in a given area of research and development (Perry 1967: 72), of which there are indeed examples. But success is never guaranteed and such roads may lead to dead ends.

More often, military requirements are adapted to the state of art, as Perry (1967: 72) has observed in a case study on the variable-sweep aircraft: The eventual requirement was more directly influenced by the progress of research than by abstract notions of what was needed in the inventory. The requirement can best be stated once the technology is reasonably well in hand. This points to the alternative, in a sense the reverse, course in which new technological possibilities emerging from the laboratories are presented to the military or the Ministry of Defense, leading to new military requirements or needs.

However, like other actors, weapon engineers cannot act wholly independently of the other actors in the process. Rather than either a simple demand-pull or a technology-push mechanism, there is a much more subtle interplay between military needs and requirements on the one hand and military technological developments on the other. This becomes most evident in the interaction process among the various organizations involved. In these negotiations, military requirements serve, so to speak, as a vehicle for interaction. The fact that these requirements themselves appear not to be rigid but negotiable is an essential feature of this process. One place in the interaction process where this feature of negotiability becomes manifest, is in the *Request for Proposals* procedure. On the one hand, such requests are guided by Mission Need Statements; on the other hand, they offer the defense industry an opportunity to present their technological options.

In practice, as may be expected, the role of needs and requirements differs from this idealized description. They are in competition with other (economic/societal) factors and are subject to change. The case study of the European Fighter Aircraft (EFA) allows us to show the role of military requirements in the initial stages of the co-evolution of a weapon system and its socio-technical network. We explore the extent to which needs and requirements direct the innovation process and we show their level of flexibility. In so doing we will focus on two episodes in the overall story of the EFA: the birth of the European Fighter Aircraft and the choice of its radar system. We will stress the emergence and the features of the socio-technical network, the role played by needs and requirements, and the changes occurring in their context and content in the course of real-life innovation processes.

3 Case Study: The European Fighter Aircraft

Plans for European cooperation in building a new fighter aircraft have been in the making since the late 1970s. A number of factors worked as incentives:

- From the late 1970s onward, the notion that U.S. security interests differed in a number of respects from European security interests began to take hold.
- There was a growing desire to become less dependent on the U.S. in the field of arms production (A),[10] while economic competition between the U.S. and Western Europe fueled the will to become an independent competitive technological innovative power. Aerospace technology was considered a key technology in this respect.
- West European countries faced very high R&D costs relative to their comparatively small domestic markets.[11]

At the start of negotiations over EFA several socio-technical networks, coupled to two different projects, already existed. The European Fighter Aircraft project had precursors in the British-led Experimental Aircraft Program (EAP) and the French Avion de Combat Experimentale (ACX). The EAP followed from an initiative in 1979 by the British Aerospace company (together with the United Kingdom electronic companies Ferranti, GEC Avionics and Smiths Industries) to develop a new advanced technology aircraft.[12] EAP was announced at the Farnborough International Air Show in 1982. The French reacted within weeks, announcing their own national demonstrator program — ACX, later to be called Rafale — to be developed by the French aerospace company, Dassault. Prototypes of both aircrafts were to fly at Farnborough in 1986.

The West-German government, favoring one common European program for a new fighter aircraft, regretted this duplication of effort. Inasmuch as it did not want to favor either of the projects, it put pressure on the German aerospace firm, MBB, to withdraw from the British EAP. Subsequently, serious negotiations started among five nations (West-Germany, United Kingdom, France, Italy and Spain) on the joint development of a European Fighter Aircraft.

We will describe two episodes in the development of the European Fighter Aircraft (EFA).[13] The first episode is the start-up phase of the project during which a socio-technical network first emerged. Throughout this phase the technological definition of the project remained controversial. This episode will show how certain characteristics of pre-existing networks constrained the formation of the new socio-technical network of the EFA.

3 Coordination in Military Socio-technical Networks 83

The second episode is the controversy that arose over the EFA radar some years later. By this time, the network had matured and had become strong enough to (co-) determine its own further development. We will use this part of the analysis to show how the dynamics of the network as a whole can help us understand how one of the technological options was finally realized. It illustrates how the resilience of an existing network represents a force for coordination. In its turn, the resilience of a socio-technical network is continuously re-established through coordinated actions within the network: resilience and coordination are mutually reinforcing.

A Episode 1: The Birth of the European Fighter Aircraft

1 Military Requirements for EFA

At the instigation of their national governments, the Air Force staffs of Germany, France, Italy, Spain and the United Kingdom discussed the possibilities for a shared new generation fighter aircraft. In December, 1983, the Chiefs of Staff reached agreement on a preliminary statement of operational characteristics for what thenceforth became known as the European Fighter Aircraft (EFA). The outline agreement, the "Outline European Staff Target" (OEST), emphasized air-to-air combat capability as EFA's primary requirement while air-to-surface combat would be secondary. It specified that the aircraft should be a single-seat, twin-engine aircraft, highly agile and possessing short takeoff and landing capabilities. In addition, agreement was reached on the type of armament and on a "basic mass empty" of 8.5 tons. It was expected that the five partners would have a total requirement for about 800 aircraft while independent studies indicated that at least another 300 could be sold, primarily in the Middle East.

In July, 1984, the defense ministers of the five nations met in Madrid where they underscored the importance of the joint development of EFA. They signed an agreement for a formal feasibility study on EFA and committed themselves to financing a study of the technology and to work-sharing for the airframe, engine and avionics systems. The national armaments directors were to present a final recommendation on the project to the defense ministers in Rome the following March. The ministers would then decide whether to go ahead with the project definition phase. The countries compromised over an aircraft of about 9.5 tons, optimized for air superiority but with built-in ground attack capability. In the meantime the Chiefs of Air Staff had continued joint Air Force studies and by October, 1984, had arrived at a more definitive agreement — the "European Staff Target" (EST). Up to this point it had been only the military requirements, specified by the Air Force staffs, that defined expectations about the type of technology that would be used. Soon, however, the plot would thicken as a number of other factors gained importance, like development and operational costs, possibilities for export, national policies on industrial development and employment, the striving to become independent of the U.S., commonality with existing programs, etc.

2 Industrial Requirements — Heading for Trouble

The next step in the EFA program was to get the aerospace industries of the five collaborating nations to agree on work and design sharing and on program funding. Deeply rooted differences of opinion soon became apparent. First, there was the problem that French industries demanded design leadership of both airframe and engine. German and British industries wanted equal shares.

The French position was rooted in the guiding principles of its national armament policy. After World War II, France had developed a national armaments industry for all major categories of weapons, including advanced fighters. For a medium-sized country, however, sustaining a high-tech defense industry solely for domestic purposes proved too great a burden on the national budget. This made French industry and political leaders very keen on export as a means to amortize development and production costs over larger volumes of production.[14] For example, with reference to the EFA, the French favored a future fighter that would be as light as possible — and therefore cheaper, it was believed — in order to facilitate export.

Hence, in France, long-standing weapon innovation networks had developed with strong ties between the government and a number of branches of industry that were considered strategic. In regard to technology, the main issues in these networks were (1) maintaining innovative power and (2) amortizing development costs over larger volumes of production through export.

In the United Kingdom, the same issues played a role, but the defense technology networks that had developed there exhibited some important differences. To a greater extent than the French, British firms had developed relationships with foreign companies in the field of arms development and production. For instance, West Germany, Italy, and the United Kingdom had previously cooperated in the development and production of the Tornado fighter. To a greater extent than the French, the British military and government relied on foreign technology for their armaments, especially on American technology. Although the British also wanted to be less dependent on the U.S., they preferred to realize this through European cooperation.

Although the Chiefs-of-Air-Staff had reached an agreement early on, this did not prevent differences of opinion between France and the United Kingdom on the military requirements for the fighter. Given the distance between Britain and central Europe[15] and its standing long-range overseas responsibilities, the Royal Air Force (RAF) believed that a heavier, longer-range aircraft would be needed. The French, being closer to central Europe, did not need this long-range capacity. These different ideas about the tasks EFA should fulfil, motivated the British to favor an aircraft some 20% heavier than that desired by the French.

In November, 1984, the defense ministers again convened and asked industry to submit a design with two variants: one weighing 9.5 tons plus 250 kg; the other weighing 9.5 tons minus 250 kg. The French firm Dassault started working on the lighter version and presented a design which was derived from the Rafale demonstrator; Aeritalia, British Aerospace, Casa (Spain), and MBB (West Germany) worked on the upper limit and presented a design very similar to that of the British EAP.

Early in 1985, British Aerospace announced that it was prepared to give France design leadership in air frame and engine segments in return for French agreement on an international development organization in which all members would be equal partners. This position reflected a certain willingness to compromise but, on the other hand, it also reflected some strong differences with the French position. This was underscored when a British industrial consortium put forth a six-point program for the United Kingdom government position in negotiations with other EFA partners. This program included a call for equality of management authority and responsibility among the five nations with no overt leadership.

In February 1985, industry presented two designs, neither of which completely fulfilled the European Staff Target, although the Dassault design was considered to be further afield than the proposal made by industries from the other countries. The proposals were discussed by the national armaments directors but they failed to reach an agreement on design leadership, engine type and size. The defense ministers were equally unable to resolve the conflict and called for more studies on the air-frame and power plant configuration. The studies ordered included:

- Determinations of performance characteristics of an aircraft with a "basic mass empty" of 9.5 metric tons and 9.75 metric tons.
- Studies of performance capabilities of an aircraft powered by an engine sized between 88-92 kN (kilo Newtons) thrust and a second aircraft in the 80-85 kN thrust range.
- Studies of possible interim engines.

This evaluation of different engine sizes underscores the different meaning that the EFA, i.e. the function of the potential EFA-network, had for the French and British participants. French industry and government preferred the French Snecma M–88 engine that was already under development. Their preference for the M–88 was, at least in part, inspired by the guarantee it offered against the aircraft becoming too heavy given the M–88's "low" 80 kN thrust. The competing option was being advanced by the Turbo Union consortium (that had earlier built an engine for the Tornado fighter in a joint program involving West Germany, Italy, and the United Kingdom). This consortium preferred an all new engine for the EFA. On the British side at least, the anticipated long-range mission capability implied a preference for a powerful engine. Additionally, there was also a plan in the offing to upgrade the Tornado fighters somewhere in the mid-1990s with a new engine that would have a thrust of over 90 kN.

In early June, 1985, industry presented 16 different, more refined, designs. These plans were discussed by the ministers but again they could not reach a decision. Once more they asked industry to submit new proposals for a design of 9.5 tons plus 250 kg for potential additional mass. Three engines were to be considered of 84, 88, and 91 kN thrust respectively. When these plans were presented in mid-July and discussed by the national armaments directors, it appeared that large differences of opinion still remained. Points of dissensus included:

- the type of political and industrial organization.
- who would have design leadership.
- where to locate headquarters. The United Kingdom did not want it to be in Paris while the French did not want it in London.

3 Breaking the Deadlock

At this point West Germany, grown impatient because of the apparent deadlock in negotiations, started pushing for a decision. The German defense minister, Wörner, made a number of compromise proposals to the armaments directors regarding the technology to be used, the type of organization and the staffing of various organizations. To the French, however, these proposals remained unacceptable.

During a marathon meeting of the armaments directors on 1 August, 1985, the Germans broke the deadlock by choosing the British side. West Germany, Italy, and the United Kingdom signed a Memorandum of Understanding for the development of the EFA. France and Spain were given until mid-August to make a commitment. Under this agreement, Germany and the

United Kingdom would buy 250 aircraft each, Italy at least 100. The total number of 600 aircraft was expected to cost $20 billion. Shortly thereafter, Spain decided to participate.

France continued on its own and pursued the development of the "Rafale" fighter. Like the EFA, Rafale would be canard-shaped and double delta-winged, but its mass would only be 9 tons while the EFA's would be 9.75 tons. The Rafale would be powered by 2 Snecma M88-2 engines, each developing 75 kN dry thrust. Three years of negotiation had not produced sufficient overlap in the meanings attributed by the different actors to the function of the EFA-network.

4 Evaluation: The Birth of a New Network

The initiative for international cooperation on a new fighter was taken by the governments of (initially) five European countries. Although governments are in a favorable position to initiate international cooperation, the case underscores the fact that they cannot simply instruct their defense industries to proceed with the necessary arrangements. From their own perspectives, the various national industries perceived no need for international cooperation. Industry preferred national solutions to international ones because of the uncertainties the latter would invariably introduce as to who would do what, who would get access to new technology, etc. Within the context of established relationships with their own national governments they knew what to expect and they understandably preferred to remain within those national frameworks. Those frameworks were the *pre-existing networks* from which, among other things, the new network would have to be built. The reluctance of national defense industry to embrace multinational networks illustrates the resilience of established networks against alteration of existing patterns.

Still, the new EFA network did emerge thanks to the continuous pursuit of international cooperation by various governments. We see that especially where pre-existing networks are robust, as is often the case in military technology, one or more actors will have to invest heavily in building a new network which will at least partly co-opt them. We call these actors — who for one reason or another, do not take the existing patterns of interaction for granted — *dedicated network builders*. These actors have to work against the odds on two counts. First, they act counter to existing patterns of interaction (i.e. socio-technical networks) of which they themselves are a part and they have to overcome pressure from other actors trying to prevent them from doing so. Second, they have to enroll new actors who also already belong to existing networks. Therefore, they have to invest a lot of effort to overcome the resilience of those other networks as well.

This raises the question why some actors turn into (partial) dissidents in a pre-existing network and what makes them seek to create a new network. The obvious answer is that they encounter major problems in the existing networks. In the case of advanced fighters the European governments faced three related problems, notably (1) the increasing costs of new generations of fighters, (2) increasing budgetary constraints, and (3) the declining harmony of interests with the United States as guarantor of security. Increasing budgetary constraints are interesting from the perspective of a theory of socio-technical networks, as they show how the government is an actor not only in the weapon innovation network but in a variety of other networks as well. It illustrates the importance of inter-network interactions (i.e. interactions with the environment) and shows that actors can carry issues (such as shrinking budgets) from one network to another.

Looking at the fighter networks, we see that some actors, e.g., a company like Dassault, interact almost exclusively within the framework of those networks. Others, like the governments, have a *low inclusion* (Bijker 1987) in the network, i.e. only a small proportion of their total interactions are framed exclusively within the network under analysis. The paradox is then that actors with a relatively low inclusion in pre-existing networks are most likely to become dedicated network builders seeking to construct a new socio-technial network. In that role they have to work to link the pre-existing networks to each other by coordinating their respective activities. Such coordination includes the mutual adaptation of internal network routines, and hence encroachment on the resilience of these pre-existing networks.

At face value one might have expected that the establishment of the 1983 Outline European Staff Target by the air-force staffs would have created sufficient coherence to enable the development of a technology to fulfill those requirements. However, on the basis of network dynamics and the role of actors it could be expected that this was unlikely to happen.

The (partial) coordination of the Air Force staffs by means of military needs and requirements might well have been necessary, but it turned out to be insufficient. The attributed — and yet to be negotiated — functions of the new network were broader than that. The resilience of pre-existing networks implies that the new network will either have to adapt or not be able to enroll actors from all of those networks.

B Episode 2: The Battle over the EFA-radar

1 The Continuing Story of EFA

Before going into the details of the disputes over the radar we first sketch some further general developments. The development phase of the EFA program was conducted under supervision of an International Project Office called the NATO European Fighter Management Agency (NEFMA) which, *de facto*, represented the four Defense Ministries of the participating countries. This management agency was to have the final say on certain items such as the radar and engine. Industries from the four nations formed a consortium called Eurofighter, based in Munich, to carry out the actual development work. Another four-nation consortium, called Eurojet, was to manage production of the EFA engines.

In May, 1986, the Air Force staffs of the four remaining participants accepted basic design parameters that called for an aircraft with a mass of 9.75 metric ton, two engines of 92 kN thrust each, and a wing area of 50 sq. meters. This design met the operational requirements of all 4 Air Forces. By September, 1987, work had progressed to a point that a further freeze of EFA's design was possible and the Air Force Chiefs-of-Staff were able to sign the EFA European Staff Requirement for Development.

Eurofighter expected to place slightly under 300 Requests for Proposals (RFPs) covering various EFA subsystems. Before releasing the Requests for Proposals, the four partner companies within Eurofighter had committed themselves to producing a technical specification of every subsystem and component that would go into EFA. These specifications would then be approved by Eurofighter. It would also be up to the partners to send RFPs out to firms within the four nations, which would then have six weeks to respond. The bidder was required to submit five copies of his tender: one to each of the four companies with design responsibility, plus an additional copy to Eurofighter, the overall arbiter. The company with system design responsibility would then close the contract on Eurofighter's behalf, following a commercial and technical appraisal of the subsystem, in which Eurofighter would have the final say in the event of any disagreement. NEFMA, i.e. actually the four Ministries of Defense, would retain some discretion over certain items such as the radar and engine.

2 The Radar — EFA's Eyes and Brains

Radar fulfills a critical role in most fighter missions. Except for close air combat, where performance (e.g., agility) of the fighter is a crucial factor, the fighter's usefulness depends largely on the radar and the weapons carried. Requirements for the radar depend upon the envisaged military missions of the aircraft, as detailed in the military need formulated by the Air Forces. A radar should enable search, detection, identification, and tracking of enemy aircraft. Subsequently it has to guide missiles to these targets or to provide initial data for launching weapons that have their own homing devices. Increasing demands are put on the detection range capability. In addition the radar system has to warn the pilot of enemy missiles and to provide electronic countermeasures.

The operational requirements of the fighter as stipulated by the Air Force staff indicated that the aircraft be usable in close combat as well as in combat beyond visual range. This implied arming it with an Advanced Medium Range Air–to–Air Missile (AMRAAM) as well as an Advanced Short Range Air–to–Air Missile (ASRAAM). The use of these missiles defined in part the type of radar that would be needed. A further operational requirement was an all-weather and nighttime combat capability and adaptation to the increasing importance of electronic warfare. Although originally conceived as an air superiority fighter, the EFA gained an additional attack responsibility as its mission was redefined by the partner nations.

3 Two Radar Consortia: ECR–90 Contra MSD–2000

In the period that France was still part of the EFA consortium, the French electronic giant Thomson–CSF had approached Ferranti (United Kingdom) and AEG–Telefunken (Germany) about collaborating on radar and avionics systems for the fighter. In early 1985, in spite of the threat of a breakdown in the five-country EFA collaboration, Thomson was still exploring possible cooperation with these companies as well as with the Italian FIAR and Spanish EESA companies. After the break with France, however, Thomson–CSF dropped out.

This event clearly illustrates the additional value of an approach in terms of socio-technical networks rather than one that considers only individual actors (like companies). Within the latter approach, Thomson–CSF would in an analytic sense have been equal to other electronics firms, like Ferranti, AEG, etc. In the socio-technical network approach, however, actors building up a new network may also be actors in pre-existing networks. This makes us aware that, to get EFA off the ground, pre-existing national arms-production networks (including national government and industry) had to be accommodated. That explains why Thomson had to drop out when the French government terminated its participation in EFA.

Subsequently, two consortia emerged in the competition over the radar contract. The British company Ferranti, leader of a consortium including the Italian FIAR and Spanish INISEL, proposed a radar, called ECR-90 (European Collaborative Radar for the 1990s) based on the development of Ferranti's Blue Vixen multimode radar. Since 1983, the Blue Vixen had been developed as part of the Mid-Life Update Program of the British strike/attack aircraft Sea Harrier. The ECR-90 would thus resemble radar technology used elsewhere in the British armed forces. The ECR-90 would be particularly software intensive and could therefore be tailored to the needs of different users. Work within the ECR-90 consortium would be divided according to the number of aircraft ordered by the EFA member nations. An important condition for the participating countries was that each of the consortium members would assemble and test complete radar systems and would have access to all technology developed.

The competing radar candidate, called MSD-2000 or Emerald, was a derivative of a radar developed by the American company Hughes for the U.S.-Navy F/A-18 Hornet fighter. This radar had a proven performance record for a complete set of air-to-air and air-to-surface modes and had growth potential to meet EFA radar requirements. A derivative of this radar was proposed by the U.S./German consortium of Hughes and AEG. This radar would also be used in the update program of the F-4 fighter, which would allow Germany to use spares bought for its F-4s and also to use the radars themselves when the F-4s would eventually be retired.

In the contest over the radar the two consortia as well as the individual companies tried to strengthen their position by linking up with other companies. Thus, they tried either to increase their technological capacity for developing more capable radar systems by adding new know-how, or to strengthen their position within particular EFA countries. They emphasized those issues that appeared to be important for decision makers, like exportability, technological capability, flexibility, reliability, growth-potential, etc.

The United Kingdom, seconded by Italy, had the most demanding radar requirements. In the competition between domestic electronic firms for participation in the EFA radar, the United Kingdom had opted for Ferranti (leaving out its main competitor the British electronics company GEC) and subsequently kept pushing Ferranti in the international competition.

On the other side, the German government strongly supported the MSD-2000 (Emerald) radar. Its main objection against the Ferranti radar was the risks entailed in its development and thus possibly prohibitive costs. Another reason for Germany's preference for the MSD-2000 was compatibility with the Hughes radar that would be used in upgrading its F-4Fs — which would then also be equipped with the same weapons as scheduled for the EFA.

Consequently, to German pilots the EFA radar would appear virtually identical to the radars in their upgraded F–4Fs. The same applies to the upgrading of the Blue Vixen in the British case.

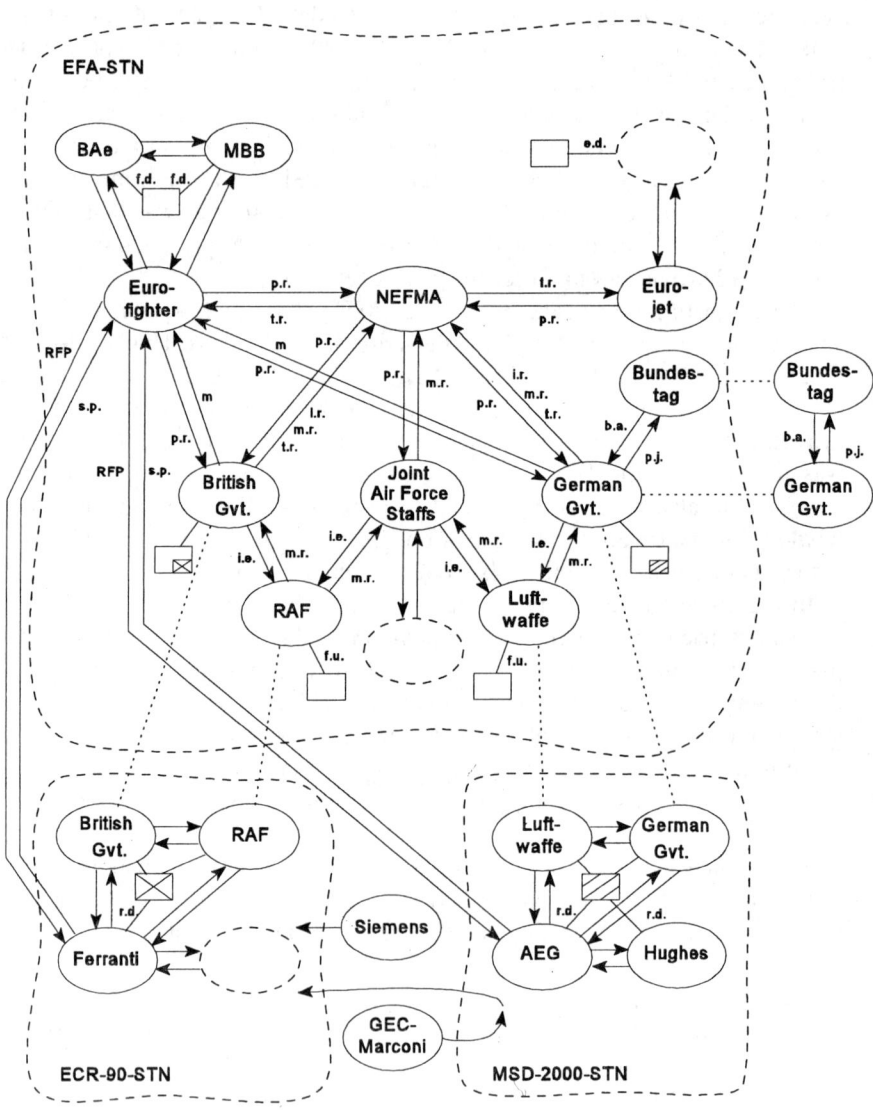

Figure 1: Sociotechnical Networks of the EFA, MSD-200, and ECR-90

3 Coordination in Military Socio-technical Networks

STN and Environment

- The dashed line surrounding the various STNs indicates their respective boundaries. Beyond this closed loop is each of the networks' environment.

Actors

- The ovals denote the actors. Actors can be part of different STNs; a dashed line between two STNs indicates the ovals connected denote one and the same actor that belongs to different STNs. These dashed lines characterize that actors can carry issues from one network to another.
- Acronyms used for actors:
AEG = AEG (German electronics company); BAe = British Aerospace; MBB = MBB (German aerospace company); NEFMA = NATO European Fighter Management Agency; RAF = (British) Royal Air Force
- To reduce complexity only the German and British actors are represented. A dashed oval indicates that several other actors are also part of the network.
- Some actors (outside the STNs) are indicated that will be discussed later in the text.

Intermediaries

- The lines with arrows indicate the intermediaries that go from one actor to another. Typically, there are two lines between two actors indicating that intermediaries go both ways. The intermediaries are labeled according to what is going in the direction of the arrow.
- Acronyms used for intermediaries:
b.a. = budget appropriations; i.e. = information exchange; i.r. = industrial requirements; m = money; m.r. = military requirements; p.r. = progress reports; p.j. = policy justifications; RFP = Request for proposals; s.p. = submitted proposals; t.r. = technical requirements
- To keep the picture simple not all of the intermediaries are labeled.

The Technology

- The rectangular boxes connected to the actors denote the actors' version of the artifact. These denote what characteristics of the artifact are of specific relevance to that actor. As this may vary across actors all these artifacts are different.
- In case a subset of actors would find the same characteristics of relevance this can be indicated by suspending the artifact between these actors as is done in the case of MBB and BAe and in the cases of the two radar-STNs
- The artifacts suspended from the British and German governments in the EFA-STN are drawn with different subsections to indicate they favored different radars.
- Acronyms used to characterize relationship between actor and artifact:
e.d. = engine development; f.d. = fighter development; f.u. = fighter use; r.d. = radar development
- In a complete STN each of the actors involved should be connected to (its version of) the artifact. Here we have only indicated a few that highlight some of the issues discussed in the text.

What is interesting here in terms of socio-technical networks is that we are not simply dealing with developing technology on the basis of functional demand but that we are dealing with two different sets of actors connected to two different technologies, i.e. two socio-technical networks: the ECR-90 and the MSD-2000 network, of which only one could eventually become part of the EFA socio-technical network. Again, the development of the radar appears to be a co-evolution of technology and socio-technical networks. What makes the analysis somewhat complicated is that some of the actors involved are part of more than one socio-technical network. Figure 1 presents a schematic representation of the various relevant socio-technical networks.

Figure 1 illustrates that we are dealing with two different radar-networks bounded along national lines. This is comparable to the situation when France was still a candidate for participation in the EFA-network-in-the-making. In that situation the deadlock was broken in such a way that all the French actors dropped out, including the radar company Thomson-CSF. In the case of the radar, however, breaking the deadlock turned out to be far less easy. It became difficult to take actions uncongenial to other actors in the network: the EFA socio-technical network had started to manifest its resilience.

4 Fierce Competition

Initially, disagreement over the radar was formulated chiefly in terms of differences in the (required) capabilities. The EFA-partners even considered a split buy: West Germany and Spain[16] the MSD-2000; Italy and the United Kingdom the ECR-90.

In July 1986 the Hughes/AEG consortium strengthened its position by linking up with the United Kingdom company GEC Avionics (later GEC-Marconi).

The addition of the GEC signal processor would allow two versions of the Hughes radar to be offered — one with only relatively minor modifications for Germany and Spain and a more sophisticated one equipped with the GEC processor for the United Kingdom and Italy.[17] Like the competing ECR-90 consortium, GEC now emphasized the importance of flexibility and growth potential in the Hughes radar, arguing that the military threat might have changed substantially by the time that the radar would go into service, i.e. the mid-1990s.

The "official" battle for the radar started in December 1986 with the Requests for Proposals (RFPs) issued by Eurofighter to 15 radar manufacturers in the 4 member countries and the U.S. The winner was to be selected by

Eurofighter (i.e. the industrial consortium) in the spring of 1987. The choice would be made primarily on the basis of technical and economic criteria.

In the event, however, it became apparent that the decision had strong political overtones. As a result, Eurofighter started changing the specifications and extended the bidding several times. The growing discord encouraged Hughes/AEG to submit two proposals, one offering a radar that would be fully compliant with the latest specifications, and another offering a shorter-term version that would not be fully compliant but would cost significantly less.

In late 1987, Eurofighter, in order to break the deadlock, asked for refined bids for a less expensive radar. To this end some of the initial requirements, chiefly those of a technical and commercial nature and, to a lesser extent, those related to performance, were relaxed. Like the AEG/Hughes consortium, Ferranti now offered a less expensive radar, notably a variant of the Blue Vixen. This manoeuvre brought no relief, however, and in the summer of 1988, Eurofighter submitted a split recommendation to the EFA management agency (NEFMA, i.e. the political organization).[18]

In late 1988, negotiations on the radar became tougher. The issues that dominated the negotiations and interactions were (1) uncertainties in estimating development and production costs and (2) export restrictions by the U.S. The technological capabilities of the radar were no longer an issue. Thus, at this stage the military and technologists had shifted to the background.

At NEFMA-meetings (basically representing the Ministers of Defense) in December 1988 and early 1989, West-Germany took an uncompromising position and warned that it would proceed with the MSD-2000, thus threatening a split decision. A breach of consensus over the radar, however, threatened the EFA consortium as a whole, since selection of one common radar had been part of the basic Memorandum of Understanding.[19] In this clash the United Kingdom, supported by Italy, stood firm on the ECR-90.[20]

It is illuminating to evaluate this episode in terms of socio-technical networks. At first sight it seems odd that the military requirements and technological capabilities of the radars lose their significance as evaluation criteria. However, as Figure 1 illustrates, we are not simply dealing with requirements and the technologies to satisfy them; we are dealing with socio-technical networks that have a broader function for the various actors and which exhibit resilience. On the one hand we have the two radar socio-technical networks, whose functions include national jobs, national prestige and the preservation of a national technology-base; on the other hand, some of the national actors are also committed to the emerging EFA network, which they do not want to put at stake. Under these conditions the situation appears to be a complete deadlock.

Whereas Eurofighter had left the decision to NEFMA because the matter was too politicized, the latter agency apparently hoped that a decision could still be made on technical grounds by Eurofighter. Therefore, in February 1989, NEFMA instructed the Eurofighter consortium — for the third time — to make a recommendation on the radar after re-evaluating the two competing bids. Again, this led nowhere and only made it plain that the conflict would have to be resolved at senior government levels.

In West Germany, the Ministry of Defense was under heavy pressure from the *Bundestag* to limit the costs of EFA and its radar.[21] The negotiations on the financial and technological risks, so strongly pushed by the Germans, entailed several moves on both sides. To defuse a crisis, it was agreed at high level consultations between the United Kingdom and German Secretaries of State for Defense that the EFA radar would adhere to current budget and time schedules and that a single shared design would be pursued — thus defusing West Germany's threatening stance. The United Kingdom Ministry of Defense further tried to lay German fears about the costs and risks of the ECR-90 to rest by offering to underwrite the risk on the ECR-90. However, the offer to Bonn by the United Kingdom Chief of Defense Procurement to transfer DM150-300 million of EFA work of its choice to British Aerospace, thus allowing Germany to retain liquid assets that would otherwise have to be frozen in a fund to cover possible budget overruns, was turned down. Such a transfer would not only take jobs away from MBB (the German aerospace firm), but would also deny the Germans valuable technological know-how.

West Germany responded with a mirror-image proposal to underwrite the risk of the Ferranti-led ECR-90, offering to pay almost all of the costs of optimizing the AEG-led MSD-2000 radar to satisfy the British requirements for EFA, and to underwrite development risk as well.

In the meantime, the company-actors tried to fortify their positions. Ferranti strengthened its hand by concluding an agreement with Siemens, giving ECR-90 an active West German partner. In addition, both Ferranti and the German electronics firm AEG were working on reducing the price tag on their radar bids to present best and final offers by early April 1989, in accordance with Eurofighter's latest instructions. In the summer of 1989, NEFMA and government officials again met, after which Eurofighter asked the two bidding teams to submit yet another round of "best and final offers."[22]

During 1989, Ferranti ran into financial difficulties due to a bad takeover in 1987 of the International Signal & Control Group. A number of companies considered buying Ferranti because of its high-tech base and in early 1990 GEC in fact bought Ferranti's radar division. Although not a direct cause, the EFA radar controversy did play a role in this strategic take-

3   Coordination in Military Socio-technical Networks   97

over, as the British Government had encouraged GEC to buy Ferranti in order to reassure the EFA partners that the ECR-90 radar would remain a viable option.[23]

As a result, the ECR-90 had substantially strengthened its position. First, teaming up with Siemens in 1989 had given the ECR-90 an active proponent in Germany. Second, the GEC take-over of Ferranti, combined with GECs strong financial position, implied that the financial risk for the development of the ECR-90 would be lower.[24] In addition, GEC now became a strong advocate of the ECR-90. All these elements helped strengthen the relative position of the ECR-90 with the result that when the final decision was made in May 1990, the ECR-90 was selected to become EFA's radar.

The above episode can be retold in terms of Figure 1 as follows: at a certain point, GEC became a member of the MSD-2000 network, whereas Siemens became an actor in the ECR-90 network. With GEC's take-over of Ferranti, however, the former decided to drop out of the MSD-2000 network and to become an actor in the ECR-90 network. This satisfied the German government that possible financial risks would remain acceptable while the participation of Siemens insured that Germany would also benefit from radar development. Thus, the German government also became an actor in the ECR-90 network which tipped the balance and ushered in the demise of the MSD-2000.

5   Compromise Found: Increased Resilience Against New Threats

Reaching a unified decision on the radar was important for the EFA project as a whole: a time bomb under the project had been removed. In Figure 1 the difference in meanings attributed to the function of the EFA-network by respectively the German and the British Government had disappeared. As a result, the EFA network was reinforced by the commitment to a specific radar. New actors had been connected and had increased the network's resilience.

New threats emerged, however. The changes in Eastern Europe in the late 1980s and the subsequent defense budget cuts in Western Europe, threatened the EFA's future once more. In Germany, especially after the German re-unification, both the German Social Democrats and the Liberal Party questioned the need for the EFA and called for its termination. This serious threat to the EFA project caused consternation in the EFA network. The United Kingdom Air Vice-Marshal J.S. Allison reiterated Britain's support for the EFA program "despite the lessening of tensions around the world." Rather than a reduced need for EFA, he argued that "[m]ulti-purpose weapons like the EFA will become increasingly important." Also

Schaffler, president and chief executive officer of Germany's MBB (with a 33% workshare in the EFA's development) suggested that reduced threat did not necessarily mean more stability and that NATO and the Warsaw Treaty Organization might acquire new roles in guaranteeing stability.[25]

Nonetheless, in the West–German *Bundestag*, opposition mounted against EFA, especially in view of the high costs of the re–unification with Eastern Germany. This even led the German Minister of Defence, Volker Rühe, to declare the EFA dead repeatedly in early 1992. The EFA network demonstrated its resilience, however, as various actors exerted themselves to search for compromises which could take the wind out of German opposition. In late 1992, this led to a new compromise in which EFA was redefined as the New European Fighter Aircraft (NEFA; called Jäger Light in German–English) that would cost DM 90 million rather than the DM 130 million the EFA was originally to have cost. Ironically, the requirements for NEFA very much resemble the original French requirements for the EFA that were rejected out of hand by the Germans and the British in 1985, causing France to withdraw and to pursue its own fighter project.

C Evaluation: Surviving Adolescence

From the moment that the EFA network was formed through the establishment of the NATO European Fighter Management Agency (NEFMA) and the industrial consortia Eurofighter and Eurojet, it had begun to build up resilience of its own.

As during the birth of the EFA, military requirements were a primary vehicle for getting interactions over radar off the ground. One might have expected the technology chosen to have been the one that best fitted these requirements. However, it appeared that the main stumbling blocks were not the translation of the requirements into technology but rather costs and what might be called the national technology base. The resilience of the pre–existing national industry–government networks prevailed once again when both Germany and the United Kingdom proved to favor a radar that conformed to the national update programs already being pursued by their national defense industries. Instead of breaking the deadlock (as had happened prior to France dropping out in an earlier phase), the dedicated network builders tried for more than three years to find a compromise that would not cause a split. For the actors committed to the EFA network, keeping the interactions going became a major goal in and of itself, thus demonstrating the network's resilience.

It is striking that in the later discussions over the radar, arguments hardly touched on the linking of technologies to requirements. Instead the debate was about which technology (or better: which combination of companies) might enable the EFA-network to persist. The technical differences between the two radars shifted to the background. Aware of this change, the competing companies teamed up strategically with companies from the hostile countries. The team that finally won was the one that was quickest in making the proper strategic moves, i.e. the consortium that was quickest in allaying fears that national industry would be benched and that costs would be overrun.

Although compromises were found over various issues in the development of the EFA, thus further augmenting the resilience of the EFA-network, its continued existence remained imperiled. The latest threat was the rising opposition to the EFA in Germany, triggered by what might be called external events. This warns us that actors in a network should never be considered in isolation from their environment, and that network dynamics should not be analyzed exclusively in terms of the network itself. In this respect, the German *Bundestag* functioned as a bridge for introducing external issues. It is part of the socio-technical network because its interactions in relation to EFA have a recurring and patternlike nature. In its (infrequent) debates on the fighter, the *Bundestag* judges it from the perspective of various national issues, including budgetary, military and industrial considerations.

By far the bulk of the *Bundestag's* activities, however, deal with issues other than the EFA. In Figure 1, therefore, the *Bundestag* is also represented as part of the EFA socio-technical network's environment. Its interactions with the German government on budget appropriations more generally may also influence the interactions in the EFA network as is indicated in the figure.

This dual positioning of the *Bundestag* means that it has *low inclusion* in the EFA-network. This is typical for any parliament, which is concerned with a broad range of political, economic and budgetary issues and is, via its members, actor in a host of other networks as well.[26] At the same time its support is (considered) essential for the network to survive, because it may cut or withhold the necessary budgets. Because of this special status in the network, parliament may be called a *critical actor*. The support of critical actors is considered essential but it is not guaranteed. The combination of its status of critical actor and its low level of commitment to the development of any given specific weapon system, may turn such an actor into a destabilizing factor in a socio-technical network. The lower their inclusion in the network, the less they are committed to what has been achieved and the less they will be inclined to adapt.[27]

4 Conclusions

It is evident from our case material that in the analysis of technological development one has to pay attention to the shaping of the social dimensions of the associated network and, in particular, to the functions attributed by the actors to the latter. It appears that a specific historically emerging pattern of social relations around technologies time and again appeared to be a major factor in directing the subsequent course of technological innovation.

For the international collaboration on military technology that we have been considering, the following dynamics in the development of the socio-technical network can be discerned. Initially, the dedicated network builders play a prominent role. In the network-in-the-making the dynamics of the pre-existing networks in which they function and the problems they face there will, to a large extent, determine the type of technology they try to construct. Initially, a great deal of effort is put into what amounts to negotiating the function of the network. Whereas different meanings will be attributed to this function, sufficient overlap is nonetheless necessary in order to get the network off the ground.

In due time a shared function emerges to which the actors can commit themselves and resilience is built up.[28] Especially when the new network is still relatively small and the dedicated network builders are fighting for its survival, the technology remains quite malleable. In order to thrive, the network builders subsequently have to enroll new groups of actors, especially critical actors, which in some cases requires adapting the artifact — within limits — to their wishes. Enrolling new actors subsequently introduces new themes, originating from the pre-existing networks in which these actors participate. These new themes co-determine the further course of technological development in the emergent network.

While the shared component of the attributed functions is no longer under discussion, negotiations will still continue on elements not shared by the various actors. Success in these negotiations is important to the actors, because they stand to gain or lose by them. In this process, the socio-technical network also builds up resilience. Thus, one may say that resilience results both from past experiences and from future expectations. In some cases specific characteristics of the technology can be "the glue" that binds various actors together. This makes these characteristics into a constitutive element of resilience. They have become a starting point for further development and are no longer (easily) questioned. The EFA-case shows that once new (groups of) actors are committed to the network, the technological features of the artifact that accompanied their enrollment (like deciding for a specific radar) become more determinate and more difficult to change.

3   Coordination in Military Socio-technical Networks   101

The preceding analysis suggests the following conclusions:

a)   When attempting to understand how actors behave in relation to the development and use of technology it is *not* sufficient to look at them as a single entity with specific characteristics. Existing patterns of interaction with other actors and with technology (i.e. existing socio-technical networks) co-determine their behavior.
b)   Existing socio-technical networks show resilience. This implies that within a socio-technical network those technology developments that enable the interactions to continue are most likely: that is, developments that are congenial to the shared part of the imputed functions of the network and which do not threaten the essentials of the non-shared component of the functions attributed by the different actors. Developments that are problematic for one or more actors are less likely.
c)   The more complex a socio-technical network is (by which we mean the participation of *different types* of actors whose interests in the course of technology stem from different backgrounds) the more likely it is that a proposed change in the agreed technology development will be problematic for one of them. In combination with conclusion b) this implies that in such complex socio-technical networks, technological change is likely to take place in incremental steps.
d)   Dedicated network builders play an essential role in getting new socio-technical networks off the ground. These may be producers of new technology, but also customers, as is often the case in international collaboration on military technology.
e)   Critical actors with low inclusion in the network may initiate the decline or disintegration of the network by forcefully introducing outside issues unacceptable to other network actors.

A   The Limits of Needs and Requirements as Coordination Mechanisms

From the case it is evident that much effort went into negotiating military requirements for the EFA. At the outset, requirements were negotiated among the Air Forces of the countries involved. These requirements were subsequently a vehicle for interactions between the Ministers of Defense (or NEFMA) and industry on the EFA technology and design, e.g., when formulating requests for proposals. The initial focus on military requirements (like the European Staff Target) might have warranted the expectation that consensus on requirements would provide the framework for binding together the various countries, including their defense industries, and would further coordinate their activities. In the event, however, the binding and coordinating power of the military requirements proved insufficient both for linking the pre-existing networks of the five original countries (France dropped out) and for deciding on the radar issue. The point is not that military requirements have no coordinating power — they clearly do, as argued in the first part of this chapter — but rather that this power depends on the circumstances. These may vary. In our case, the function attributed to the EFA network, in particular the shared part of the meanings attributed by the actors to this function, provides a coordinating force. In the EFA case it was actually this (shared) function rather than the mere military requirements that had to be negotiated in order to link a number of pre-existing networks — in the event unsuccessfully, as we saw (France dropped out). To think that military requirements could do this job is to equate them with the (shared) function of the network-in-the-making. Military requirements, however, are too limited in scope to cover all aspects important to the actors involved, thus limiting their coordinating power. By contrast, in well-established networks, where military requirements may to a great extent reflect the function of the network, their coordinating power may be substantial, though here also they will have to compete with other (economic and societal) factors. The requirements themselves may still be the subject of negotiations between the network actors, but even then they function as an important vehicle of interaction and play a coordinating role.

B Afterthought on Influencing Technological Development

In this chapter we have analyzed technology development basically as an inter-organizational process, in which many actors are involved. Being relatively autonomous, but at the same time mutually dependent in developing a new technology or technological system, these actors become tied together in a socio-technical network. This implies that technology development cannot be determined from one position in the network (though the potential power to influence may differ). Influencing technology development is a multi-actor process.

Still, a specific actor can attempt to influence specific outcomes of the process by supplying new input and taking into account how other actors are likely to react. One process through which the course of military technological development might change, is a change in guiding principles. Guiding principle was introduced as an analytical concept that somehow governs the direction of military technological innovation and its accompanying military R&D in the defense community. Guiding principles are a representation of military needs and requirements at a higher level of abstraction. Turning guiding principles into an instrument of influencing technological innovation implies giving the concept a normative, rather than an analytical meaning. A change of guiding principles would imply that new orientations become important in the interactions between the actors, leading to different heuristics in the innovation process.

Whereas dedicated actors are needed to try to initiate such a shift in guiding principles, this type of influencing is not a one way street, but rather a multi-actor steering process, because of the existing interdependencies between the actors. The new guiding principles would then gradually materialize in new military needs and requirements. Critical actors are in the best position to stimulate the development of such a process because their "support" is considered essential by the other actors in the network while they are relatively weakly committed (because they have a low inclusion) to the existing state of affairs.

Notes

1. For a review of the relevance of policy network approaches to technology development, and for further references, see Smit (1995).
2. Actually, as Johanson and Mattson (1987) have remarked: networks are stable *and* changing. At the empirical level, nothing is stable in any absolute sense. Human actors change (in terms of what they do and in terms of the networks they operate in), non-humans change (like technologies), interactions change. Stability, therefore, is always relative. It expresses the idea that while everything can change, not everything can change at the same time (at least not drastically).
3. The constraining doesn't occur in a rigid way because networks themselves may change or may be the target of actions for change.
4. In the empirical case of the European Fighter Aircraft, for instance, the function at the network level is the production of a new advanced fighter at reasonable costs. For some countries elements of the function are also the survival of their national defense industry (United Kingdom, Germany), for others to get, via collaboration, access to modern advanced technology (Spain), and for France to be able to maintain a national security policy based on a viable national defense industry.
5. We are even tempted to claim that perpetuating existing patterns of interaction is more important to actors than the actual content of interaction on the condition that the content reflects possibilities for continuing other interactions. The French stepping out of EFA (see case-description below) might be a case in point because choosing the British concept would have been threatening to certain French network relations. This perpetuation of existing patterns is similar to phenomena described by Giddens (1984) in his analyses of bureaucratization and institutionalization.
6. An actor's inclusion in a socio-technical network refers to the degree to which his thinking and acting are adapted to the (rules and function of) this network.
7. As a dynamic concept, guiding principle may be compared with the concept of democracy, which also, for instance, is not simply equivalent to the majority decides, but also implies that the interests of the minority should be taken into account. In that sense, democracy must also be operationalized in concrete situations. Though

8. The following may illustrate the distinction between needs and requirements. In the 1980s the issue of defense against tactical ballistic missiles became an issue within NATO. It was formulated as the need for extended air defense, that is an air defense supplemental to the traditional one against aircraft. One option for satisfying this need was to develop an Anti Tactical Ballistic Missile (ATBM) system. The technical specifications of the ATBMs (e.g., as to radar range and resolution power, speed, manoeuvrability, intercepting warheads, and so on) are then part of the operational requirements (Borg and Smit 1987; Hafner and Roper 1988).

9. In practice the distinction between needs and requirements is somewhat ambiguous. The interplay, and even fuzzy boundaries, between military needs and requirements were noted in the RAND study as follows: "[...] there is no universally accepted distinction in the Air Force between needs and requirements" (Stanley and Birkler 1986: 4-5).

10. Moreover, in Western Europe opposition against the American export restrictions on key technologies was growing, which made advanced apparatus produced in Western Europe subject to prior American consent in cases of intent to export.

11. In 1984, for example, in order to promote European cooperation on military technology, the governments of European NATO countries (including France), had revived the Independent European Programme Group (IEPG), an informal governmental consulting association (A: 237).

12. Their initiative reflected concern about the termination of the Jaguar, Tornado, and Harrier aircraft, while the British government had no firm plans for a next generation aircraft. An excellent overview of the "prehistory" of the European Aircraft has been given by Ivan Yates, Deputy Chief Executive [Engineering] of British Aerospace and Chairman of Eurofighter GmbH (C).

13. A more detailed case description and references can be found in Elzen et al. (1990).

14. France's very active weapon export policy has moved this country to third place on the world list of arms-exporting countries, at some distance behind the U.S. and the USSR (B: 199; see also Kolodziej [1987]).

15. The same argument made Spain and Italy incline to the British position.

16. For Spain the MSD-2000 radar is attractive for reasons similar to those of West Germany, namely its resemblance to the Hughes radar that is to be installed in Spain's McDonnell Douglas F-18s.

17. The new signal processor would be highly flexible, allowing major updates with only software changes. Its development had been started by GEC three years before and had been company-funded.

18. The Eurofighter consortium itself became split along national lines. British Aerospace supported the ECR-90 submission on the grounds that it believed it to be technically superior, whereas the German MBB was backing the MSD 2000, because it offered a less risky approach to full-scale development.

19. The German Defense Ministry and the Luftwaffe (with backing from MBB), warned its EFA partners that it would install the enhanced AN/APG-65, that was already under development for the F-4 upgrading. The budget would be significantly overrun if two radars were to be developed and integrated separately into the EFA airframe.

20. From an industry/technology point of view the governments of Italy and Spain had little preference one way or another, for, no matter which radar project would be chosen, both FIAR and INISEL would participate in its development and production.

3 Coordination in Military Socio-technical Networks 105

21 The Ministry of Defense in Bonn claimed that adoption of the ECR 90 would result in DM 300-500 million in additional costs.
22 In the meantime, the United Kingdom Ministry of Defense was working on an optimization study of the MSD-2000 to see if it could be worked up to satisfy the Royal Air Force's demanding radar requirement. At the same time, AEG had prepared a study of the MSD-2000 Plus which sought to ascertain whether the improved radar could meet RAF requirements.
23 Another reason was to preserve Ferranti's technological know-how in regard to radar.
24 This was an important factor in overcoming the West-German opposition while, at the same time, the British Ministry of Finance became less hesitant to guarantee a limited financial risk for West Germany.
25 "Defense as a means for preserving national sovereignty will not become superfluous" according to Schaffler, who referred to Sweden's national procurement program "which even includes the development of a new fighter aircraft" (Schaffler 1990: 91).
26 So Parliament is a nodal point in which very diverse issues can become connected to each other. This may vary across different nations. In France, for instance, there is very little parliamentary interference in matters of arms production and procurement.
27 Many similar examples can also be found in the development of the American B-1 bomber which we also studied (Enserink 1993; Elzen et al. 1990).
28 Elements in the shared component of the function of the EFA network included the development of a fighter aircraft of which: (1) performance capabilities were important as long as costs would not become too high; (2) commonality of the radar choice was important to limit the costs of systems integration; (3) companies from all four countries should participate so that every country would benefit in gaining know-how; (4) the chosen solution should contribute to the continuity of the four-country EFA-network.

Sources

A. SIPRI (1985): *World Armaments and Disarmament* — SIPRI Yearbook 1985. London: Taylor and Francis.
B. SIPRI (1989): *World Armaments and Disarmament* — SIPRI Yearbook 1989. Oxford: Oxford University Press.
C. Yates, I. (1988): *Evolution of the New European Fighter; A British Industrial Perspective*. London: British Aerospace Ltd.

Chapter 4
Getting an Experiment Together in High Energy Physics: Big Plans, Big Machines, and Bricolage*

Cornelis Disco

1 Introduction

Experiments in high energy physics have become such massive projects that coordinating them in space and time is now a major challenge in itself. It is not surprising that high energy physics is the field which originally inspired the idea of "big science" (Weinberg 1961).[1] "Bigness" refers in the first place to the size, complexity, and expense of experimental apparatus. In high energy physics, the machines needed to give the elementary particles their high energies (accelerators) as well as the machines needed to observe their interactions (detectors) are huge, complex, and getting more so all the time.

Colossal particle accelerators like those at the *Conseil Europeen pour la Recherche Nucleaire (CERN)* or at Fermilab in Illinois (or the terminated SSC in Texas) have captured the bulk of media and public attention. Nevertheless, detectors, which are unique to a specific experiment or sequence of experiments and which have short lifetimes (relative to the accelerators) are large, complex, and delicate machines in their own right and it requires many physicists over considerable spans of time to design them, build them, and get them to produce robust data. There are never enough physicsts, enough money, and enough resources in any one location to produce such a detector, so they are invariably constructed in the context of large collaborations affiliated with a particular accelerator lab. Recent collaborations have begun to assume truly phenomenal proportions. The "Atlas" collaboration at CERN now numbers some 1,500 physicists and is expected to run for the next twenty years. This single collaboration alone compriscs a noticeable portion of the field and any physicist who faithfully sticks to his Atlas guns can expect the experiment to fill much of his or her career. In comparison with Atlas, the experiment studied in this chapter (the "Muon g–2" experiment now under construction at Brookhaven National Laboratories at Upton, New York) is a much more modest affair: only 60 or so physicists participate and it was expected to start producing data in January, 1997, a mere ten years after its approval.

But even so modest a collaboration as this faces major problems of coordination. In the first place the collaborators are physically distributed at a number of sites about the globe. And, even though high energy physicists are highly peripatetic, especially the senior collaborators must also spend considerable time at their home institutions teaching and generally holding the fort. So, while the g–2 detector itself is being assembled and will be run at Brookhaven, collaborating groups have good reason to do much of the basic R&D "at home." As a result there is a manifest problem of coordinating work at local sites so that it adds up to a coherent experiment at Brookhaven. At Brookhaven itself, there are different project groups working on different subsystems of the detector (cryogenics, magnet shimming, beamline, etc.) and their efforts must be tightly coordinated. In the second place, the collaboration is a long–term process and measures must be taken to coordinate the design cycles of various components and to ensure, for example, that earlier decisions do not needlessly jeopardize degrees of freedom, accessibility, or performances of other subsystems in later stages.

But just how are these big and lengthy high energy physics collaborations coordinated? The most obvious, but inadequate, answer is that from their inception they are well–planned missions. Karin Knorr-Cetina (1992, 1995, 1996) has approached this question from the perspective of a comparative socio–epistemology of science. Instead of coordination, she speaks of consensus, emphasizing the epistemological dimension of coordination. She notes that in the "little" sciences favored by laboratory ethnographers, consensus is normally achieved late in the experiment, in fact as a possible outcome of contestation about the validity and meaning of data, i.e. contests about the results of experiments. In high energy physics, she argues, consensus is achieved quite early in the experiment, that is, in the phase when the collaboration is being formed and the participants are working out a division of labor and basic parameters for the experiment:

> Yet contestation of results in experimental high energy physics appears to be rare — rare not just because surprise results are unlikely, but because 'consensus' has been relocated into the beginning of an experiment, when the technology is still unsettled, the physics processes within reach unclear, and the protoexperiment is still in a fluid, unincorporated state (Knorr-Cetina 1995: 139).

Once the groups are in place, the basic plan for a detector established, the necessary R&D trajectory mapped out, and access to an accelerator regulated, the experiment simply unfolds itself in time. This image of early consolidation and a subsequent process of articulation and unfolding of the already achieved consensus, leads Knorr-Cetina to speak of the post-formative collaboration as a "superorganism," working out its destiny through structurally and normatively anchored programs analogous to the genetic programming of a termite colony or a beehive:

> The script of the machine to be built and of the results to be gained from it is written not only through some actors winning out, but through different forms of order and practice playing against each other. One has to do with the components of an experiment, the groups and individuals who have enormous stakes and interests in these matters; the other is the order of the emerging superorganism, in which the goals and means are set, the stakes are aligned and what remains to be done is 'working things out' (not a trivial matter!) (Knorr-Cetina 1995: 141).

This raises the following question: If "working things out" is "not a trivial matter," how is it in fact accomplished? What do physicists have to do to "work things out" and, in particular, how do they coordinate this activity across space, through time, and across professional-discursive boundaries? In particular, is the image of linear unfolding which Knorr-Cetina seems to suggest with a "superorganism" metaphor, adequate to denote what happens in high energy physics collaborations? My own findings suggest not and in particular argue against a progressive and linear "unfolding" from some primal consensus. Rather, there is repeated convergence to semblances of consensus and determinacy which function as temporary frames for the ongoing work of getting the experiment together. That work consists in "unpacking" the frames, exposing the repressed ambiguities and solving the associated problems. Eventually this leads to a new frame (already present as the mission of the previous phase of unpacking and repair) and again to a subsequent phase of unpacking and problem-solving.

Concretely, "working things out," once the collaboration's goals and means are set, will at the very least involve accomplishing the much researched phase of constructing data from ambiguous "inscriptions" — as is emphasized in lab studies like those of Knorr-Cetina (1977, 1981) herself, Latour and Woolgar (1979) and Lynch (1985). But in high energy physics (and mutatis mutandis in other big sciences) it will also require an intermediate phase of getting a complicated detector together and getting it ostensibly "working." Prior to this, of course, there is the difficult and "politically" challenging problem of getting the collaboration together in the first place. A typical high energy physics experiment must therefore accomplish at least three phases of problem-solving in order to "work things out." I argue that these phases converge in pragmatic "frames" which coordinate the subsequent phase of "unpacking" and problem solving. The pragmatically constructed frames, then, are the backbone of "working things out" in high energy physics experiments and the key to how they can be coordinated over such considerable ranges of space and time.

2 Building Experiments: Ethnomethods and Frames

Frames are a particular type of "constructs," i.e. material or textual objects which are products of coordinated action and which help to coordinate subsequent action. "Constructs" are generic features of social life. Actors are always buttressing evanescent and precarious agency with artifices, constructs, which stand apart from agency but also reflect, memorialize, and orient it. These constructs are narratively or functionally organized memorials to past action which allow actors to orient themselves in a collective process and which, for this and other reasons, are also means to coordinate the future. Constructs, whether textual or material, are ways actors freeze action, consolidate gains, mark progress, and attempt to establish predictable patterns of interaction on the basis of which they can proceed. Mark Twain gives a marvelous example. Twain, a resident of Carson City, Nevada, for a brief period in the late 1860s, describes how he and a friend "staked a claim" to a parcel of land on the then still pristine shores of nearby Lake Tahoe:

> We liked the appearance of the place, and so we claimed some three hundred acres of it and stuck our 'notices' on a tree [...]. Next day we came back to build a house — for a house was also necessary, in order to hold the property. We decided to build a substantial log house and excite the envy of the brigade boys; but by the time we had cut and trimmed the first log it seemed unnecessary to be so elaborate, and so we concluded to build it of saplings. However, two saplings, duly cut and trimmed, compelled recognition of the fact that a still modester architecture would satisfy the law, and so we concluded to build a 'brush' house. We devoted the next day to this work, but we did so much 'sitting around' and discussing that by the middle of the afternoon we had achieved only a halfway sort of affair which one of us had to watch while the other cut brush, lest if both turned our backs we might not be able to find it again, it had such a strong family resemblance to the surrounding vegetation. But we were satisfied with it (Twain 1962: 135-136).

For the purpose at hand, which was not actual occupancy but signifying a claim to possession of the surrounding land, the "house" did not really have to be a house at all — but only the sign of a house. To work as the sign of a house it had to resemble a house but not necessarily to function as one. This awareness, feeding on an aversion to work, not only produced a progressively less functional house, but in the end even imperiled the semiotic function of the construct — simply because it had become nearly invisible. But for all its tongue-in-cheek drollness, Twain's example precisely illuminates how constructs function as mnemonic placeholders: as summations of activity and as frames for intended activity — in this case as a claim to ownership and a promise of future homesteading and occupancy.

Constructs in high energy physics are very similar but as a rule neither ironic, slipshod, nor ineffective. They perform the same function as Twain's house. They are placeholders, markers, claims to ownership, efforts to sum up and fix agency at a point in time, devices of summation and memory which guide and constrain future action. Knorr–Cetina (1995) provides a pertinent catalogue of ethnomethods which describes how particle physicists are constantly assembling heterogeneous and hybrid constructs to remind themselves where they are, to show others where they are, to demarcate ownership, to plan a future course of action, to implement such a course.

- *Pervasive discourse* as the basis of the collective consciousness of high energy physics experiments. Knorr–Cetina describes this, in a delightful turn of phrase, as a "constant humming of the experiment to itself, about itself"(Knorr–Cetina 1995: 130). Participants in the experiment seize upon any encounter, planned or by chance, to communicate about the experiment. "Experiments are mapped into a fine grid of discourse spaces created by intersections between participants. Its existence is perhaps the most important vehicle of experimental coordination and integration"(Knorr–Cetina 1995: 131). In more formal gatherings the primary mode of communication is the status report. Such reports provide accounts to the collaboration of the interactions of some group with the specific objects for which it claims responsibility. Such reports are a key part of the gradual unfolding of the object of the entire collaboration — the detector and ultimately the data. "The importance of the report derives from what it unfolds: reports are called upon to exhibit all relevant wrinkles of a problem and to convey the message that all relevant wrinkles have been aptly identified and investigated" (Knorr–Cetina 1995: 131). Via status reports, then, the collaboration is kept in touch with itself — its various groups/objects are kept aware of one another's doings and can adjust their own practices accordingly. The ongoing story which the collaboration tells itself about itself consists of the continual summing up of reports. In the muon g–2 project a series of *G–2 Notes* have resulted which now number in excess of 250 and are distributed to all members of the collaboration. In addition, the collaboration has maintained an internet bulletin board, a list server, and now sports a Web page for internal communication and publicity purposes.[2]
- *Unfolding the objects (and subjects)*. Knorr–Cetina notes that in high energy physics collaborations, overt conflicts are avoided as much as possible and, when unavoidable, papered over. There is an ideology that optimal solutions will emerge after due deliberation and that conflicts are an outcome of premature decisions. Hence, explicit decisions for or against some particular contribution or line of R&D are avoided or delayed as

long as possible. However, it is clear that not all things are possible at once; for example, particular lines of development must sometimes be curtailed in favor of others. Time and resources are not infinite. Under these constraints, decisions are allowed to emerge as inevitabilities rather than being taken as arguably political acts. How is this possible? Knorr-Cetina argues that an ethnomethod called the "unfolding of the object" allows the collaboration to achieve such thorough knowledge of its technological objects and options that particular choices emerge of themselves as the most advantageous (and others simply drop by the wayside as inferior). High energy physics collaborations rigorously unpack any and all black boxes that might possibly "unfold" spontaneously at some point in the future and cause unexpected problems. The detailed assessment of performances, reliabilities, costs, etc. allow rational decisions simply to emerge as self-evident. Time always presses, but never to such an extent that inadequately founded decisions are authoritatively imposed only to meet deadlines. Even when unfolding doesn't immediately indicate obvious paths to the future, it always opens up possibilities for further unfolding so that decisions can be delayed until further evidence comes in. "[...] one could say that decisions emerge through processing becoming reflexive; and through the paradoxical pathway of avoiding making the decisions which need to be made" (Knorr-Cetina 1995: 137).

- *Back to the Future*. The unfolding of the object is an exploration of properties of a construct (the detector and its component systems) which already exists in the form of more or less detailed plans and, particularly, in simulations. So the unfolding is in the first place a double check to see if the assumed functionalities are in fact available — or can be made forthcoming. As Knorr-Cetina says: "Reflexivity turns forward, then, when the experiment fictionally anticipates itself in partial, full, and finessed variants whose actual construct, if they were built, would take many years. The future is captured in these simulation stories, and then slowly caught up with through being fed into the ongoing stories of actual instrument construction and use" (Knorr-Cetina 1995: 138).

In short, through their ethnomethods physicists build constructs that work as waystations and as coordinating signposts and which by dint of summation eventually become the very road itself: physicists communicate and report, they unfold their objects by modeling and physically reproducing portions of their fine-structures, they go back to the future by filling in projected agency and simulated futures with increasingly palpable constructs.

If it is true that a high energy physics experiment is gotten together by dint of constant textual and artifactual "summing up," it still remains to be seen what steers and coordinates this "summatology." Knorr-Cetina (1995) argues, though with evident vacillation, that the "summing up" is coordinated through space and time by dint of adherence to a basic moral order in high energy physics and by scientific and technological constraints that are implicit in the tacit founding agreements of each collaboration. This might be called a "big bang" theory of coordinated summation.

Evidence from the Muon g–2 experiment suggests that this is an unsatisfactory metaphor and that there is no progressive and linear "unfolding" from some founding consensus. While the g–2 experiment was initially also defined by a founding consensus, this only applied to the basic goals and core experimental parameters. Unfolding these goals and parameters into a ramified experiment has been accompanied by a resurgence of disorder and by efforts to achieve a new consensus at more fine-grained levels of detail. Hence the experiment appears to be proceeding not only via the small cycles of incremental summatology (the gradual unfolding of the early consensus) but also through several large cycles of indeterminacy and the reduction of indeterminacy in a new consensus. The gradual summation of the experiment is, in other words, not linear but punctuated at several points by a rapid increase of indeterminacy which is then subsequently resolved in new sequences of "summatological construction." The intermediate constructs which emerge from these cycles of construction then become frames for the summatological construction of the following phase. The intermediate constructs summarize what has gone before into a coherent and robust material or textual object which represents the standing consensus on the nature and construction of the experiment. Such a construct-frame can be an authorized text describing how the experiment is put together and how it will produce the desired data, or it can be a combination of machine and text (and software) which provides the physical and logical environment in which and with which the experiment (that is, the actual data-taking and processing) is actually realized. Frames therefore serve as a point of departure, as a set of coordinates, for a subsequent phase of pragmatic construction. Because frames are never the experiment but only a consensus on how to carry the experiment out, they necessarily contain indeterminacies which emerge and become dramatic (in the sense of generating new storylines) when they are acted on.

As far as can be ascertained from this point in time, the Brookhaven g–2 collaboration will have generated two major construct-frames before its completion. Hence the experiment has three major phases. The first phase entailed the formation of the collaboration and the achievement of a consensus about experimental aims and parameters. The social and textual achievements of this phase are summarized in the text of the final project proposal. This proposal is the first and currently (1997) active frame of the experiment. It is a kind of charter which guides and constrains the current technological effort of designing and building the detector. However, in the g-2 experiment, the robust-seeming proposal has proved a black box whose initial unfolding produced a spate of new difficulties and problems to be resolved. These had been foreseen in only an abstract, generic, kind of way. This second phase of design and construction will, if all continues to go well, be resolved by the achievement of a second construct-frame, namely the "working" detector. As a careful summation of all the physics considerations, design reports, prototypes, simulations, drawings, etc. which went into its making, the detector is projected as a carefree instrument whose precise working has been foreseen long before. Presumably, if past experience is any guide (Galison 1987; Knorr-Cetina 1996; Traweek 1988) it will turn out to be a worrisome thing.[3] The detector, despite the most profound care in its construction, will require extensive testing, adjustment, and probably the most imaginative kind of bricolage in order to produce the data which is sought. What might at first sight seem an incontrovertibly robust frame will again generate its own new agenda of indeterminacies which have to be resolved before the experiment can proceed to the production of robust data. Nonetheless, the very existence of the frame as a kind of theatre in which work proceeds, coordinates the pragmatic building of the evolving landscape of performances and data.

This is schematically represented in Figure 1. The horizontal axis represents linear time and the vertical axis the degree of indeterminacy. Moments of minimal indeterminacy coincide with the achievement of global constructs, basically, a well-defined theoretical problem, publishable results, and the two "summatological frames" sandwiched into the experiment itself, a proposal and a detector.[4] Immediately subsequent to these achievements, uncertainty typically increases dramatically as the repressed ambiguities are first explored. In the course of the ensuing construction, bits and pieces of new construct are generated and as these are aggregated into overarching new constructs, uncertainty again decreases. When there is consensus that additional bricolage will lack sufficient pay-off, loose ends are summarily tied up, remaining bits of awkward dust get swept under the rug, and a new construct-frame is finalized and authorized.

4 Getting an Experiment Together in High Energy Physics

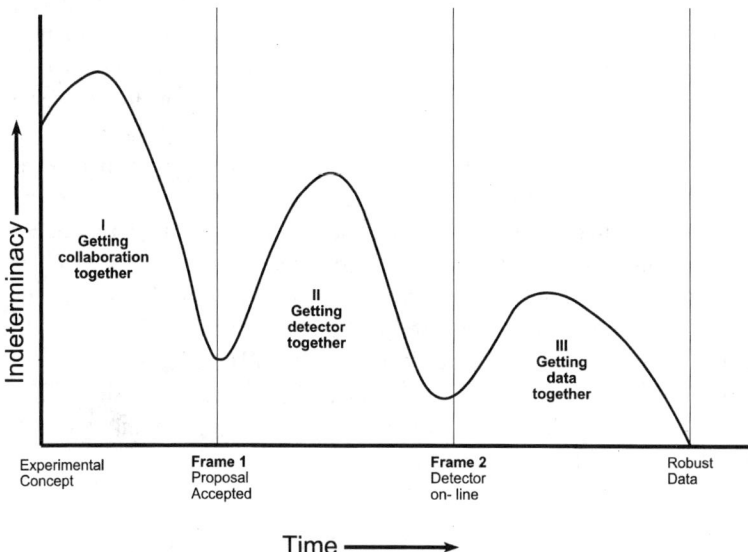

Figure 1: Frames and "Summatological" Construction in High Energy Physics Experiments

Two movements are entertained in Figure 1:

1) pragmatic construction condensing into frames;
2) frames initiating and coordinating new construction.

I will discuss these in turn. Initially, as scientists begin to work within a new formal structure which engenders and enables work leading to some subsequent structure, they tend to flounder in a sea of indeterminacy and uncertainty. No one knows exactly what the problems are, which resources will turn out to be useful, or which of the many possible solutions might work and at what costs. Successful construction produces partial and tentative islands of order in this sea. This is neither necessarily a linear nor for that matter always a successful process. Uncertainties may simply increase for a time or even remain intractable — if they cannot be eliminated the project may fail. Successful construction, however, is marked by the achievement of more and more of these islands plus their gradual linking together into a coherent structure.

This suggests that in practice, and in line with Knorr-Cetina's catalogue of ethnomethods, summatological construction is a layered, a hierarchically organized, mode of action. Preliminary and halting attempts at constructs which might become pieces in the puzzle are assembled from available resources. These bits of structure then become possible resources for more aggregated constructions. Finally, the available results of earlier and partial bricolage are assembled into a coherent frame. The proposal is written or the detector is "commissioned," and a new phase of summatological construction ensues. The "demo-bureaucratic" organization of high energy physics collaborations — which is formally, albeit tentatively, laid down in the proposal — typically mirrors this layered organization of bricolage. There is both a horizontal division into specialized groups working on different sub-systems and individuals responsible for different components, i.e. groups and sub-groups getting the different islands together, as well as a vertical division of responsibility in regard to coordinating the overall aggregation. So, as the experiment is unfolded it is also reflexively steered by those formally responsible for aggregating the overall structure into something that will plausibly perform the functionalities premeditated in the earlier formalization — which is now no longer a product but a point of departure, a coordinating frame for action.

3 Building Constructs and Unfolding Frames in the Muon g–2 Experiment

In this section I describe some of the summatological assembly of constructs and the unfolding of frames in the Muon g–2 experiment. Because the experiment is still very much under construction, I have been able to study only the first two phases, the original assembly of the collaboration (in retrospect via documents, mock ups and interviews) and the design and construction of the physical detector (by direct observation and interpellation). These phases are separated by the first construct-frame, namely the approved text of the experiment proposal. I will first describe getting the collaboration together as a process of assembling humans and resources together under the purview of a proposal. Then I will show how this proposal becomes a frame for the subsequent phase of designing and building the physical detector, i.e. through the process of unfolding its repressed lacunae and finding appropriate solutions.

A Getting a Collaboration Together: From Notional to Approved Experiment

Collaborations in high energy physics are organized around a good idea for an experiment. This invariably means the investigation of some theoretically interesting problem. All high energy physics collaborations are sensible and

4 Getting an Experiment Together in High Energy Physics

legitimate (in the eyes of high energy physicists) because they propose to demonstrate effects or make measurements that test or confirm theoretical predictions. This is how particle physicists justify ideas for experiments and it is a necessary condition for assembling the people, skills, equipment and texts necessary to carry the experiment out. Hence, initial statements of intent, draft proposals, and the final proposal itself all emphasize and feature the anchorage in high energy physics theory. Figure 2, for example, is a transparency from an early presentation arguing for the "sensibility" of a new Muon g–2 measurement.

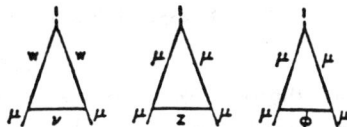

SINGLE LOOP DIAGRAMS CONTRIBUTING TO THE MUON g FACTOR

$$\Delta a_\mu (W) = \frac{G_F m_\mu^2}{8\pi^2 \sqrt{2}} \times \frac{10}{3} = +3.89 \times 10^{-9}$$

$$\Delta a_\mu (Z) = \frac{G_F m_\mu^2}{8\pi^2 \sqrt{2}} \times \frac{1}{3}\left[(3-4\cos^2\theta_W)^2 - 5\right] = -1.94 \times 10^{-9}$$

$$\Delta a_\mu (\phi) = \frac{G_F m_\mu^2}{2\pi^2 \sqrt{2}} \int_0^1 \frac{y^2(2-y)\,dy}{y^2 + (1-y)(m_\phi/m_\mu)^2}$$

$$= \frac{G_F m_\mu^2}{2\pi^2 \sqrt{2}} \left(\frac{m_\mu}{m_\phi}\right)^2 \ln\left[\left(\frac{m_\phi}{m_\mu}\right)^2\right] \le 0.01 \times 10^{-9}, \quad \text{IF } m_\phi \gg m_\mu$$

$$= \frac{3 G_F m_\mu^2}{4\pi^2 \sqrt{2}} \qquad m_\phi \ll m_\mu$$

$$\sin^2 \theta_W = 0.217 \pm 0.014; \quad m_\phi > 7 \text{ GeV}$$

$$a_\mu(\text{WEAK}) = (1.95 \pm 0.01) \times 10^{-9}$$

Transparency for talk at Columbia University by Vernon Hughes, 25-4-84. This diagram reappeared at the 1984 summer workshop and featured in the proposal as well as subsequent articles on muon g–2.

Figure 2: Feynman Diagrams Explaining the Theoretical Aims of the Project

Theoretical relevance is not enough. The experiment must also be practicable (in the basic sense that it can be built, that it will work, and that it not be extravagantly expensive or time consuming, relative to the expected payoff). In the phase of assembling a collaboration and writing a proposal it is basically a question of convincing the high energy physics community that this is the case and that the practical challenges can be met.

The proposal must show:

1) that the notional experiment is theoretically interesting;
2) that it is not prohibitively expensive;
3) that it is technologically feasible, given the state of the art;
4) that this collaboration, i.e. this particular assemblage of people and facilities could plausibly succeed.

In order to produce such a report, the organizing spokesman must have a sense of which groups are critical, which experts must be recruited, which facilities and resources must be mobilized. Producing a convincing proposal is not merely a rhetorical challenge, it is a partly after-the-fact description of plans and commitments which have already been brought together, or are being brought together under the aegis of producing the proposal. As the text acquires form, it may itself attract new collaborators who see opportunities to make specific contributions and stake a claim in the results. In this way the text of the proposal and the arrangement of people and facilities are co-produced.

What criteria steered this joint "socio-textual" aggregation in the present case? What resources were necessary? Basically, anything that would increase measurement precision. This experiment is not the first measurement of muon g-2. In fact, the value of this constant is already known very precisely. The most immediate ancestor of the present attempt was a series of measurements at CERN in Geneva in the early to mid-seventies.[5] The Brookhaven effort aims at reducing CERN's margin of error by a factor of 20 (to a mere 0.35 parts per million). It is claimed that the sharply increased precision of the proposed Brookhaven measurement will allow determination of the electroweak contribution to g-2. This could have implications for physics theory beyond those of the older CERN measurement. So increased precision, and technological means for increasing precision, were the *Leitmotif* of assembling the g-2 collaboration. In the summatological process of gathering people, things, and texts, the spokesman looked for elements that could contribute to that aim.

The first place to look is always improved statistics. A major key to precision in high energy physics measurements is simply massive data gathering. Subatomic events, being profoundly indeterminate and intrinsically error-ridden, provide data which can only be interpreted as contributions to a statistical average. In consequence it is always desirable to record as many events as possible within constraints of time and money. This places a premium on experimental setups in which the rate of events is high. In the early 1980s, Vernon Hughes, the initiator of the present muon g-2 collaboration, saw congenial statistical possibilities in plans then afoot to upgrade Brookhaven's main accelerator, the Alternating Gradient Synchrotron (AGS), such that its proton intensity (flux) would be some 50-100 times that of the CERN machine on the occasion of the previous muon g-2 measurement.[6] This meant a spectacular increase in the number of events which could be measured in the necessarily limited beam time which might be made available and hence a significant improvement in statistical error (CERN = 7 ppm, Brookhaven = 0.3 ppm). The energies at which the CERN experiment had operated were not spectacularly high (relative to the present state of the art) and well within the range of the older and more modest Brookhaven machine. In sum, the proposed measurement seemed almost tailor-made for the beam parameters of an upgraded Brookhaven AGS. This was, however, merely fortuitous. Other experimental configurations had also been considered by Hughes. These would have linked the experiment with other accelerator facilities, for example LAMPF at Los Alamos. With the choice for the CERN configuration the AGS became the obvious choice for the experiment.

In addition to statistical errors, high-energy physicists also distinguish systematic errors. These are errors due to the instability of systems components coupled with ignorance of their real-time variations. They can be managed only by imaginative and careful design. Hughes identified the principal systematic error in the CERN measurement as excessive instability and non-homogeneity in the magnetic field ('bumps" and secular variations in the field introduce non-uniformities in rates of muon precession and hence inaccuracies in the determination of g-2). This was largely due to CERNs use of conventional water-cooled magnets. The use of superconducting magnet coils, argued Hughes, would alleviate most of these problems. Remaining instabilities could be caught on the fly by much improved real-time monitoring of the magnetic field. Results of such monitoring could be reinserted either as real-time feedback or subsequently as corrections to data.[7]

Superconducting magnets, although carrying a higher risk of catastrophic failure, (so-called "quench") were known to have inherently more stable and uniform fields than the conventional magnets used in earlier generations of accelerators and detectors. Hughes, who had followed the CERN experiments closely (though not as a member of the collaboration) had in fact suggested superconducting magnets for the 1976 CERN measurement, but at that point in time the technical difficulties seemed insurmountable and the consensus favored conventional practice. Ten years down the road in the development of superconductor technology, Hughes' suggestion appeared much more feasible. By then, of course, other aspects of experimental high energy physics technology had advanced as well, particularly in the field of detectors, NMR measurements, control electronics, and data acquisition systems. These were strong allies in the emerging campaign to launch the next precision measurement of muon g–2.

By the summer of 1984, the time seemed ripe for a preliminary conference on the possibilities for a new precision measurement based on the following parameters: improved statistics at the upgraded AGS, superconducting magnets with full-field NMR monitoring, and the incorporation of whatever other technological advances might seem likely to reduce systematic errors. Hughes and several close collaborators presented preliminary ideas on how these design parameters could be realized in concrete conceptual designs. The conference, held at Brookhaven, deliberated on whether the proposed improvements to the CERN experiment could be realized and, if so, whether they would in fact provide sufficient improvements in accuracy to make the proposed measurement theoretically relevant. The *Brookhaven Bulletin* reported on the meeting in part as follows:

'The precision required for the magnetic field is about 0.1 parts per million,' says Hughes 'which has never been achieved before and is well beyond the present, obvious state-of-the-art.' Essentially the motivation behind the workshop is to examine whether the idea seems sensible for an experiment that would take four or five years to build (B: 1).

This conference was of course not only about assessing "sensibility," but also about making it. Achieving "sensibility" meant assembling a public consensus on the worth of the experiment and on the desirability of actually carrying it out given existing commitments of time and money in experimental high energy physics at large and Brookhaven in particular. Nearly 50 physicists attended the conference, including 6 veterans of the 1976 CERN measurement and 13 resident Brookhaven physicists. Copies of the conference agenda were sent to the managerial level of Brookhaven and to the Department of Energy (DOE), the main funder of Brookhaven and a prospective source of grant-money for the g–2 project.

The conference's primary achievement was a working consensus on the nature and configuration of the magnets. Hughes and a close collaborator, Gordon Danby, went into the conference proposing a 5 tesla field to be produced by 4 helium-cooled superconducting coils of ± 4.2 m diameter arranged around the toroidal beam chamber — in other words, adding superconductivity, increasing the flux, reducing the diameter, and eliminating the heavy iron magnet yokes of the CERN design.

The reduced diameter was apparently to circumvent the formidable challenge of making superconducting coils at the original CERN diameter, i.e. 14 m. With the stability guaranteed by superconductivity, Hughes and Danby considered the addition of iron mass to shape and stabilize the field to be superfluous. With the iron gone and with it the problem of physical stress and saturation at high magnetic flux, the intensity of the magnetic field could be increased without penalty. However, at the meeting, magnet experts from MIT's Magnet Lab and from the Brookhaven Magnet Division seriously questioned the wisdom of leaving out the iron yokes. Although in the context of the workshop this was a technical criticism (of potential threats to field stability and uniformity) it had to be interpreted as a political sign as well. Continuing to push the small, strong, non-ferric magnet design would encounter opposition in the high energy physics community and would reduce chances of approval. By the time of the submission of a formal proposal to the Brookhaven National Laboratories program committee in October 1985, the iron yokes had reappeared and the field strength and dimensions of the magnet coils had reverted to those of the CERN design: 1.5 tesla and 14 m diameter. The upshot was that the g–2 collaboration was now proposing to design and build the largest superconducting magnet coils ever made. The emerging storyline of the experiment was shaping up as a major technological challenge but, at the same time, the contours of the required expertise were also emerging clearly.

This 1984 meeting can be seen as a major "construct" in the fine-grained process of assembling the collaboration. It assembled previously distributed experts. Its collection of overhead sheets served as a rough draft of the storyline which would become the text of the proposal. Subsequent "politicking" took place within the temporary frame of this meeting. The meeting could provide this forceful focus because it was in fact a public assessment of the acceptability of the original proposals. These were found wanting on certain points, and collectively cast into a more acceptable and robust version. The meeting was a preliminary and informal gantlet, somewhat less demanding perhaps than the "eye of the needle" through which the proposal would eventually have to pass, but indicative nonetheless of the latter's general shape and tolerance.

In proceeding beyond the meeting it was first of all necessary to work out a proposal that would be attractive to Brookhaven, relative to its other possible commitments of funds and beamtime. In the mid-eighties, superconductivity was all the rage in high energy physics magnet design and it is likely that the opportunity to do hands-on design work on very large superconducting magnets made g–2 especially attractive for Brookhaven physicists and accelerator engineers. Particle accelerators like Brookhaven's AGS are basically concatenations of many electromagnets and their effective design and upgrading depends on the continuing development of expertise in magnet physics. Hughes and his initial collaborators seem to have converged on an experimental configuration, especially of the magnet assemblies, which fit into Brookhaven design traditions and so could provide an enticing site for collaboration. R&D at various local sites and a series of meetings over the next year established enough intermediate constructs (texts, simulations, prototypes, drawings) so that by September, 1985, these could be assembled into a detailed 115 page proposal and circulated to the members of what could by now be called a "collaboration."

While the proposal was thus clearly a summary of accomplished politics, preliminary designs, and public criticism, it was also a plan, i.e. a storyline stipulating what was to be done in the coming years to achieve the next "summatological frame," a real physical detector. The manifest technological gaps which feature in any interesting proposal now became salient here as well. The exploration, subjugation, and cultivation of these pieces of technological *terra incognita* were already anticipated and specified in the form of conceptual designs and proposed R&D trajectories. However, as a rhetorical ploy, the profundity and obscurity of these gaps were routinely downplayed by referring to the positive results of preliminary studies, by arguing that new technological goals were in fact only extrapolations from existing practice, and by emphasizing the technological prowess of the collaborators. This rhetorical politicking temporarily bridged the gaps in the proposal which allowed all those already involved (and all those who still needed to become involved, like accelerator program committees and funding agencies) to proceed with the confidence that, in the end, the experiment could be achieved. Consider the following passage from the g-2 proposal:

> The conceptual design presented herein has evolved from the earlier CERN design which received considerable developmental effort and has subsequently provided considerable operating experience. The problems are known and expensive modeling would provide little or no new understanding. The design principles by which previous errors can be reduced are also known and were implemented in the present concept. The focus of the development will therefore be on the manufacturing problems and processes which will result in the required precision, employing design techniques that use available procedures rather than on the development of major new manufacturing processes. The effort will thus consist primarily of an analysis to determine how performance varies with the choice of reasonable, available material and the establishment of fabrication tolerances. In this way the material specifications, design details, and corrections systems will be optimized for cost effectiveness (K: 43).

In retrospect (see e.g., the following section) this seems rather optimistic. No doubt everyone associated with the project at the time understood this to be the case. However, such rhetoric is a way of constructing expectations of risk which allow actors to proceed in spite of misgivings. In high energy physics too wars are won on morale, and the rhetorical strategy of choice seems to be to define a possible best case which allows actors to commit resources in spite of inevitable uncertainty. Texts like these fuse the interests of program committees and funding agencies (investing in progressive and likely-to-be-successful experiments) with the interests of the collaborators (getting the green light for a potentially risky — but precisely by virtue thereof interesting and challenging — experiment).

On November 13, 1986, after a year's deferral on its initial application, the muon g-2 group acquired a so-called stage 1 approval; i.e. "scientific approval with final approval contingent upon assembly of major resources" and was accordingly re-baptized as experiment AGS 821.[8] By early December a joint g-2/Brookhaven proposal was being submitted to the high energy physics Division of DOE asking for 470 K$ over 1987 to support R&D, primarily on the magnet ring. With the exception of 45 K$ for computer time, equipment, and travel funds, the funds were allocated for engineering and design manpower: 1 full time senior mechanical engineer, 1 full time associate engineer, 1 physicist, 3 designers, 0.5 computer programmer, and 25 K$ for shop labor. Clearly AGS 821 was being outfitted for the first steps on the long and involuted road which would lead to a second frame, the muon storage ring and detectors (J).

B Getting the Detector Together: From Proposal to Working Machine

What is the storyline of the g-2 proposal? First, it defines the theoretical relevance of the experiment in terms of the data which will be sought. Second, it describes an operating principle and basic parameters for the required detector. Third, it describes plausible conceptual designs for the detector and its major subsystems. Fourth, it presents a schedule and cost estimate. As a proposal for an experiment seeking access to accelerator beamtime and funds for construction of the necessary apparatus, it has first and foremost been a political and therefore rhetorical document. However, politics in high energy physics is embedded in stabilized canons of evidentiality, authority, and plausibility. In order to be effectively political, the proposal has had to show in detail both that its aims were worthwhile at the field level and that they could be achieved with "reasonable" effort and costs. This is the sense in which the proposal is a textual summary of the preliminary R&D and politicking that went into its making.

However, to understand how the proposal frames the current phase of designing and building the detector, it is important to distinguish the non-negotiable hard core of the proposal from its more fluid and negotiable outer layers. The most utterly non-negotiable element of the proposal is the specific physics data which will be pursued. These anticipated data are the promissory notes against which "the field" has extended credit to the collaboration and continues to do so. In principle it will be up to the collaboration to figure out how to deliver the data; in the event, any method which can be shown — even in retrospect — to yield valid and reliable data will be acceptable. Of course, "the field" is very interested in knowing beforehand just how the collaboration intends to go about producing robust data; this provides at least a ballpark indication of the likelihood of returns on the investment. So an important part of the hard core of the proposal is a specification of what might be called the "operating principle" of the experiment, i.e. the kinds of interactions which are to take place in the detector and the means of making these visible. Then "the field" wants to know in considerable detail how and at what cost a detector can be built which would incorporate this "operating principle" and so yield the desired data. This part of the proposal is, as I have argued, firm in only a rhetorical sense. To be sure, the collaboration has to exert itself to demonstrate that at least one design is possible, that the human and material resources are available, and that the technologies to realize it exist (or can be developed with reasonable effort). However, certainly in view of the unpredictable dynamics of high energy physics-related technologies, no one ultimately expects the collaboration to hew to the letter of its design proposals. The important thing is to establish a concrete point of departure and the conviction that the road ahead will be technologically manageable: not to specify this route in unswerving detail.

In sum, as a frame for coordinating detector design and construction, the proposal is no more than a point of departure and a performative "design envelope." This envelope is stipulated as a set of dimensional and performance parameters which will have to be respected whatever specific design options are ultimately worked out. So, as the very possibility of measuring g-2 depends on the operational principle and the configuration (borrowed from the CERN measurement) the required precision, which was the *raison d'etre* of the current Brookhaven measurement, will depend utterly on achieving the stipulated performance and dimensional parameters. All these elements have therefore become part of the non-negotiable hard core of the proposal and hence now "frame" the R&D efforts of the current phase of the experiment.

The proposal also coordinates current R&D by specifying a work structure and a corresponding division of labor. The notional detector, like any complex apparatus, is easily broken down into various subsystems and components. It clearly exhibits what Clark (1985) calls a "design hierarchy."[9] Recruitment of collaborators took place chiefly along the lines of this design hierarchy — different groups assuming responsibility for a particular subsystem or component of the proposed detector. This means that the R&D takes place at widely distant sites like Brookhaven, Tokyo, New Haven, Heidelberg etc. This places a premium on active coordination, so that on assembly the different components will fit and function together. Again, the proposal provides general sorts of guidelines and points of departure, but the specific details have to be hammered out in the course of lengthy meetings and frequent communications (see for an example Duncker and Disco, this volume).

The design hierarchy articulated in the proposal not only describes the architecture of the detector and suggests a division of labor, it also suggests a sequencing of design and construction steps. A basic element of the design hierarchy of detectors is the subsystem which acts as a support structure for other subsystems. This support structure tends to determine the physical shape of the detector and is the mechanical matrix onto or within which other subsystems are mounted — much like, for example, the chassis of a truck or the plastic exo-shell of a coffee maker or hand drill. In the case of the Gargamelle detector described by Galison (1987) the large pressure chamber together with the field magnet constituted the frame. In the muon g–2 detector, the iron magnet yokes serve the same function. The logic of assembly dictates that the support structure be present before the detector can be physically assembled — generally by "bolting-on" or "inserting" the various other subsystems. Each subsystem, in turn, also has a "support structure" component which serves as its skeleton and whose architecture determines the order of assembly.

However, the sequencing of R&D in the early stages is not strictly tied to the logic of assembly. Because of the relative autonomy of subsystems, parallel R&D trajectories are in fact the rule. Different groups can thus be working on different parts of the detector — including the support structure — at the same time without waiting for each other to finish. This compresses the total effort into a shorter time span. The result is that the detector emerges in a parallel fashion, bit by bit, in pieces. This emergence must be carefully coordinated — in terms of dimensional and functional compatibility and sometimes in terms of sequence. Figure 3 shows a flow chart from the proposal for the Gargamelle detector which graphically illustrates the parallel scheduling of R&D trajectories, their coordination, and their gradual assembly into the working detector.

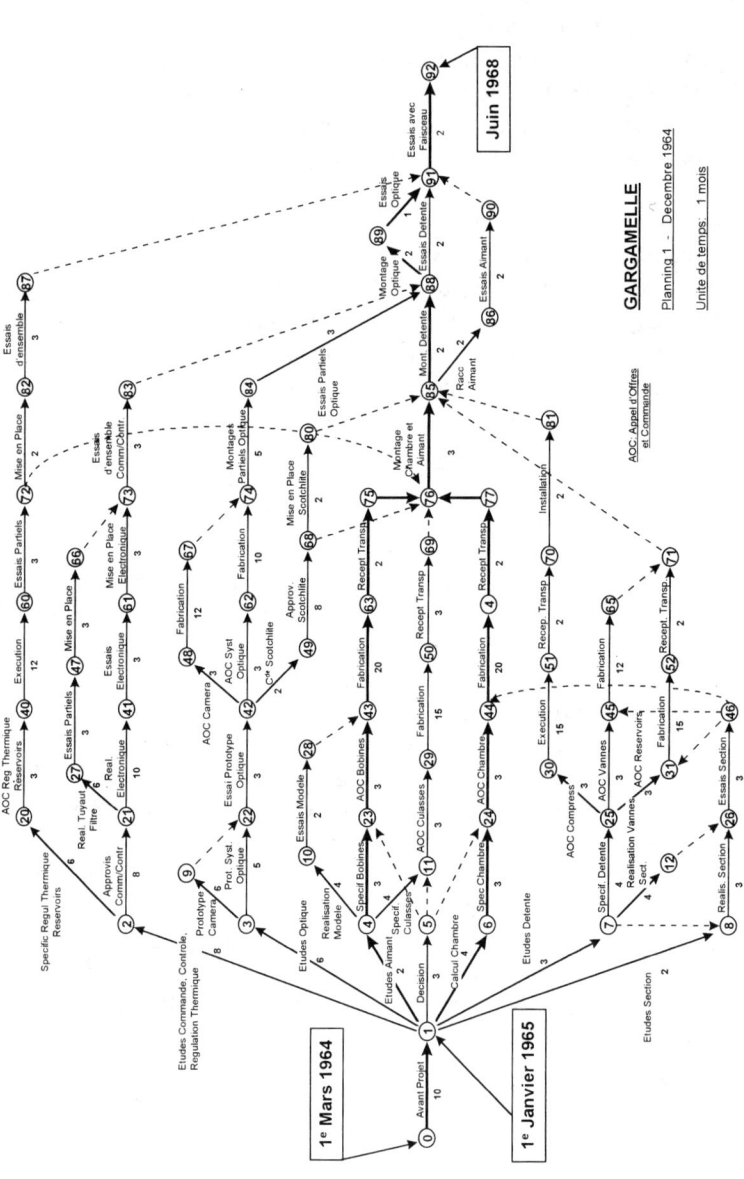

Note the seven parallel tracks along which the design and construction is planned to take place, once the "avant project" to define them has been completed. The boldfaced tracks beginning at tasks 4 and 6 together comprise the support structure of the detector, in this case initially composed of the two subsystems of the bobines (magnets) and the actual chamber. Task 76 consists of joining them together. Subsequent tasks along this axis consists in assembling (assemblies of) other subsystems onto or into the "support structure." Although the g–2 design and construction follows the same kind of logic no diagram as complex and detailed as this was included in the proposal.

Figure 3: Flow Chart for Gargamelle R&D   Source: Galison (1987: 145).

In the ambiguous flux of dimensions and functionalities which mark the start of detector design and construction, there is a premium on fixing basic configurations and dimensions, so that each project group has at least a ballpark notion of the features and dimensions of its environment. Without such parameters none of the projects can proceed much further than the investigation of basic concepts and generic simulations. The proposal itself provides a number of basic parameters, chiefly derived from physics considerations. In the case of g-2 this included dimensions and tolerances for the beam cross-section, for the magnet diameter and field strength, for data-handling capacity and response times. These fundamental parameters provide fixed points of departure for the progressive articulation of other dimensions and functionalities in the course of the R&D process. This articulation tends to begin with the dimensioning of the support structure for two reasons. In the first place, the support structure intersects mechanically with almost all of the other subsystems, so that once its dimensions are fixed the mechanical design envelopes for many of the parallel projects are defined as well. In the second place, the completed support structure is necessary in order to proceed with the actual assembly of the detector as more and more subsystems are completed and get attached or inserted.[10] For these reasons, the g-2 project began with the dimensions and construction of the magnet assemblies. Working groups were established both for the superconducting coils and for the iron yokes. A third working group was established to locate and prepare an experimental site at Brookhaven. With these organizational measures the phase of detector design and construction took off in earnest. As we shall see, this summatological search process was framed by the basic parameters of the proposal, but not in the sense that it provided a simple recipe for making the detector.

In the wake of the 1984 Brookhaven meeting, the experimental concept of a "superferric" magnet with CERN dimensions and field strength (14 m diameter and 1.47 tesla) had become hegemonic and thenceforth binding. The official proposal further articulated this concept in a design which was represented as being definitive and relatively unproblematic. One of its elements was a cost-effective measure to use superconducting stock left over from a terminated Brookhaven accelerator project. The following reassurances about the choice of this so-called CBA conductor betrays the technological optimism of the proposal:

The use of the CBA conductor, with stabilizer added, results on the one hand in an extremely stable design, which consequently permits excellent electrical and mechanical integrity, and on the other hand in a design that is comparatively easy to protect. The cryogenic system is simple and straightforward, representing technology readily available at Brookhaven. Technology currently under development at SSC may further increase the efficiency and reliability of the system (K: 36).[11]

Photo of magnet yokes assembled as muon storage ring support structure taken in January, 1995. The storage ring beam chamber fits into the azimuthal groove visible just under the electric fan. The four superconducting magnet coils run parallel to it in appropriate channels in the magnet yokes.

Figure 4: Basic Support Structure of the g–2 Detector

The coil design advanced in the proposal clearly exploited Brookhaven's long-standing expertise in building magnets, both conventional and superconducting. Moreover, it incorporated superconducting cable which Brookhaven was prepared to provide for free, and mimicked the overall parameters of the successful CERN measurement. It was in effect a conservative, more or less tried and true, cost-effective extrapolation from a local version of the current state of the art. In the proposal, clarifying drawings were supplied, although they understandably left much to the imagination.

This was the point of departure for the Superconducting Coil Group which was installed at the general g-2 meeting of February 5, 1987. The group was chaired by a Brookhaven engineer (EK) and included a changing complement of physicists and engineers in the course of its career. At the first meeting of the coil group on February 12, 1987, it appeared that the rhetorical optimism of the proposal had been premature. While there was clear consensus on how the coils ought to behave, i.e. on their overall dimensional and performance parameters, there was no consensus at all on how to achieve these parameters. There was basic disagreement on how to wind the coils, what material to wind them on, and how to package them. Considering that the original proposal had been written in the fall of 1985 and that the various local collaborating groups, each with its own particular practices and prejudices, had had at least a year in which to meditate on its possible shortcomings, this is not surprising. The meeting served to inventory the various positions which the physicist FK (CERN Muon g-2 veteran, acting resident coordinator, and recorder at the initial meetings) reconstructed in his minutes as specific combinations of binary choices along four dimensions:

A) Winding configuration: [1] "double pancake" vs. [2] "zig-zag";
B) Mandrel material: [1] Aluminum vs. [2] Stainless Steel;
C) Winding method: [1] horizontal vs. [2] vertical "ferris-wheel";
D) Cooling method: [1] Epoxied package in contact with liquid He cooling tubes vs. [2] Immersion of coil package in liquid He cryostat.

These options defined a matrix of possible design variations within the constraints of the proposal envelope. This envelope fixed the key performance criteria with which any design for the coils would have to comply. The disagreement was not about the criteria but about how best to meet the criteria.

FK's minutes for February 12 indicate that three positions had emerged, corresponding to the backgrounds of the major participants in the Coil Group, i.e. Brookhaven, Yale, and a representative of the 1976 CERN measurement. From some "objective" perspective like that of the collaboration spokesman or funding agencies, the problem now was to forge a careful consensus on the basis of open debate on the options — without premature closure on any of them, yet within realistic constraints of time-budgeting. For each of the participants in the group the problem was to demonstrate why his particular proposal for realizing this bit of the new frame would deliver "the biggest bang for the buck."

At the next meeting of the coil subgroup on February 26th, the Brookhaven physicist, RS, argued for a stainless steel mandrel/cryostat because "the pitfalls are best known." This seemed to have the provisional assent of the meeting. FK, fearing carelessness in premature closure, nonetheless took the opportunity of touting the virtues of aluminum in his minutes "This does not rule out aluminum cryostats! Any proposition, reasonably worked out, will be thoroughly discussed, since we are all aware that aluminum allows extrusion, incorporation of helium cooling duct, vastly superior heat conduction and absence of magnetic disturbance."

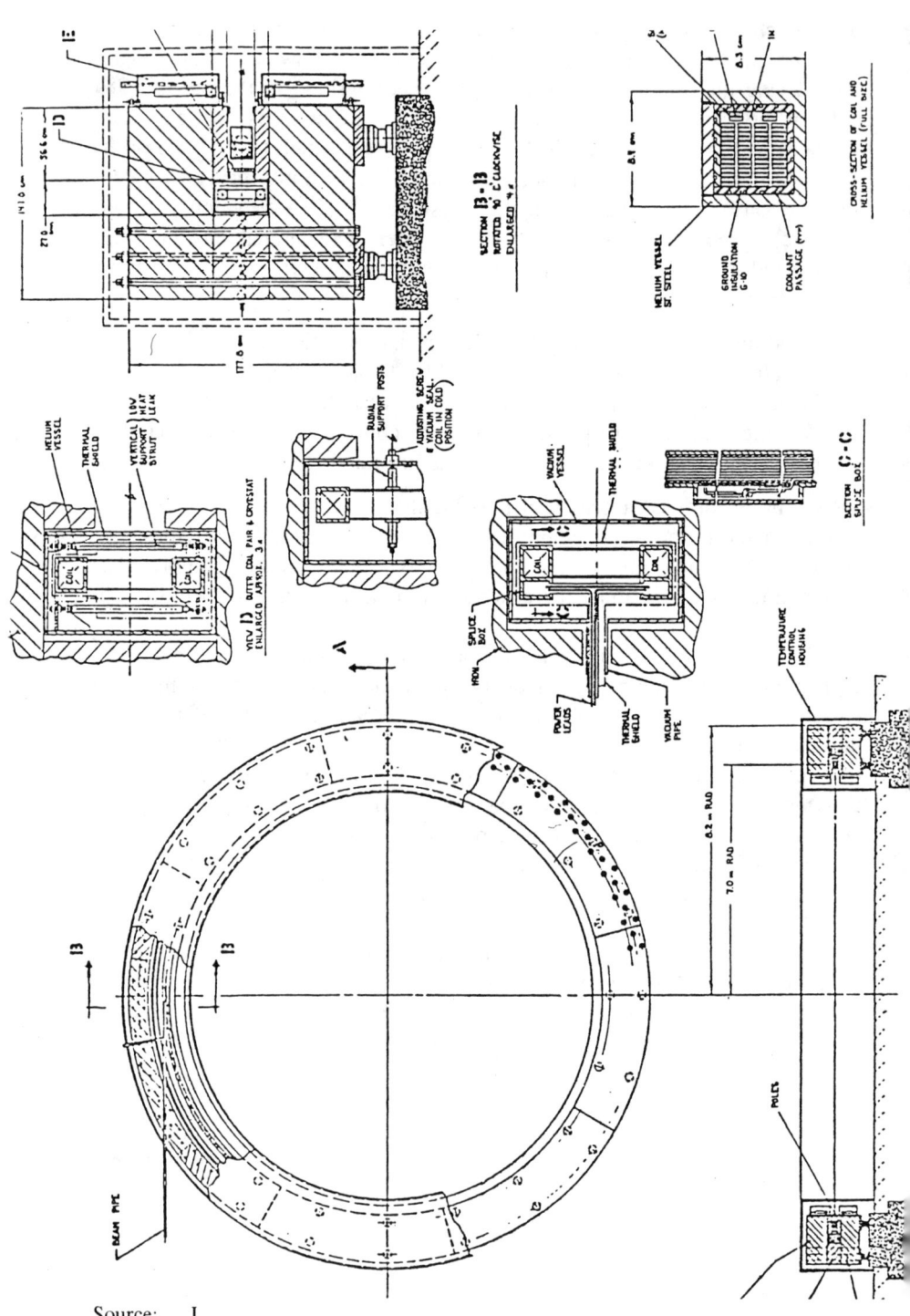

Source: L.

Figure 5: Basic Configuration of Muon g–2 Storage Ring and Magnet Details.

4   Getting an Experiment Together in High Energy Physics   131

By March 26th, although many options and strategies had been aired, a decision did not appear to be forthcoming. In a joint meeting of all three existing subgroups, the chairman of the coil group (EK) described the situation as follows (notes by FK):

EK reports on the coil design. The group struggles with the material for the helium vessel: stainless steel or aluminum, see previous coil minutes. In short, stainless steel is more conservative, but the extreme slenderness of the coil section with respect to the coil radius made us look for other material combinations in which an aluminum tank or aluminum strips wound together with the conductor are possible alternatives. The conductor fabrication is under control. The leads are, as always, difficult to realize, in particular for the outer coils, where they have to pass through a slot in the back leg of the magnet. The coil winding is still under discussion. The coil support is tricky because of tight space, large forces, and low heat loss (D).

Trench warfare continued through the end of summer. The deadlock was partly the outcome of risk management by the individual groups. Rather than approve design options employing unfamiliar constructions and materials, each of the groups favored solutions based on their own conventional practice. They justified their own conservatism as seasoned wisdom, while impugning that of their opponents as blind traditionalism or recalcitrance. A remark made by one of the Yale engineers, AD, in a design note of April 28, 1987, illustrates the contours of the stalemate:

We find that during several meetings of this subgroup the general tendency was mostly in the direction of pancake coil construction. As a matter of fact, continuous winding coils were not seriously considered and the strongest argument was that pancake was "the way it was always done" (A).

Although the coil group itself remained stalemated, its technical and organizational environment was in flux. At the end of April a Brookhaven mechanical engineer (JC) was hired to expedite coil design over the next three months. In the wake of a trip to Japan, Hughes had succeeded in interesting a group from the large Japanese accelerator facility, KEK, and the University of Tokyo in participating in the collaboration. Japanese financial and technological resources were later to prove crucial in breaching the impasse in coil-design. At the time, however, a paper by A. Yamamoto, entitled "Development of Fabrication Technique in Large Scale Solenoid in these 10 Years," submitted to the May 7th coil meeting by Hughes, passed unnoticed. This paper described the considerable Japanese experience in producing superconducting coils like those proposed for g-2 (albeit significantly smaller) and in fact contained a cross-section drawing (see Figure 6) which adumbrated the final g-2 coil design in essential outline — based on the principle of winding the coils inside the mandrel rather than outside it.

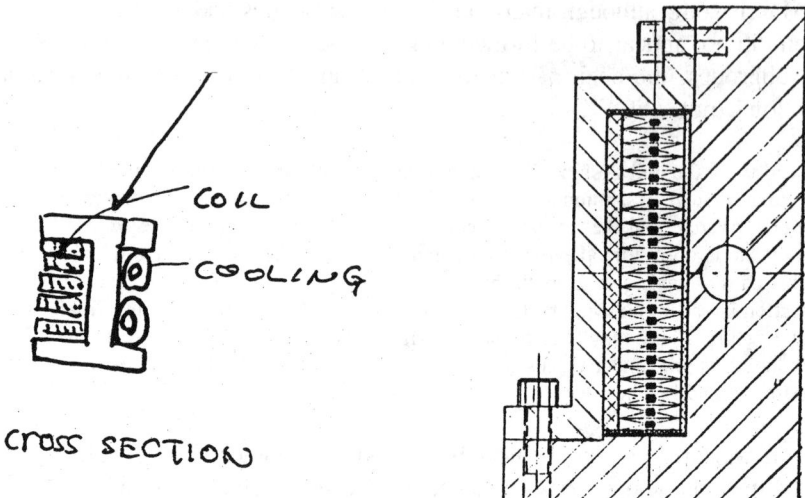

Figure 6: Coil Drawing by Yamamoto Submitted May 7, 1987 (left) Compared with Early Rendition of Final g-2 Design by JC, March 17, 1988 (right)

By June it was becoming clear to everyone that the coil subgroup had bogged down. In a radical departure from the usual consensual procedure, the chairman twice attempted to achieve closure by calling for a vote. Both efforts stranded in stalemates along established lines and only firmed up partisan commitments to the various options. In some desperation, the Brookhaven physicist, RS, proposed to adopt a compromise consensus (FK's aluminum mandrels with Brookhaven's and FK's pancake windings) as a point of departure for concerted work. As FK formulated it in his minutes of the meeting of June 10th:

RS proposes to choose aluminum and pancakes, and work from thereon: more optimization, calculation to make it stick, adapting details of the design as more information is gathered. Having thus set the scenario, JC, AD, and FK (participants in the debate from, respectively, Brookhaven, Yale, ex-CERN) could produce conceptual drawings of the coils, their space requirements for the suspension and the vacuum tank (E).

Ironically, at the very same meeting, Hughes distributed a Japanese proposal to wind the g-2 coils on the inside of the mandrels (so-called "outside in" winding and a radically different approach from the one RS had just proposed as hegemonic).

The next meeting on June 19th was attended by the Japanese team leader HH. EK opened by stating that "nothing has been frozen and [...] the Japanese input might possibly change Brookhaven's view on the matter" (F). He

4 Getting an Experiment Together in High Energy Physics

then went on to detail the current state of that view and its problems. As FK reported:

EK gives full account of proposed aluminum cryostat. The voids (i.e. between turns of the winding, CD) will be a problem. The supports of inner- and outer coils will be entirely different and very difficult to make. Our ideas on the gas-cooled leads need to be worked out. The welding of the enclosure should be done whilst the coils are firmly clamped on the winding machine. The stability is a big problem: mounting concentric at room temperature, expecting it to be concentric at cryo-temp and last but not least cope with the expansion when the current is ramped up, while the outer coil is intrinsically unstable — all this demands superior engineering (F).

Those present responded with detailed solutions to a number of these problems. In the course of this discussion RS proposed taking what was in effect a first step away from the "Brookhaven package": i.e. abandoning the requirement to design on the basis of the cross-section of the surplus CBA superconducting cable. RS argued that if a superior design could be achieved with other dimensions, then free conductor would be a "small bonus." HH then described the Japanese approach of "outside-in" winding. EK queried him about the costs of this approach. At the conclusion of the meeting EK announced his intention to finalize a Brookhaven conceptual design before September and invited HH to do the same. FK noted: "HH agrees to do this home work."

At the beginning of July, JC (the Brookhaven mechanical engineer appointed in April) assumed chairmanship of the Coil Subgroup. JC pursued EK's agenda and worked toward finalization of the Brookhaven-dominated conceptual design, including cost calculation, time scheduling, and a functional division of responsibility. By the end of August, a paper for a Boston Magnet Conference (September 21-25, 1987) was being circulated among 22 authors for approval, which was presumably the active population of the (still very small) collaboration at that point. The paper, entitled "Ultraprecise Superferric Storage Ring Magnet for the Muon G-2 Experiment" described the basic Brookhaven conceptual design: aluminum cryostat and vacuum vessel, copper-stabilized superconductor, double-pancake inside-out winding, and a heavy mandrel/dewar construction to limit radial movement of the coil. By the September 10th meeting, JC could note:

It is now believed that the conceptual design of the coil is well enough advanced to start with the preparation of engineering drawings. However, we will wait until the September 28th meeting when the complete g-2 assembly has the opportunity to discuss the design and to propose improvements (G).

At this plenary meeting the Japanese option for the coil design was once again placed forcefully on the agenda (by Hughes among others). On September 28th, JC, as chairman of the Coil Design Subgroup, presented the multilayer "pancake" design as the consensus of his group. But his files show that by September 30th he was already making preliminary drawings on the basis of the Japanese system. However, transparencies by JC dated October 2 (for another presentation) again revert to the double pancake. The apparently precarious balance was ultimately tipped in favor of the Japanese option by the visible support of prestigious members of the collaboration, in this case especially by the collaboration spokesman.

Given the already impressive size of previous Japanese coils wound by this method, the option clearly had the advantage of a sound basis in experience. At least the degree of extrapolation from past superconducting practice appeared to be far less than for the Brookhaven-Yale option. Moreover, the coil subgroup itself had already liberated itself from the design constraints involved in using the surplus CBA conductor. While this increased the design flexibility, it unfortunately also hiked the projected costs up by some one to several hundred thousand dollars. Hence, when the Japanese offered to donate aluminum-stabilized superconductor material left over from the KEK-TOPAZ detector, their design became financially attractive as well. In any case, by October 22 the Coil Subgroup was critically reviewing the design details of the Japanese proposal and the minutes show absolutely no mention of the original double-pancake copper-stabilized superconductor design. In JC's terse retrospective formulation: "Stop talking is the way decisions are made" (H).

RS, as protagonist of the Brookhaven design but also loyal member of the collaboration, contributed to the critical review of the KEK option by questioning the self-evident superiority of aluminum conductor from the viewpoint of managing stresses (I). RS sums up, carefully, skeptically: "Overall, we conclude that the use of aluminum-stabilized superconductor for the Muon g-2 magnet needs careful consideration of all stresses and motions with respect to allowed values. The advantages of aluminum stabilization over copper stabilization for the present application are not all obvious" (I).

Over the next few meetings, the Japanese proposal gradually gained credibility within the coil subgroup. In this process, JC, as formal expediter of design and construction, persisted in presenting schedules and advocating "do" agendas. The physicists clearly accepted the need to make decisions but were nonetheless reluctant to do so. Notwithstanding, on November 19, JC presented an R&D agenda for winding the coils outside-in.

At the January 14th meeting JC presented engineering details of coil/vacuum chamber construction based on the KEK approach. A collaboration memorandum by Brookhaven's magnet experts, Danby and Jackson, affirmed the collaboration's commitment to KEK's proposal on the basis of calculations of the forces on the coils (which turn out to differ little from those in the "pancake" design). The authors state:

> The use of "thin" coil construction techniques, using aluminum stabilized conductor wound from the outside in utilizing an aluminum support ring, was proposed by Hirabayashi and co-workers at KEK.
> The KEK members of the g-2 collaboration have extensive experience with this technology. This approach has been accepted by the entire g-2 group. As was pointed out by the authors, a single layer coil design was practical: ... the design presently proposed by the magnet coil subgroup uses single layer coils (C).

Danby and Jackson's note announced consensus on the KEK proposal to the collaboration and thereby effectively quashed any remaining resistance to adopting the Japanese design strategy. It was in effect a public rebuttal of Shutt's *Note 014*, which still questioned the wisdom of adopting the KEK proposal as an integral package.

The superconducting magnet coil controversy, which was first a tripartite struggle among Brookhaven, Yale, and ex-CERN designs, and later a battle between Brookhaven's hegemonic design and the new Japanese option, can be seen simply as a power struggle, i.e. as "political" strife in which the player with the best cards won out. But if this is politics, it is certainly politics within a constitutional framework. The point is that all the options were tested (by means of simulations, calculations, mock-ups, and prototypes) for their conformity to the physics and technological parameters set down in the proposal — including parameters like reliability and costs. AD's complaint, therefore, that Brookhaven's design was being adopted simply because "that's the way it was always done," seems to have been given the lie by the adoption of the Japanese option. Careful scrutiny seemed to have convinced everyone of the likelihood that it could embody the proposal's design requirements in a much superior — and cheaper — fashion. In the final analysis the text of the proposal — and not the contingent interests, strategies and conventions of the actors — coordinated the process of summatological construction of the magnet coils. The manifest design struggle was framed by an ironclad set of specifications and parameters to which all the players ultimately had to accede.

With the conceptual design hurdle taken, the engineers were free to move on to design-prototype-testing cycles to familiarize themselves with the problems of winding aluminum-base superconductor on large mandrels outside-in. These intermediate constructs were still only simulations of the real assemblages which would later be designed and assembled into the next active "frame" of the experiment, the working detector itself.

At the time of this writing, the basic support structure of the detector is in place and, with the commissioning of power supply and cryogenic subsystems, initial tests of the superconducting magnets and the resulting magnetic field are underway.[12] The next step is an elaborate process of "shimming" the magnetic field, i.e. fine tuning the field so that it meets parameters set down in the proposal in terms of uniformity and stability. In the meantime other subsystems are in various stages of design and construction and will be added as they are needed. Hence, although I have postulated a third phase of summatological construction involving the production of robust data within the frame of the working detector, I cannot yet report on this phase for g-2.[13] I have been told, however, that given the contentious nature of the acceptance of experimental findings in high energy physics, a strategy will be developed to incorporate a controversy within the collaboration itself about the significance of the detector's traces. One group will argue the case for interpreting the detector's traces as indicative of a real measurement of muon g-2 while another group will argue that the traces are background or an artifact of the operation of the detector itself. In the ensuing controversy the architecture and performance of the detector will inevitably be the main point of reference for both parties (Knorr-Cetina 1996). In other words, one can expect the assembly of data to be coordinated by the previously assembled frame of the physical detector in the same way as designing the detector was coordinated by the text of the proposal. The construction of data will clearly be just as contestatious, although here the contest will be contrived and have the dramaturgical aspect of a dress-rehearsal for the real controversy which may well take place — but at the field level.

4   Conclusion

In this chapter I have presented findings which suggest that the coordination of experiments in high energy physics depends on the production of summatological frames which are constructed in the course of the experiment itself. These frames are the goals of ongoing construction and in their final form are robust and coherent summaries of that activity. I argued that in the g-2 experiment, and quite likely in high energy physics experiments in general, two such frames are especially prominent, namely the formal proposal and the accomplished detector. This implies that the life cycle of high energy physics experiments exhibits three distinct phases of construction: getting the collaboration together, getting the detector together, and getting the data together. Each frame implies a subsequent trajectory of summatological construction which it coordinates by defining possible outcomes and establishing points of departure and constraints for action. Nonetheless, each phase of construction retains important degrees of freedom so that its accomplishment also incorporates decidedly non-trivial actions and decisions.

We can interpret this as an instance of what is generically known as "reversal" whereby agency, in becoming routinized as predictable patterns and expectations, recoils upon subsequent agency as a constraint. The g-2 experiment shows that this phenomenon is not confined to macro-sociological formations over the *longue durée* but can also play a role in relatively short-term organizational and inter-organizational processes (see also van Lente and Rip, and Duncker and Disco, this volume). It may be that this generation of intermediary frames which summarize a phase of action and carry its implications over into the coordination of a subsequent phase is a necessary feature of complex processes taking place over medium-range time spans of the order of several years and more. Such "frames" are like Wittgenstein's ladder, ad hoc structures which allow coordinated action to proceed in spite of the capriciousness of human memory and intentionality and which can then be cast aside when no longer needed. Countless obsolete detector-hulks and subsystems, mere way-stations to the production of data and publications, now lie scattered about the accelerator laboratories of the world.

Recent work in STS, particularly laboratory studies, allow us to approach this from a slightly different angle. Social studies of science and technology have taken pains to deconstruct science and technology as "bricolage" or "tinkering," rather than as premeditated, methodically planned, search processes. As Levi-Strauss originally conceived it, bricoleurs work intuitively, achieving results by assembly of materials given by the environment or willy-nilly accumulated in the course of previous projects, but not made specifically for the task at hand (Levi-Strauss 1972). Engineers, according to Levi-Strauss, typically consider the specific requirements of the problem at hand and, on the basis of reflexive assessments of what must be done and how to do it, acquire or design the necessary materials and components. Engineers, in short, make big plans and carry them out methodically, while bricoleurs have less clearly defined aims but greater flexibility due to a pragmatic and improvisatory style of work. Sherry Turkle (1984) in attaching a gender meaning to the distinction, speaks of the "soft mastery" of bricolage, as opposed to the "hard mastery" of engineering.

Levi-Strauss intended his distinction to highlight a contrast between two kinds of science, that of the primitive — involving magic but also the impressive practical knowledge and technologies of the Neolithic Revolution — versus that of the modern scientist and engineer. Barry Barnes' (1974) ironic commentary on Levi Strauss argued that, appearances to the contrary notwithstanding, modern science and technology were just as "neolithic" as the bricolage of Levi-Strauss' "sauvages." In subsequent studies of science and technology, scientists and engineers were portrayed as pragmatic actors making use of particular local assets to achieve "credit" in cosmopolitan scientific fields. Karin Knorr, arguing in this vein, described the aim as "success" rather than "truth" (Knorr-Cetina 1979). To "cash in" on bricolage, i.e. to transform a local product into scientific succes, it is necessary to "work up" the fragile and ambivalent local results into credible and objective data. Bruno Latour (1987) has described a number of canonical ways this is accomplished.

I would argue that Knorr-Cetina's (1995) inventory of ethnomethods i.e. the reflexive, communicative, and summatory agency which gradually produces textual and material constructs of increasing complexity and functionality as the experiment proceeds, is exactly how Levi-Strauss would have portrayed collective bricolage. Imagine a group of neolithic bricoleurs, single-minded in the pursuit of some future construct (some vision of wood, of signs, of knowledge) but differing in experience and abilities and constantly skirting disorder when it came to actually going about the work of assembling materials and putting them together. Such a collective bricoleur would constantly have to be, as Knorr-Cetina puts it, "humming to itself — about itself." It would constantly have to be reviewing what materials were available, it would have to "unfold the available objects" to see how they could be used and what they were good for, it would constantly have to be "making reports" to its several selves so that they could coordinate the production of the developing construct over time. Such a collective bricoleur would be very like a high energy physics collaboration seen from up close.

The evidence in this chapter (and in Duncker and Disco, this volume) support the widespread view in STS that scientists are tinkerers and bricoleurs. If you look closely enough you always see scientists engaged in Turkle's (1984) soft mastery. In their efforts to get collaborations together, detectors together, and data together, particle physicists manifestly follow novel and improvised routes in search of materials for the constructs they are working on. If you look closely enough all you see is the agency of bricolage and in this process "anything goes."

At the same time, it is utterly impossible to imagine how high energy physics experiments can be so complex, massive, and extended in space and time without massive coordination. How could a large, yet complex and delicate piece of apparatus like a detector simply emerge from "soft mastery" — at least on time scales of a mere 5–20 years? It would be preposterous to think that massive funding could be commanded simply on the basis of promises that collaborators intend to do their level best to produce the desired data. The very scale and complexity of high energy physics, as opposed to the benchtop sciences which have been the province of most laboratory studies, prompt a search for means by which bricolage is constrained and coordinated over space and time.

I have argued that while the state of the field provides the initial coordination for choosing problems and framing collaborations, two additional means of coordination are assembled in the course of the experiment itself: a proposal and an experimental machine. The two frames, I can now argue, are both summations of specific kinds of bricolage as well as frames for subsequent (and different) kinds of bricolage. In this way, engineering, in the sense Levi-Strauss used it, returns in the guise of constrained and well-oriented bricolage which repeatedly generates the constructs which are its own successive incarnations: experimental concept, formal plan, functioning detector, and finally compelling data. In this way, high energy physics experiments combine engineering and bricolage in a coherent process of premeditated exploration and construction on the fly.

Frames, to put it yet another way, are a kind of temporal "boundary objects" (Star and Griesemer 1989) which allow actors performing a specific type of bricolage to communicate with their future and other selves as performers of distinctly different experimental roles. Proposals, for example, mediate between physicists as politicians and physicists as engineers. Detectors mediate between physicists as engineers and physicists as scientists (if we understand the latter in the questionably restrictive sense of producers of empirical data).

High energy physics is not fundamentally different from other sciences, little or big, in this regard. Experimental sciences all share a similar fundamental structure. In every field, experimenters must do the same basic three things: organize participants and resources, assemble experimental apparatus, and reduce or produce compelling data. The differences — and hence in part the differences in visible "epistemic cultures" (Knorr-Cetina 1996) — derive from the relative prominence of the different phases of bricolage in each specific field. This relative prominence, in turn, seems to depend most immediately on the scale of the technologies needed to perform experiments and on the concomitant competition for access to scarce experimental facilities.

Most benchtop sciences seem to exist as many relatively small experimental sites with only a modest interchange of personnel among them. Moreover, the actual experimentation often depends on no more than relatively standard configurations of off-the-shelf components. Hence, in these types of sciences, the bricolage involved in producing data is the most indeterminate and hence labor-intensive portion of the entire trajectory and consequently, perhaps, the most visible and the most often studied. In high energy physics, and to varying degrees in other "big" sciences, it is the other two phases, i.e. defining and organizing the experiment and constructing the experimental apparatus, that immediately strike us as the fulcrum around which the scientific practice is organized. While it is important to recognize the "epistemic" differences associated with the relative preponderance of earlier phases of experimentation, it is also important not to lose sight of the basic similarities in the organization of experimental sciences. On this view, studies of high energy physics and other "big sciences" reveal generic aspects of experimental science which are also present, but atrophied in "normal" benchtop sciences. Having understood how high energy physics is gotten together — simply because we cannot ignore its massive organizational and technological aspects — we can start noticing things about benchtop sciences that may not have been so apparent before. For example how the seemingly decontextual bricolage of producing scientific success is in fact coordinated and constrained by textual and artifactual frames produced in earlier phases of the scientific work itself.

Notes

[*] Thanks to Wim Smit, Arie Rip, William Morse, Stuart Blume, Michael Lynch, Barend van der Meulen, and Richard Rogers for very helpful comments on earlier drafts. My appreciation also to the members of the Muon g–2 collaboration for allowing me to stick my sociological nose into their business.

[1] This concept was used by Derek L. de Solla Price as the *Leitmotif* of a series of lectures at Brookhaven National Laboratories, at the time one of the premier high energy physics facilities in the world. The lectures eventually became the chapters of his influential (little) book *Little Science, Big Science* (Price 1963). The experiment on which the research for this chapter is based is now, some 30 years later, under construction at the same facility.

[2] The g–2 WWW homepage is: http://www.phy.bnl.gov/g2muon/home.html.

[3] As a matter of fact even what little there yet is of the detector has already acted up.

[4] In a sense, the publishable results too are a kind of frame, only not of the experiment that produces them, but of a future experiment. Insofar as the results raise further questions or point to new directions for research, they shape the subsequent formation of new collaborations. This is another way of formulating a "genealogical" mode of progress in high energy physics (Knorr-Cetina 1995).

[5] Knorr-Cetina (1995) argues that high energy physics progresses "genealogically," i.e. via "lineages" and "generations" of experimentation. I am not sure if the unfolding of experiments follows such an autonomous logic in general, but the ancestry of the Brookhaven National Laboratories g–2 measurement (going back a number of "generations") certainly supports the thesis. Ironically, one of the underlying causes of successive generations of experiments, the continual upgrading of accelerator energies, is not a factor in the muon measurements. Here the issue is increased precision, for which beam luminosity is more important than energy. Quite aside from the opportunities offered by the AGS there are independent theoretical reasons to want to specify muon g–2 more precisely.

[6] Hughes is by now one of the "elder statesmen" of the high energy physics community. He and his contemporaries constitute a coordinative network tying together locations, resources and research agendas in the field. They can do this by utilizing

elaborate networks of personal contacts, impressive reputations, and well-developed talents for both physics and for organizing physics (Traweek 1988; Nothnagel 1992).

7 Knorr-Cetina (1996) notes how one of the varieties of "care of the self," namely "self-monitoring," produces knowledge of detector deviations which can be introduced as ex post corrections to data.

8 The list of institutional collaborators had grown again, specifically including a Japanese group and a "strong American group" as recommended by the Director of Brookhaven National Laboratories in the wake of the 1985 deferment. Participating at the time of the second application were: Boston University, Brookhaven National Laboratories, CCNY, Columbia University, Cornell University, University of Heidelberg, Los Alamos National Labs, University of Michigan, University of Mississippi, Sheffield University, University of Tokyo, and Yale University.

9 One of the most engaging descriptions of "design hierarchies" is Robert Pirsig's (1974) memorable Kantian deconstruction of a motorcycle.

10 In practice there may be other, in particular logistic and geometric, considerations that modify the sequencing so that the support structure is not the first assembly to be built. In AGS-821, for example, the magnet coils were the first elements to be designed and fabricated because their support structure consisted of an "exoskeletal" cryogenic dewar (cryostat) within which the coils would be suspended and which had to be welded in place around the completed coil assemblies. The fabrication of the coil-cryostat assemblies required the same building space that would later house the entire detector itself. If the basic support structure of magnet yokes had been assembled first, there would have been no room to wind the coils and assemble them in the cryostats. Hence, as the three coil/cryostat assemblies were successively built, they were removed to the parking lot behind the building. When the last of them had been assembled and moved to the parking lot, the iron magnet yoke bases were assembled inside the building, the two outside coils placed in position and the upper parts of the magnet yokes bolted on. This was the assembly of the basic support structure of the detector and at the same time the irrevocably material fixation of a number of the basic dimensional parameters for other subsystems.

11 I do not want to suggest that the assertions made here are in any way devious or untruthful. The most that can be said is that they tend to repress the possibility of unexpected difficulties. The point here is that the style of the proposal can be read as suggesting (incorrectly, in the light of hindsight) that optimal designs have already been found — and that the collaboration has unanimously recognized this.

12 The magnet assemblies were first powered up in January, 1996. On powering to full field, unexpected noises and excessive voltage differentials across the two outside coils required shutting the magnets down ("extracting the energy") in order to avoid damage. The cause appeared to be physical motions of the coils in excess of what had been predicted. These motions were due to shrinkage on cooling down and expansion on powering up (Lorentz force). Both of these forces and the resulting motions had been calculated in advance. Nonetheless, analysis revealed deficiencies in the design and number of positioning stops for the coils, an error which had allowed the coils to buckle and assume an unexpected oval shape. Both the electromechanical integrity of the coils as well as field uniformity required a new design for the stops (and in the event also better monitoring possibilities for coil position). Implementing this design required warming up the coil assemblies to room temperature, dismantling the top half of the magnet yokes, welding new stop assemblies into the coil-cryostats, re-assembling and readjusting the yokes and, finally, cooling the coil assemblies down again. All told it took until mid-May 1996 before the magnet coils

were ready for their second cooldown. At that point a major leak in the cryogenic system caused a further three week delay. On cooldown, which takes about a week, a new leak developed and to repair it the storage ring had to be warmed up once again. This was the situation as of June 10, 1996. The moral of the story is that even when the dangers are known and the utmost care is taken to avoid them, physics and engineering judgements remain just that — judgements. The story gives some inkling of the complexities high energy physics collaborators are up against and the costs of errors and helps explain why deliberations about optimal designs seem so nit-picking and interminable to outsiders. At the same time, ironing out the wrinkles in the machine is revealed as a stern learning process which allows the physicists, engineers, and technicians to come to terms with the capabilities and limits of their machine and to learn to feel at home in the electro-mechanical frame in which they will finally have to conduct their experiment and construct their data.

[13] For a fascinating example of the complexities entailed in this phase of bricolage see Peter Galison's (1987: 167-241) account of the struggle to confirm the Gargamelle detection of so-called "neutral currents."

Sources

A. AD (1987): "Evaluation of superconducting coil concepts," *Muon g–2 Note 004*, April 28.
B. *Brookhaven Bulletin* (1984), June 15, 38.
C. Danby, G. and J. Jackson (1988): "Forces on g–2 Coils with proposed single layer design," *g–2 note 016*, January 20.
D. FK (1987): "Minutes of April 26th, 1987, Meeting of three subgroups," *Muon g–2 Note 007*.
E. FK (1987): "Minutes of coil subgroup meeting," *Muon g–2 Note 009*, June 10.
F. FK (1987): "Minutes of coil subgroup meeting," *Muon g–2 Note 009*, June 19.
G. FK (1987): "Minutes of coil subgroup meeting," *Muon g–2 Note 018*, September 9.
H. JC (1993): Interview with JC on November 16.
I. Shutt, R. (1987): "Muon g–2 Magnet: aluminum stabilized superconductor wound 'outside-in': stresses and deflections during winding, cooling, and powering," *Muon g–2 Note 014*, October 26.
J. Hughes, V. (1986): Design of a Muon storage ring for AGS experiment 821, Proposal to Department of Energy, high energy physics Division (Draft); December 8.
K. Hughes, V.W. et al. (1985): *A new precision measurement of the Muon anomalous magnetic moment,* Proposal submitted to the Program Committee of the Alternating Gradient Synchrotron, Brookhaven National Laboratories, Upton, New York.

Chapter 5
Why Are Chickens Housed in Battery Cages?*

Ibo van de Poel

In 1992, the Dutch Ministry of Agriculture, Nature Management and Fisheries explained the high productivity of the Dutch poultry industry in an advertizing brochure as follows:

> The Netherlands has managed to secure a place among the world's largest egg exporters thanks to its efficiently managed poultry farms. The revolution in egg production started in the early 1960s, with the introduction of slatted floors and mechanization of feeding, manure removal and egg collection. A few years later the cage system was introduced and it gained ground rapidly. Pullet rearing units were established simultaneously. At present 90% of the layers and 80% of the pullets are reared in battery cages [...] (U: 6).

In this quote, and as a matter of fact in a lot of statements by either proponents or opponents of intensive livestock farming, the pursuit of efficiency is named as *the* reason for housing chickens in cages. But where does this pursuit of efficiency come from? Is it, as poultry farmers often try to convince us, an economic necessity or is it, as some animal rights groups want us to believe, an ideology, born out of the disregard of other issues, like animal welfare? Is the pursuit of efficiency, in other words, basically an *economic* category or a *cultural* one? In this chapter, I will argue that both perspectives have their merits, but insist that they fall short of presenting a comprehensive picture of stability and change in the design of battery cages. I will argue that, although the laying battery is technically a relatively simple system, the activity of designing this system is surrounded by diversified and complex forms of coordination. The stabilized outcome of these different modes of coordination will be called a design regime. Changing this regime, I will argue, is not only a matter of economics or ideology, but a matter of re-directing different interdependent modes of coordination.

1 A Brief History of the Battery Cage

In 1911, Professor J. Hulpin of the University of Wisconsin built what were probably the first cages in which to house chickens for egg production (E; M; Q: 79). The first serious experiments with what later became known as "battery cages" or "hen batteries" were carried out in the United States in 1924 at the Ohio Agricultural Experiment Station (E; P). Starting in the thirties, battery cages were commercially produced in the United States and the United Kingdom, but at that time the system was not yet in widespread use.[1] Most authors locate the large-scale introduction of the battery cage for egg-production in California in the fifties (B). The increasing popularity of the battery cage at that time and place seems to be directly related to the emergence of large-scale poultry farms. The battery cage made it possible to keep a very large number of chickens on a small area in an efficient way.

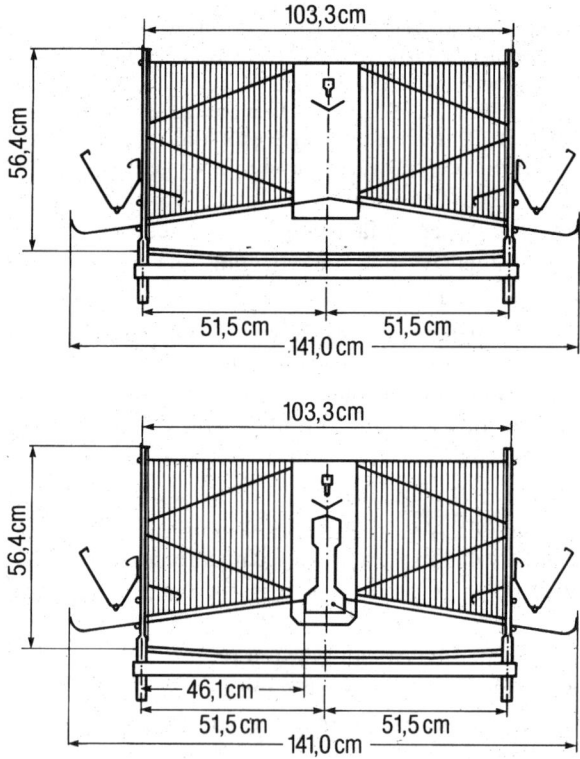

Source: Advertizement Brochure, *Salmet Deutschland*.

Figure 1: Typical Present-day Battery Cage in Cross-section

Large-scale poultry farming would have been impossible without scientific insights into the nutritional and health requirements of chickens (BB). Hence, the existence of scientific disciplines like veterinary science and animal husbandry was a precondition for the use of the battery cage on a large scale. Moreover, this essential knowledge had to be made available to farmers in a practical way.

As a matter of fact, a cognitive infrastructure around the husbandry of laying hens had already emerged long before the development of the battery system. By 1850 the first poultry organizations had been established in the United States (L). By 1870 the first specialized magazines on poultry farming had appeared. Between 1870 and 1900 more than 200 different poultry magazines were founded.[2] Although a great number of these folded before 1900, by then an institutional infrastructure had been established for the purpose of exchanging information on, and experience with, poultry husbandry. This included the exchange and dissemination of scientific findings. In 1921 the first international congress on poultry farming was held (EE). In the same year the International Association of Poultry Instructors and Investigators was established; changing its name in 1928 to World's Poultry Science Organisation. Various scientific journals on poultry and poultry husbandry were established.

In the Netherlands, a core role in the research infrastructure has been played since the 1920s by what is nowadays called the Spelderholt Centre for Poultry Research and Extension. At the Spelderholt, research is done on poultry housing, egg-quality and feeding.[3] Poultry research is also done at several Dutch universities, preeminently the Agricultural University at Wageningen and the Veterinary Faculty at the University of Utrecht. Practice-oriented research was long carried out at special experimental stations (farms), but since 1990 it has been concentrated at the Spelderholt.

The battery cage was not widely used in the Netherlands until the sixties, despite the fact that there was a lot of publicity about battery cages in the thirties and fifties. Two reasons can be given for this late adoption. In the first place, there was a lack of specialization in Dutch farming. Until the fifties, most farms in the Netherlands were mixed. As long as keeping chickens for egg production was only one of the activities of the farmer and therefore of marginal importance for his or her income, there was little reason to invest heavily in battery cages in order to produce eggs more efficiently. The second reason was the absence of farms, whether mixed or specialized, where poultry was held on a large scale. This was due chiefly to government regulations, especially the so-called *Pluimveeregeling* (Poultry Regulations), that put a legal ceiling on the number of chickens a farmer was allowed to keep. This *Pluimveeregeling* was supported by the farmers' organizations, because they feared the intrusion of outsiders into the poultry industry, i.e. non-poultry farmers looking for profitable investments. The consequent small scale of the sector, which persisted till 1961, did not provide an incentive to invest in more efficient housing systems like battery cages.

In 1961, the revocation of the *Pluimveeregeling* was followed by rapid upscaling in the poultry sector. The proportion of hens kept at farms with more than 1,000 chickens rose from 7% in 1961 to 23% in 1964 and 40% in 1966 (Q: 158).[4] The number of farms where poultry was kept fell from about 199.000 in 1960 to about 94.000 in 1967 (Q: 161). In the meantime, the number of specialized poultry farms increased. The decline in the number of farms and the rise in number of chickens kept per farm was, however, not only due to the revocation of the *Pluimveeregeling* but also to developments in the market for eggs. In the fifties, the Netherlands became one of the most important egg exporters in the world. The establishment of the European Economic Community (EEC) in 1958, seemed to offer good opportunities for a consolidation or even expansion of this position, because the EEC open market would put an end to the protectionist measures of some (EEC) countries like England. However, the formation of an open market was prefaced by a transition period of six years. During this period, EEC-taxes were imposed, disadvantaging countries like the Netherlands which had achieved a high productivity per hen. In some cases these EEC-taxes were even higher than the existing import taxes. The result was a decline in Dutch egg exports from 3,4 billion in 1961 to 1,3 billion in 1966. During this period, many of the existing farms disappeared.

In the sixties, and partly as a result of these developments, Dutch poultry farmers began to demand housing systems which could accommodate more hens per square meter and ultimately produce eggs more efficiently. Keeping chickens indoors had in any case already become common practice in the Netherlands by that time. Normally this was done in big sheds, in which large numbers of chickens were kept on the floor. Early in the sixties slatted floors were introduced in such sheds and the traditional manure area was covered with slats and enlarged to half and later two thirds of the entire floor area (see Figure 2). This made it possible to keep more chickens per square meter.

In the latter half of the sixties the battery cage became increasingly popular and the system progressively replaced the traditional housing systems and the slatted floor system. Although the housing costs per chicken in laying batteries were probably somewhat higher than in other systems, the battery cage had a number of advantages over slatted floors (and traditional housing systems); advantages like lower food conversion,[5] a smaller likelihood of diseases (due to the absence of litter in battery cages) and better egg quality. Moreover the system required less labor per hen and in particular

5 Why Are Chickens Housed in Battery Cages? 147

did not require the collection of so-called ground-eggs — i.e. eggs laid on the ground by hens.[6] In the sixties and early seventies there seemed to have been some discussion about the efficiency and advantages of the battery cage vis-à-vis slatted floors. But despite these discussions, the battery cage rapidly became popular. In 1971, 50% of the chickens was housed in battery cages; in 1976 this had risen to 80%. Presently more than 90% of the chickens is housed in hen batteries (Q: 221). An important reason for the rapid adoption of the battery cage in the sixties was probably also the fact that after 1965 — when the transitional EEC-taxes came to an end — the profitability of poultry farming started rising again. In these circumstances it was especially advantageous for farms with a large number of hens to switch to battery cages.

Source: DD.

Figure 2:   Slatted Floor or "Scratching" System

Battery cage design in the sixties and seventies was oriented chiefly to one goal: efficient systems, i.e. systems that made it possible for the farmer to produce quality eggs for as low a price as possible. Achieving efficiency demanded attention to three factors: Housing, food, and labor. Efficient housing meant as many chickens per square meter as possible. However, there was a limit to the number of chickens that could efficiently be held per square meter, because a hen's productivity is dependent on the number of chickens held in a battery cage and the size of the battery cage. Efficiency with respect to food meant that "food conversion" should be as low as possible. Food conversion was one of the parameters used to evaluate the results of tests with housing systems. Efficiency with respect to labor was generally achieved by different forms of mechanization, for example with respect to feeding, egg removal and manure removal. As a rule, these forms of mechanization were easier to implement in battery cages than in traditional housing systems (see Figure 3).

Source: Advertizing Brochure, *Salmet Deutschland.*

Figure 3:   Battery Cage Showing Feeding and Egg-removal Mechanisms

The design of battery cages in the sixties and seventies was guided by generally accepted heuristics,[7] like more chickens per square meter, lower food conversion, more mechanization. But while these kinds of heuristics were helpful in designing efficient systems, they did not in themselves guarantee maximum efficiency. The development of the battery cage was therefore accompanied by a lot of trial and error and testing, in experimental setups as well as in the field. At present, after several decades of development, the search for more efficient systems has become a matter of detail design.

5   Why Are Chickens Housed in Battery Cages?   149

Moreover, most designers already know "what will work" on the basis of experience (see Figure 4).

Source: Advertizing Brochure, *Salmet Deutschland*.

Figure 4: Overview of Present-day Battery Cage Setup

2   The Primacy of Efficiency

Supporters as well as opponents of the battery cage tend to see its emergence as the logical outcome of a process wherein efficiency was the main goal. Of course, a historically more adequate account would show that there was not a simple linear development leading to the battery cage as the one and only solution in the drive for efficiency. On the other hand, it seems quite reasonable to say that in the sixties and seventies efficiency became the prime goal in the design of battery cages. Why did efficiency become so important in these years? Two explanations will be given below. In the first explanation the drive for efficiency is considered to be the outcome of economic developments since the fifties. In the second explanation the drive for efficiency is regarded as the result of a culture or, perhaps better, an ideology of efficiency.

The economic argument is probably more popular among the proponents of the battery cage, whereas opponents will probably lean toward a more cultural explanation. Nonetheless, the arguments given below should not be understood simply as the arguments of pro-and opponents. The argumentations are my own constructs, building on insights from economics and technology studies. They deliberately stress only one line of argumentation, economic or cultural, so as to make possible a more explicit discussion of the strengths and weaknesses of these lines of thought. As we shall see, this approach will lead us to a more *comprehensive explanation* of the striving for efficiency in the design of battery cages.

A The Economic Argument

According to the economic line of argumentation, the adoption of the battery cage in the Netherlands is the outcome of economic transformations with which the users of battery cages — i.e. the poultry farmers — were faced. To understand the nature of this change, it is important to realize that the market for battery cages has a layered structure. The users of battery cages — i.e. the farmers — are the producers of eggs, supplying mostly anonymous egg-consumers. There are two important groups of egg-users: the end-consumer and the egg-based foods industry. This last group, which has to compete on specific consumer markets, is extremely sensitive to the unit price per egg. If they are unable to produce cheaply enough, they will either lose customers or face a decline in profits.

Because poultry farmers have to compete with each other on the egg market, they regard efficiency in terms of costs per egg produced as an important criterion. Due to the formation of the EEC in the fifties and sixties, this competition became so harsh that only poultry farms producing cheap high-quality eggs efficiently could survive. So the market structure that came into existence in the sixties — many large scale poultry farms competing on an anonymous market — forced farmers to produce eggs more efficiently. Those who did not do so risked going bankrupt, as indeed a good number did. This in turn meant that the only viable housing systems were those that made it possible to produce eggs efficiently. This suggests that the demand for efficient housing systems is not primarily an outcome of the convictions of users, the farmers, but of their economic position. Any poultry farm that did not produce eggs efficiently enough was likely to go bankrupt, or at least to pay a price in terms of its market share.

According to this argument, it is not the convictions of the parties involved, but rather the economic structure of the sector that determines what

5 Why Are Chickens Housed in Battery Cages?

kind of systems will survive. Therefore, housing systems that are not efficient enough in the given economic circumstances, can only survive if these economic circumstances are changed, for example by creating new types of markets.

B The Cultural Argument

According to the cultural argument, the pursuit of efficiency in the design and use of battery cages is due to a cultural disposition towards efficiency. This disposition is embedded in the rhetoric used and ideology shared by the different parties. In other words: "efficiency" is a rhetorical term and a normative goal that guides the behavior and interactions of the different actors involved.

This coordinating power of efficiency can be explored by seeing it as a "guiding principle." I use the concept "guiding principle" here in the sense developed by Wim Smit (see Smit, Elzen, and Enserink, this volume), who has applied it to the study of military technology development. Smit describes the concept as follows:

> We have defined the concept of 'guiding principle,' first, as an inter-organizational concept, that plays a role in the interactions between the various organizations in the military technology network, and, second as an interface between military doctrines or strategies [...] and the concrete shaping of weapon systems (Smit 1993: 402).

In other words a guiding principle is a form of rhetorical coordination, because the guiding principle is used as a legitimating principle by the various actors involved in technology development and guides their (inter)actions. It is also a form of normative coordination, because it restricts these (inter)actions by relating them, explicitly or implicitly, to general strategies, doctrines or, I would add, values.

As rhetorical means of coordination, guiding principles act like what Susan Leigh Star has called "boundary objects." She defines boundary objects as "objects which are both plastic enough to adapt to local needs and the constraints of several parties emplacing them, yet robust enough to maintain a common identity across sites" (Star and Griesemer 1989: 393). This means that a guiding principle, as a kind of boundary object, has a shared as well as a group-specific meaning. This is recognizable in the case of the design of battery cages, where the concept efficiency is a common denominator, while at the same time the main actors involved have different specific interpretations of what it means to strive for efficiency.

For the *users of battery cages*, i.e. the poultry farmers, efficiency means in the first place minimal costs per egg produced, that is, a *pro rata* portion of the housing and investment costs, the cost of feeding the chicken, labor costs, etc. With respect to the *design of battery cages* the general guiding principle (efficiency) is translated into more specific heuristics like more chickens per square meter, lower food conversion, more mechanization, and in methods for testing and evaluating housing systems, generally accepted lay-outs for laying batteries, etc.

It could be argued that other relevant groups, such as agricultural suppliers, researchers, policy-makers, consumers of eggs etcetera, also have specific interpretations of efficiency, which guide their day-to-day practice. The important point, however, is that these interpretations are all interpretations of one general goal, the striving for efficiency. In other words: the actions and interactions of the actors involved are all guided by the general drive for efficiency.

According to this argumentation, efficiency and its derived terms have become a kind of common language in the research, design and use of battery cages. This implies that actors will *normally* not undertake actions that cannot be explained in terms of this pervasive rhetoric or cannot be legitimated according to this reigning ideology. This does not mean that actors are unable to undertake such actions, but that deviance requires the development of new rhetoric or ideologies. In this sense, deviant actions are (strongly) discouraged by a guiding principle.

We now have an economic and a cultural explanation for the primacy of efficiency in designing battery cages. My purpose is to highlight some of the strong and weak points of these explanations by discussing their relative ability to explain the enduring primacy of efficiency despite the changes that took place in the design of battery cages from the seventies onwards. These changes will be described in the next section.

3 Attempts to Change the Design of Battery Cages

In the sixties, when the change-over to the battery cage — at least in the Netherlands — was still under way, the first criticisms of the effects of the battery cage on animal welfare began to be heard. The appearance of Ruth Harrison's *Animal Machines* in 1964 is widely regarded as the point of departure for resistance to the battery cage (M). In her book Harrison attacked the drive for efficiency in animal husbandry on the modern farm. She showed that people have a quite unrealistically romantic picture of animal life on the farm. In fact, as she argued, animals on the modern farm are reduced to production machines. Harrison's book was widely read and suc-

5 Why Are Chickens Housed in Battery Cages?

ceeded in creating a negative public image for the battery cage and other intensive housing systems (O; Q: 242).

In the early seventies two new animal rights groups were established in the Netherlands: *Lekker Dier* (Nice Animal) and *Rechten voor al wat leeft* (Rights for all that lives). These groups promptly attacked the battery cage. Later, they were joined by the venerable *Dierenbescherming* (Dutch Society for the Protection of Animals), founded in the nineteenth century. These groups succeeded in catching the attention of the media and of several politicians (Q: 242–243). However the growing public unease with the battery cage had as yet no direct impact on the design of battery cages.

A   Changes in the Market

Starting in 1972, animal rights groups tried to introduce an alternatively produced egg onto the Dutch egg-market, i.e. one not produced in battery-cages.[8] This so-called *scharrelei* (scratching egg)[9] was produced in chicken sheds with slatted floors, in which at least a third of the ground area was not covered with slats, in order to give the chickens the possibility to scratch (see Figure 2).[10] Such sheds with slatted floors were still in use on some farms in the early seventies; however, the eggs produced at these farms could not be distinguished as scratching eggs by the consumer. This spurred animal rights groups to develop a label that could be given to eggs produced in systems more benign to animals than battery cages.[11] Farmers did not like the idea of selling discernibly alternative eggs, probably because it would imply recognition of the fact that battery chickens lived in worse circumstances.

In 1974 the first scratching eggs were put on the market at a somewhat higher price than "normal" eggs. Thenceforth, it became possible for consumers to choose between "benign" and "cheap" eggs. However, authenticating the origins of the scratching eggs — which in the beginning was done by the group *Rechten voor Al wat Leeft* — was problematic, partly because there were no legal procedures for certification. Incidental fraud was reported. In an attempt to save the situation, the Dutch government issued a *Landbouwbesluit Scharreleieren* (Agricultural Decree on Scratching Eggs) in 1978, enforceable as of January, 1979.

This decree created a new certifying authority.[12] This authority, in which farmers, egg distributors, and animal rights groups participated, became responsible for authenticating the origins of scratching eggs. To make scratching eggs easily recognizable for the consumer, a special stamp (a little lion) was designed. The stamp, as well as the word *scharrelei* (scratching egg), were copyrighted.

The institutionalization of a new market for more benign eggs, which gave consumers the possibility of choosing for animal welfare, not only required legal procedures, but also the support of the egg distributors and supermarkets. In 1976, Albert Heyn, a large Dutch supermarket chain, undertook a trial with scratching eggs in ten of its stores. The trial was successful and an important channel for distribution was thereby created. The sale of scratching eggs increased until 1982, but then stagnated. The main reason was held to be the limited availability of scratching eggs. Therefore, in 1984 the *Dierenbescherming* started a national campaign to promote the sale of scratching eggs. Supermarket chains, the hotel and catering industry and the bakery sector were approached. The *Dierenbescherming* not only tried to create new distribution channels, but also developed special stickers to make the use of scratching eggs in bakery and restaurant products more visible. As a result of this campaign, the consumption of scratching eggs started rising again. The number of scratching eggs delivered to packing stations rose from 218 million in 1984 to 632 million in 1990. In 1991 approximately 10% of the laying hens in the Netherlands was kept as scratching hens (Z). In 15 years the scratching egg has captured 20% of the domestic egg market. However, as things now stand this may remain only a limited success.[13] Eggs used in the egg-based foods industry are not discernible as either battery cage or scratching eggs for the end consumer. Moreover, 75% of the eggs produced in the Netherlands are exported and the scratching egg plays only a limited role in this export.

B A (New) Role for Ethologists

Criticizing the battery cage or promoting alternative eggs required explaining why chickens suffered in battery cages and were better off in alternative systems. For some, it was quite clear why the battery cage was an inhumane system. *Lekker Dier*, for example, stated in a brochure:

> The use of laying batteries, in which 95% of the laying hens in the Netherlands are kept, radically changed the living conditions of the animals. The laying battery is practically the opposite of the farmyard. A laying battery consists of a number of small mesh cages, each with a floor area of 40 by 50 centimeters. In each cage there are four to five hens, placed there when they are seventeen to eighteen weeks old. During the laying period more than 10 percent simply die. After fourteen months the survivors are pulled out of the cages and put into crates. With broken legs and wings they go into the chicken soup (S).

For others, there was no clear proof that hens in battery cages actually suffered. Take, for example, the following commentary by a farmer on the purported humanness of an alternative system (the aviary: a system with a somewhat lower productivity): "If chickens felt better in an aviary, they would also produce better. But they don't" (A).

As in many societal controversies, scientists were called in by the contestants to buttress their respective positions. "Science" in this case, at least for the critics of the battery cage, meant ethology. Ethology is a branch of biology that studies animal behavior. Ethology had already played a role, albeit a marginal one, in the design of battery cages. In the fifties, for example, important ethological research had been done on the social behavior and status hierarchy of chickens in relation to their egg production (HH).[14] This research doubtless had an impact on the development of the battery cage.

Although ethological welfare research on chickens was not widely carried out until the sixties,[15] individual ethologists were among the first to criticize the battery cage (BB). This is not surprising. The disciplinary basis of ethology is the study of the behavior of animals in their natural environment. This "natural" behavior gives ethologists a kind of reference point with respect to which they can claim to discern "abnormality" in the behavior of, for example, chickens in battery cages. Deviant or absent behavior can then be interpreted as possible failure of the animal to adapt itself to the new environment. So, ethology as a science had a normative standard by which to judge the suffering of animals. Of course, this did not mean that all ethologists agreed on the level of animal welfare in battery cages or on possible measures that might be taken. However, ethology at least offered some instruments and concepts with which to say something about animal welfare. Moreover, as part of their professional life, most ethologists had studied animals intensively and over long periods. According to some observers this made them more sensitive to animal welfare than, for example, veterinary scientists.[16]

Governments also called in ethologists (G). In the sixties and seventies, governments were under public pressure to take measures to encourage animal welfare in intensive livestock farming. For example, the British government installed a committee on animal welfare six weeks after the appearance of Harrison's *Animal Machines*. This committee, named after its chairman Roger Brambell, a well known professor of zoology, also included the ethologist William Thorpe. The committee's report advanced several criteria for animal welfare in housing systems: "An animal should at least have sufficient freedom of movement to be able, without difficulty, to turn around, groom itself, get up, lie down and stretch its limbs" (O: 120). On the basis of these kinds of criteria and insights, ethologists developed more detailed ideas about requirements for benign housing systems. With respect to housing systems for chickens the following kinds of requirements were often named: the number of hens per square meter, the possibility for chickens to scratch and take dustbaths (presence of litter); the presence of laying nests (to lay eggs in); the presence of perches (R). These sorts of requirements also came to function as heuristics for the development of alternative systems.

C   The Development of Governmental Regulations

Most countries had generic laws for the protection of animals. What such laws implied for the keeping of chickens in cages was, however, often unclear. In a number of countries animal rights groups have sued farmers that keep chickens in cages, mostly with little success (O: 118–135). In 1979 several German court decisions, however, declared that hen batteries violated provisions of the German animal protection law of 1972. Nevertheless, the defendants were not prosecuted because it was judged that they could not have known that this *common practice was illegal*.

In different countries, as a result of public pressure, new laws were developed to protect livestock animals. In these laws ethological insights played an important role. For example, the "European Convention for the Protection of Animals kept for Farming Purposes," introduced in 1978, stated: "Animals shall be housed and provided with food, water and care in a manner which [...] is appropriate to their physiological and ethological needs [...]" (O: 128).

Because of the international structure of the egg market, most West European countries preferred to encourage EEC legislation, instead of national laws that would "punish" only their own farmers. In 1986, EEC rules with respect to laying batteries were laid down in EEC Directive 86/113/EEC.[17] This directive stipulated the minimum requirements for laying batteries coming into use after January 1, 1988. The requirements were: at least 450 cm^2 floor area per hen, 10 cm feeding trough per bird, 40 cm height over at least 65% of the area and a floor–slope of maximally 14% (see also Figure 5) (J). In most West European countries, these requirements are now mandatory. In some countries, however, requirements for floor area are more stringent. In Denmark, for example, the minimum requirement is 600 cm^2 per hen; in Germany and England the requirements are more stringent for heavier chickens or for fewer than four hens per battery cage.[18] None of these national and international requirements, however, actually compromise the legality of the battery cage as such. The only country at the moment that has such stringent legal requirements for hen housing systems that they amount to a ban on the battery cage, is Switzerland (V).

5 Why Are Chickens Housed in Battery Cages?   157

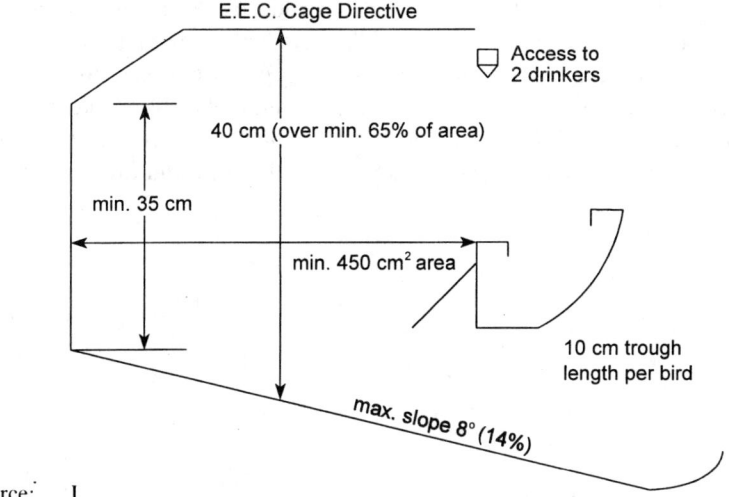

Source: 1.

Figure 5:   Effects of EEC Legislation on Battery Cage Design

D The Development of Alternative Systems

As we have seen, ethological research had pointed to some basic requirements and heuristics for more benign poultry husbandry systems. By the seventies, these insights began to encourage the development of alternative systems in the Netherlands and other West European countries. In the Netherlands, important research was carried out at the Spelderholt.

Bareham's development of the so-called get-away cage in 1976 is generally cited as a starting point for applied research on alternative systems. However, in some countries applied research had already commenced some years earlier (D; R; FF). The get-away cage is a battery cage with special areas for perches, laying nests and litter. These special areas increase the quality of the system from the point of view of animal welfare. In the beginning, most research concentrated on the get-away battery cage and other modified battery cages, because the scratching system has two major disadvantages: an increased risk of poultry diseases due to the wet litter and an increased risk of cannibalism (chickens killing each other under stress), which is more serious in a scratching system because there are more animals sharing the same space (FF: 20). In the course of time, however, it became apparent that modified hen batteries had their own disadvantages. The litter in these systems caused problems with respect to eggs laid outside nests and the chickens were more difficult to access and inspect. Moreover, modified hen batteries turned out to be labor-intensive (D: 5). Hence, other types of alternative systems were developed. One of them is the aviary or, as it is called in the Netherlands, "volière." This system is characterized by the presence of several levels on which the chickens can drink, eat, and rest (see Figure 6). Because of the different levels, a large number of chickens can be held per square meter of floor area. On the other hand, the aviary affords chickens the opportunity to scratch, take dustbaths, rest and lay eggs in laying nests.

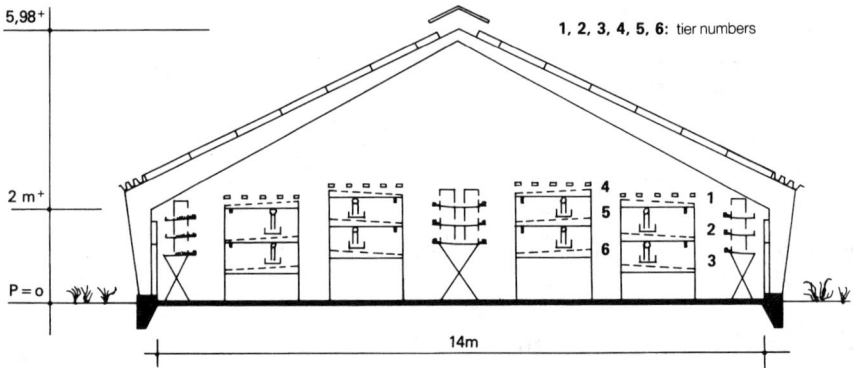

The system has three tiers, of which the upper is a resting area. The bordering mesh floor is mounted at different heights to create a staircase effect. The entire ground area is covered with litter.

Source: DD: 25.

Figure 6:   The Spelderholt Aviary System

Alternative systems would probably never have been developed without some form of active political coordination. In most countries research on alternative systems got off the ground only when government subsidies became available or as a response to government threats of a possible ban on laying batteries. Research institutes did not develop alternative systems earlier because most farmers — i.e. the users of the systems and the main clients of most research institutes — were clearly uninterested in alternative systems. Moreover, farmers in the Netherlands specifically refused to pay for research on alternative systems which were not based on the battery cage.[19] Hence, only the government remained as a source of funding for research on alternative systems.

The political coordination of the development of alternative systems also had some effects on the contents of the research. This was because governments were interested in a system that would compromise neither welfare nor economic considerations. The projects at the Spelderholt were defined in

5 Why Are Chickens Housed in Battery Cages?

those terms, i.e. they were based on the pursuit of animal welfare as well as efficiency.

The Dutch government defined the results of research on alternatives to the battery cage as a possible basis for a legal ban. Because of these implications, which were anathema to the farmers, the contending parties tended to interpret the results of research at the Spelderholt in terms of the political question: "Must the battery cage be banned?" This form of political coordination is worth describing in more detail.

In 1984 the Dutch parliament passed a so-called initiative law, proposed by representatives Tazelaar and Van Noord. In addition to specifying minimum requirements for the laying battery, the law charged the minister of Agriculture with formulating tighter requirements before January 1, 1990, to become effective as of July 1, 1994. The law aimed at nothing less than a ban on the battery cage. However, as a prerequisite it stipulated that an economically and technically feasible alternative should first be available. The Spelderholt was seen as the legitimate authority to develop and demonstrate such an alternative.[20] In the same year that the Tazelaar/Van Noord bill was passed, the Spelderholt abandoned efforts to modify the battery cage and concentrated its research on aviary systems. The development of the Spelderholt aviary was completed in 1986 and in 1988 a series of comparative tests with the battery cage were started. The first round of these comparisons was favorable for the aviary. Based on this success, the minister of Agriculture expressed his intent to forbid the battery cage as of July, 1994. Director De Wit of the Spelderholt appeared to support this decision: "The battle of the battery is lost. We need no longer entertain illusions; there will be a ban on the battery. The challenge now is to manage the transformation in such a way that the Dutch poultry industry comes out unscathed" (Y). At the same time, De Wit defined the Spelderholt system as fully developed:

> Research has done what it can do. Politics and economics must now speak. We think that the price difference of 0.8 cents [per egg, IvdP] cannot be reduced by further modifications to the floor system [aviary, IvdP]. All cost increases are now due to welfare augmenting measures. Less is impossible.

The farmers, however, firmly rejected the impending ban on the battery cage. Heated discussions between proponents and opponents of the battery cage ensued. These discussions, fired by the success of the first round of the comparative tests at the Spelderholt, encouraged further comparisons between the battery cage and the aviary. However, the subsequent second round of tests had to be terminated because of a defect in the feeding system. In an article in the magazine *Pluimveehouderij* (*Poultry Husbandry*) staff members of the Spelderholt speculated on why the second test with the aviary system failed. In so doing, they seemed to hedge on the more politically loaded statements of director De Wit:

> In the first place: staff members of 'The Spelderholt' and the IMAG have designed and tested the aviary system and have interpreted and published the results to the best of their ability. All along, we have stated that more research is needed before definitive conclusions can be drawn. Secondly: the research in question has no bearing on possible policy decisions with respect to the use of this and/or other housing systems in practice (AA).

These efforts to keep the political and technical dimensions separate were not very successful. In the press, at least, the technical and political issues were indiscriminately confused. For example, in the *Agrarisch Dagblad* (*Agricultural Daily*) of November 8, 1990, it was reported that De Wit rejected speculations that the third round of comparative tests was going to be a failure. The same article also reported that Tazelaar — former member of parliament, co-author of the 1984 law and now, ironically, chairman of the *Produktschap voor Pluimvee en Eieren* (Product Board for Poultry and Eggs) — considered a ban on the battery cage unacceptable:

> An acceptable alternative system is presently not available. If the laying battery were to be abolished in our country at this time, laying poultry farming will also be abolished. Enforcement has to take place in Europe as a whole. I do not think that likely by 1994. (A).

Van Noord, the other erstwhile proponent of the Tazelaar/Van Noord bill, had already said something similar in 1989. Indeed, parliament ultimately decided to temporize the enforcement of the law for an indefinite period, the main reason being potential economic damage to the farmers. It was therefore decided to keep in step with the European Union.

E The Spread of Alternative Systems

A number of alternative systems is now on the market. The Dutch firm LACO BV has been producing aviaries since 1981, initially for the Swiss market, where the battery cage is legally banned. At the moment, LACO BV also sells these aviaries in Great Britain, Belgium and the Netherlands. Since 1990 the aviary designed by the Spelderholt has been produced by the Dutch firm Rijvers BV. A number of foreign firms also produce aviaries.

Most farmers do not want to buy and use these aviary systems because they imply higher production costs per egg. According to the poultry farmers, the system has some other disadvantages as well: aviaries are more labor-intensive, they produce more dust which further aggravates labor conditions and entail a greater risk of poultry diseases due to the litter. Moreover, aviary systems have a higher ammonia emission. As environmental pollution standards become more stringent in the Netherlands, this is becoming an important disadvantage for the aviary. Currently, the Spelderholt is investigating how to minimize aviary ammonia emissions. Given the fact that the aviary's higher ammonia emissions are chiefly due to the presence of litter, which is seen as a prerequisite for animal welfare, the (new) design criteria of animal welfare appear to conflict with the criteria for low levels of ammonia emissions.

4 Discussion

The efforts undertaken by various actors to change the design of battery cages have, as we have seen, led to some results. Nevertheless, the drive for efficiency is still dominant in the design of battery cages. Why? For an answer, let us reconsider the two arguments for the primacy of efficiency given before.

A The Economic Approach, New Markets for Eggs?

According to the economic approach, a change in the design of battery cages can only follow on a change in the economic position of farmers. This also seems to have been the reason for animal rights groups to create a market for scratching eggs. Before these specially stamped eggs were introduced, consumers could not opt for benign eggs. This change in the structure of the egg-market required three things. In the first place eggs had to become distinguishable as either battery cage or alternative eggs. In the second place institutions and legal procedures had to be called into existence to monitor the origins of eggs distinguished as "alternative." Third, arrangements had to be made to actually sell scratching eggs through existing distribution and selling channels. Under these conditions, scratching eggs became an interesting market niche for farmers, especially for those already using the system, albeit under a different name. And, although the production of scratching eggs was somewhat more labor-intensive, the farmers were compensated by receiving a higher price per egg.

Economic arguments can explain why aviaries and other alternative systems are as yet unsuccessful. They are too expensive in terms of cost per egg, because only a small proportion of consumers is prepared to pay more for benign eggs. This line of reasoning also explains in part why the Dutch government does not yet want to forbid the laying battery: this would mean exporting poultry farms instead of eggs. However, the economic argument falls short in at least two important respects. In the first place, it *tends* to present the economic situation as given. But as the story of the scratching egg shows, the economics of the egg-market can be changed by creating new institutional structures which offer both producers and consumers of eggs new possibilities. Switzerland offers interesting examples of such openings. In this country, there are a large number of small poultry farms (T).[21] The internal price of eggs is relatively high. For every Franc they spend on imported eggs, Swiss egg importers must by law spend a certain amount on eggs produced in Switzerland by small farms — the so-called "percentage rule."[22] Moreover, imported eggs intended for direct consumption have to be recognizably stamped. Despite these measures, more than 50% of the eggs consumed in Switzerland — those for direct consumption as well as those destined for egg-products — is imported. Nonetheless, the "percentage rule" protects existing Swiss farms from a further decline in their market-share. This rule combined with the relatively high internal price level for eggs, explains how the Swiss government could ban the battery cage with relative ease and without fear of major adverse effects on Swiss poultry firms. There was simply little terrain to be lost to egg producers in other countries.

The second shortcoming of the economic explanation is fundamental. Actors never act on the basis of the "real" economic situation — whatever that may mean — but on their perceptions of it.[23] Farmers, for example, believe that consumers in foreign countries and the egg-based foods industry are not prepared to pay more for benign eggs. But they are not sure of this fact because it has never been tested. Animal rights groups have quite different perceptions of the willingness of people to pay more for benign eggs. The point is that actors tend to over- or underestimate the possibilities for economic change — thereby enlarging or reducing these possibilities — on the basis of their perceptions. In other words, the possibilities for economic change are not only due to some real economic situation but also to the perceptions of the actors involved. And these perceptions can be influenced by reigning cultures and ideologies. So, the adherence of farmers to battery cages cannot be explained solely in economic terms.

B   The Cultural Approach: Towards a New Guiding Principle?

If the cultural argumentation is right, re-directing the design of battery cages involves changing the guiding principle. As a matter of fact, efforts to influence the design of battery cages started out by criticizing the existing guiding principle, although, it is doubtful whether such criticism had its roots in an elaborate (cultural) analysis of the situation.

There seem to be two reasons why the attack on the laying battery started with criticisms of its legitimacy. The first reason is the protesters' heartfelt rejection of the simple fact that efficiency instead of considerations of animal welfare "guided" the design and use of battery cages. The other, related, reason seems to be that it is relatively easy to mobilize the general public on an issue like animal welfare. Humane values, at least as values, are deeply rooted in Western civilization. Few would contest that it is inhumane to let animals suffer unnecessarily. (The question, of course, then becomes: "What is unnecessary?"). Animal rights groups adroitly exploited the fact that people do not like to see animals suffer (see Figure 7). This dramatized the tension between the guiding principle in the design of battery cages and more generally accepted humane values.

Figure 7:   Picture Used by the *Dierenbescherming* to Demonstrate That Chickens Suffer in Battery Cages (The text reads "Put an end to this chicken-egg situation").

The technical language used by the designers of battery cages tends, by contrast, to veil this tension. "Drop-out rate" sounds better than "percentage of chickens killed" or even "mortality rate." And a term like "food conversion" sounds more neutral than giving chickens just enough food to lay good eggs and to survive. In most advertizing brochures the battery cage looks not only like an efficient system, but also like a very clean one (see Figures 3 and 4). Some battery cage producers have even presented "traditional" battery cages with a slightly larger floor area per hen as "humane" (see Figure 8).

The parties involved seem to be quite aware of the ideological bearings of their words and pictures. At the most general level, for example, farmers and their organizations like to talk about "intensive livestock farming," whereas opponents prefer the word "bio-industry."

The headline reads: "Five-star hotel for laying hens. An enormous improvement for man and beast." The advertisement praises the excellent food conversion, the reduction in ammoniac emission, the fact that the chicken has 35% more room compared to 1985; and the fact that the eggs cost 2 cents less apiece than aviary eggs. It should be noted that the system is in most respects identical to existing battery cages.

Source: II.

Figure 8: Advertizement by *Zonne-Ei-Farm BV* (a firm that sells German battery cages on the Dutch market)

In the Netherlands, criticism of the existing guiding principle was in certain respects successful. The general public was quickly persuaded that animals suffered unnecessarily in battery cages. Governments felt obliged to take measures in regard to animal welfare in laying systems. Ethologists translated the general preference for benign chicken husbandry into more detailed requirements and heuristics. These design heuristics in fact came to direct the search for alternative systems. And the heuristics for alternative housing systems were reinforced by legal norms for alternative eggs, like the *Landbouwbesluit Scharreleieren*. Such legal norms defined in what respects a scratching system is more benign for animals than a laying battery. Hence, these heuristics became more compelling as more people supported the stipulated legal rules by buying scratching eggs. So at least some actions of some actors came to be coordinated by the guiding principle animal welfare. As we have seen, the pursuit of animal welfare has led to some rhetorical re-coordination. Battery cage producers have advertised laying batteries with incrementally larger surface area as "humane." And some farmers have argued that chickens feel well in battery cages because they are so productive. Especially in the case of the farmers, however, the re-coordination has remained merely rhetorical and has not become normative. The actions of farmers are in most cases not (yet) guided by a striving for animal welfare.

The fact that farmers were not so easily persuaded is rooted in part, no doubt, in an unreflective commitment to a "normal" way of doing things. This is probably why poultry farmers claimed to be misunderstood and reacted angrily when the outside world started criticizing their methods. The specific resistance of this group, however, must also be understood as an effect of encapsulation in the existing guiding principle. As we have seen, according to the cultural argumentation the adherence of farmers to efficiency must be seen as the result of the fact that interactions between them and other actors were organized around this guiding principle. For example, farmers are dependent for credit on good relations with financiers, who expect them to produce efficiently. It is therefore far too easy to accuse chicken farmers of simply being traditionalists or conservatives with an aversion to new technology. Nor are they *necessarily* unreflective adherents of efficiency. It may even be the case that their personal attitudes conflict with the guiding principle, but that they nevertheless act according to this principle because it organizes their relations with relevant actors.

This does not, however, explain why *some* relations between actors were re-organized according to a new guiding principle, whereas other relations were not. Why did farmers object more vehemently to aviaries and other alternative systems than did other actors like the *Spelderholt* and the producers of laying batteries? To explain this, we need to take other coordinating mechanisms besides the guiding principle into account. One of these mechanisms is the existing market coordination as explained in the economic argumentation. Poultry farmers have to sell their eggs at a price that at least compensates the costs of production. Of course, this can be done by trying to create a new market for benign eggs, but this entails an economic risk for the farmer, unless the feasibility of such a market can first be proven.

The existing market coordination brings about an investment structure, in which departure from the efficient production of eggs threatens serious risks. Clearly, this is an incentive for farmers to stick to the existing guiding principle. Moreover, as the economic argumentation shows, the selection of farmers and technical systems takes place partly behind the backs of these actors. For example, in the fifties, the economic restructuring of the Dutch laying poultry sector selected for those farmers who were more successful in striving for efficiency. *As a result* of this selection, the remaining farmers were also encouraged to developed a disposition to efficiency.

Of course, at the *individual* level, market structures do not automatically produce a bias toward efficiency. As said before, actors always react to their perceptions of the market, not to some real market situation. But it should be noted that market selection can lead to a bias for efficiency on a *collective* level, since farmers not producing efficiently enough have a greater chance of going bankrupt. Moreover, bankrupt farmers provide a frightening example for the survivors. This will reinforce the *conviction* that it is very risky to switch to husbandry systems that *may* be not as efficient as the laying battery. This conviction may cause (some) farmers to start producing more efficiently or encourage the development of more efficient laying batteries. Subsequently, additional businesses may go bankrupt, providing new evidence for the conviction that only those producing efficiently, i.e. using laying batteries, have a chance to survive, etc.

Hence, it is clear that in the case of the poultry farmers the guiding principle and the prevailing market coordination are mutually reinforcing. If we take this mutual reinforcement into account, it becomes clear that the question whether the economic *or* the cultural argumentation is right is an empty one. There is no *prime mover* causing adherence to laying batteries (or efficiency), there are only, in this case at least two, mutually reinforcing, coordination mechanisms.[24]

C Other Relevant Modes of Coordination

As argued, farmers' commitment to laying batteries can best be understood as the result of at least two interacting modes of coordination, i.e. market coordination and coordination by guiding principles. As we have seen, however, governmental regulations and political coordination also played an important role. Recall the *Pluimveeregeling* and its revocation, the *Landbouwbesluit Scharreleieren* and the law Tazelaar/Van Noord which threatened to ban the battery cage. In most EU countries, EU legislation sets the minimum standards for laying batteries. And the Swiss example showed that governmental rules and regulations can play a decisive role in the development and use of poultry husbandry systems.[25]

The pursuit of animal welfare has also fired discussions about the paradigms used in animal research. As Kuhn has argued, paradigms are an important mode of coordination in science (Kuhn 1962). According to some animal welfare advocates, the dominant paradigm in poultry research reduces the animal to an input–output model, a mechanistic device, and hence degrades animal well-being.[26] Some critical ethologists have blamed classical ethology for objectifying animal behavior and ignoring animals' subjective feelings (BB: 45).[27] Nevertheless, there are ethologists who have developed definitions of animal welfare, based on a notion of animal consciousness or emotions.[28]

The underlying arguments for the existence of animal consciousness or emotions may suggest a break with the existing paradigm in animal research. This is, however, not necessarily the case. For example, the new "normative stance" in ethology discussed earlier does not imply a change in paradigm. Hence, arguing that chickens suffer in battery cages does not necessarily imply a break with the existing paradigm. Of course, new paradigms in animal research may suggest changes in the design of chicken husbandry systems, and hence animal research paradigms are an important factor in understanding the how and why of specific designs in this field. But, although some cognitive re-coordination was required to develop alternative systems like the aviary, new scientific paradigms were not at issue. Nor would they have been a sufficient condition for redirecting the design of chicken husbandry systems, given the preponderance of other modes of coordination.[29]

Political and scientific modes of coordination, no more than economic and cultural ones, are sufficient in and of themselves to explain why chickens are housed in battery cages. Nevertheless, each modality provides part of the answer. It is now time to pursue this further by shifting the discussion away from the various modes of coordination taken separately and to consider them in their mutual coherence, as a (possibly contradictory) framework of coordination.

D Towards a More Comprehensive Approach: Design Regimes

While different modes of coordination are interdependent, they are not therefore indistinguishable. The market mechanism is clearly a mechanism which differs from coordination by guiding principles or by paradigms.[30] In general, no mode of coordination can be held to be the prime mover, at least not in any *a priori* sense. Thus, the design of a technology is subject to different but interdependent modes of coordination. I will use the term *design regime* to refer to the stabilized outcome of different forms of coordination impinging on the design of a specific technology.[31] I choose the term "regime" because the different forms of coordination not only pose procedural, but also substantive, limitations on the actions of and the interactions among the pertinent actors. The existence of a design regime implies a certain normality in these interactions, leading to generally accepted design criteria, outcomes, heuristics, etcetera. Hence, a design regime is operative on the level of the interactions among the main actors involved. This means that at specific locations deviant ideas of technological propriety may well exist. As long as these deviant ideas do not play a role in the interactions between the main actors, they are not part of the regime. Therefore, by speaking of a regime I do not mean that actors *must* act in some specific way or another, but that they will *normally* act in a particular way because they expect others to act in predictable ways too. In acting, actors implicitly or explicitly, and more or less creatively, take their situation into account. So there is no direct causal link between the interaction situation and the behavior of actors. Nevertheless the interaction situation may provide a strong incentive for actors to act in one way or another.

It should be noted that while I defined a design regime as the *outcome* of the interactions among the actors involved in a design regime, this outcome also constitutes the *context* for actors' actions at some later moment in time. Therefore, a design regime may become self-reinforcing if it produces outcomes that are an incentive for the actors involved to keep on playing the game as they have been playing it. Moreover, if the outcomes of the interactions fit the expectations of the actors involved, these outcomes will assure the actors of the rightness of their role performances and of the predictability of the role performances as played by other actors. Of course, the outcomes of the interactions at the regime level may also invite actors to change their behavior. If this is the case and actors do change their behavior — once again it should be noted that this is not necessarily the case — it *may* result in new stabilization and, hence, a change in the design regime. It may, however, also lead to a run-away regime, which keeps on producing outcomes that result in changing interactions. Or, if only a limited group of actors changes its behavior, it may have no result at all at the regime level. The battery cage case provides an example in which one element of a design regime, i.e. the striving for efficiency, is strongly stabilized as a result of mutually reinforcing modes of coordination.

The existence of a design regime implies not only that the activities of (often very different) actors are coordinated, it also means that certain actors normally do not play a role in important interactions. In other words: A design regime also presumes certain boundaries. The definition of what is in-

5 Why Are Chickens Housed in Battery Cages?   169

side and what is outside is often related to what is seen as the normal way to proceed within a design regime. For example, as long as animal welfare is not an issue in the design regime of the battery cage, specialists on animal welfare will be seen as outsiders. (To a certain extent, the argument also works the other way around: as long as they are outsiders, animal welfare is not likely to be an issue in the design of battery cages.)[32]

The fact that some actors can be seen as marginal to the design regime means that insiders can, to a certain extent, act relatively autonomously from the rest of society with respect to the core activities of the regime (designing). This autonomy is the outcome of a historical process through which the design regime has emancipated itself from contiguous practical domains. There are two reasons why this autonomy is not absolute. On the one hand, autonomy exists thanks to the fact that the products produced by the design regime are seen as useful.[33] This obviously buttresses the legitimacy of the design regime. This is a necessary requirement in order for outsiders to respect the autonomy of the design regime. The other important restriction is that the design regime must command sufficient resources (financial, cognitive and otherwise) to continue with its activities. In the case of established design regimes this stream of resources is part of the normal interactions of (actors within) the design regime with the environment.

Thus, a design regime is never fully autonomous; it has to interact with the outside world. Instead of autonomy, I therefore prefer to speak of the *socio-cognitive space* of a design regime, a term which more clearly indicates that the autonomy of a design regime is always bounded.

With the help of the concept design regime, a more comprehensive description of the change process can be given, a description that does not stress only one form of coordination, but exhibits the interdependence of different forms of coordination. For the present example this produces a narrative along the following lines: the change process was initiated by outsiders, animal rights groups, criticizing the legitimacy of the pursuit of efficiency in the design regime of hen batteries (and in intensive livestock farming in general). The reason the attack on the battery cage started in this way was that criticizing the guiding principle was both a cheap and effective strategy for animal rights groups. The primacy of efficiency in the design and use of battery cage could easily be made visible and discredited. At least in late 20th century Northern Europe, public opinion could readily be mobilized against the needless suffering of animals. The effect of this criticism was not a change in guiding principle, at least not an *immediate* change, but an erosion of the legitimacy of the design regime. In other words: the socio-cognitive space of the design regime was threatened. The institutionalization of a market for discernibly different, alternative, eggs can be seen in the same light. Because the scratching system was already in use as a production system, the development of a new market-niche did not have much direct impact on the design of laying systems. The relative commercial success of the scratching egg may be said to have saved the slatted floor system, but it could not prevent the rise in the percentage of chickens housed in battery cages from 50% in 1971 to more than 90% nowadays. However, it did have some *indirect* effects on the hegemony of the battery cage system: it showed that there was a market for the products of more benign systems and created a space and a motive for the development of alternative systems. Neither development, the ideological critique nor the formation of a new market niche for alternative eggs, had a very direct impact. But by threatening the socio-cognitive space of the regime, they had an important indirect effect and they gave insiders a motive to change the regime.

In addition to "adamant outsiders" like the animal rights groups, merely relative outsiders like ethologists and governments also played a role in the change process. Ethology initially functioned in arenas where issues of animal welfare were discussed. But when ethologists developed more detailed requirements and heuristics for the design of benign systems, i.e. when they became experts in the actual formulation of requirements for chicken housing, they became more central actors in the regime.

Before the change process started, ethologists as well as the government had already been playing minor roles in the design regime. Ethologists were cognitively integrated into the regime, contributing their insights on the social behavior of chickens. Governments were integrated because of their intrinsic capacity to interfere in the regime by imposing rules and regulations (which in actuality had only occurred sporadically). Hence, both groups were well-situated to play more central roles in the design regime. Ethologists possessed important cognitive resources, like methods to study the well-being of chickens in housing systems. And the government had the legal capacity, as in all design regimes, to impose compulsory regulations. Finally, both groups — as relative outsiders — also had good reason to be interested in animal welfare: government because the public was upset about the suffering of animals; ethologists because they were, as a matter of professional method, concerned animal watchers. The roles that governments and ethologists already played, as insiders as well as outsiders, clarify why they came to play such an important role in the change process.

In the end, even insiders were mobilized for change. However, research groups and firms did not start developing alternative systems until governments subsidized the research or threatened with a battery ban. The development of alternative systems was prepared within the existing cognitive infrastructure of the regime. The cognitive role of organizations like the

Spelderholt and the producers of housing systems was no different in the process of developing alternative systems than it had been in the existing regime. However, the development of alternative systems was also politically coordinated. While facilitating the development of alternative systems, this political coordination also increased antagonism among farmers, expressed as strong opposition to the aviary and a refusal to provide financial support for the development of alternative systems.

The development of alternative systems within the regime, in combination with the creation of a market niche for alternative eggs, and hence for alternative systems, created a protected socio–cognitive space within the regime. In this niche, animal welfare, translated into heuristics like more space per chicken and the presence of litter and into alternative systems like the scratching system and the aviary, has to a certain extent become the guiding principle. This niche is supported by governmental subsidies and by consumers buying alternative eggs. Its legitimacy is, understandably, not at stake for most outsider actors, but some insiders are seriously asking if it is legitimate to develop alternative systems, especially if they result in a higher price per egg, or in a legal ban on the battery cage. This has put some insider actors on the spot. For example, at a certain moment the farmers suspected the Spelderholt of supporting a ban on the battery cage.

Of course, farmers could not legitimately criticize the Spelderholt for doing *research*;[34] this explains why researchers from the Spelderholt tried to keep technical and political issues apart. The creation of a niche in which alternative systems are designed, produced and used, offers an important opportunity to optimize alternative systems and to create new markets, as the *Dierenbescherming* recognizes:

> The development of alternative "scratching" products has very far reaching consequences for animal livestock farming. As a matter of fact, it creates separate fields with their own techniques and interests, which, in part, differ significantly from those of the bio–industry. The rise of these fields has, on the one hand, led to an increase in practical knowledge of these alternative forms of housing; on the other hand, it has spurred scientific investigations into optimization of these alternatives. Psychologically and politically it is very important that alternative housing systems show that it is now possible to keep animals in more humane ways than in the bio–industry and that this can be done in an economically viable way (X: 4).

Meanwhile, the rhetoric within the regime itself is changing. Existing, or slightly different, "alternative," battery cage systems are presented as humane. The striving for animal welfare has thus been "internalized" in the design regime, but surely not in a way that commands the approval of outside protesters.

For alternative systems to become successful on a routine basis, i.e. to become the hegemonic standard in a new design regime, a new socio–

cognitive space has to be built up. In principle, this might be achieved in a number of ways. It is conceivable that some genius may invent a benign housing system with ever-lower production costs per egg. The EU may categorically ban the battery cage. Consumers and the egg-based foods industry might refuse to buy battery cage eggs any more. Farmers could voluntarily ban the battery cage and represent themselves to the general public, and in particular the consumer, as social innovators taking economic risks in the pursuit of humane goals. Of course, not all these possibilities are probable, but they show that there are different roads to a design regime wherein animal welfare is taken seriously. It is too simplistic to say that such a design regime is a pipe-dream simply due to the economic situation or an ideological inclination to efficiency.

5 Conclusions

There is no such thing as a prime mover in technological development. Stability and change in technological design can only be poorly understood as outcomes of only one mode of coordination, be it market mechanisms, guiding principles or paradigms. This does not necessarily mean that approaches or theories stressing only one (or two) modes of coordination, are worthless. On the contrary, these theories are often indispensable in understanding certain aspects of technological development, but they simply fall short in providing comprehensive explanations and hence are not proof against unpleasant surprises.

In this chapter, the concept of design regime was presented as a strategy for achieving a more comprehensive approach, an approach that sees technological development as the outcome of different interdependent modes of coordination. This approach can profit from insights provided by other theories regarding the details of different modes of coordination, but has the advantage of integrating these particular explanations into a comprehensive explanation of sociotechnical change.

Notes

* I would like to thank Nil Disco, Johan Schot, and Wim Smit for comments on earlier versions of this chapter. Part of this chapter appeared earlier in an article in Dutch (Van de Poel 1994). The case of laying batteries is analyzed from a different angle in Ibo van de Poel and Cornelis Disco (1995).

1 See for the history of battery cages (E; M; P; Q).

5   Why Are Chickens Housed in Battery Cages?   173

2   In the Netherlands the first magazine devoted especially to poultry farming appeared in 1923 (Q: 104).
3   Recently also themes like animal welfare, biotechnology and environmental issues have been placed on the agenda.
4   In the text Ketelaars speaks about the percentage of farmers having more than 1,000 chickens. However, from the table he uses and from the other numbers he gives, it becomes clear that he must have meant the percentage of chickens kept on farms with more than 1,000 chickens.
5   Food conversion is the ratio between the weight of the (specialized) food that farmers feed their chickens and the weight of (sellable) eggs that these chickens produce.
6   In laying batteries there are no ground-eggs, because the eggs when laid roll automatically from the floor (which is at a slope) onto an egg conveyor, which collects the eggs. In housing systems with slatted floors the chickens are supposed to lay their eggs in so-called laying-nests, which distribute the eggs to egg conveyors in order to collect the eggs. However, some small percentage of the eggs is not laid in laying nests but on the "ground," these eggs have to be collected by the farmer by hand. This work is much disliked by farmers because they have to bend down to pick up each egg.
7   Heuristics are intersubjectively sanctioned rules which indicate the direction in which a good solution to a design problem can be found (Disco et al. 1992).
8   The description below is based mainly on (F).
9   The Dutch verb "scharrelen" means something like "messing about" or "scratching." The later term is used as translation by the Dierenbescherming and will be used here too.
10 This type of chicken housing is also sometimes denoted as "deep litter houses with slats."
11 It is contestable whether "scratching systems" can really be seen as benign for animals. In such systems, chickens are housed at a density of seven per square metre and the chickens are not allowed outdoors. The Dutch animal rights group, Lekker Dier, therefore long considered this system to be unacceptable. In the early seventies, this organisation distanced itself from efforts by the group Rechten voor Al wat Leeft to introduce the scratching egg onto the market. Today, Lekker Dier is willing to accept the scratching egg as "a first step on a long road to go." For a critique of the idea that scratching systems are benign for chickens, see (N).
12 These were the Stichting Scharreleieren Controle (Foundation for the Certification of Scratching Eggs), later followed by the Stichting Nederlands Eiercontrole Bureau (Foundation Netherlands Egg Certification Office).
13 As a matter of fact, at this moment there are also other types of eggs that promote chicken welfare (besides scratching eggs) on the consumer market, like eggs from aviaries and eggs from free range farms (where the chickens are also allowed outdoors). These different types of eggs are also legally protected, being controlled by the Stichting Nederlands Eiercontrole Bureau. However, the market share of these eggs is very small.
14 Ethological research on chickens was already being carried out in the 19th century. In 1873 Spalding published the first ethological study on chickens.
15 The number of institutions doing ethological research on animal welfare rose worldwide from 6 immediately after World War II to about 30 at the end of the eighties (HH). In the Netherlands the number of ethologists active primarily in animal welfare rose from 5 in 1972 to 12 in 1988 (R).

16 Schenk spells out this argument (BB).
17 Council Directive 86/113/EEC was, however, later annulled on procedural grounds. Unauthorised changes had been made by the General Secretariat in the preamble of the directive. The annulment was the result of a protest by the United Kingdom against these changes. The stipulations of the directive as such, however, were not challenged and the measures were re-adopted as Council Directive 88/166/EEC (K).
18 (K; W). In Southern Europe and the United States there are no or less stringent legal requirements. In the United States for example a floor area of 315 cm2 per chicken is common (FF).
19 Currently the laying poultry farmers are willing to pay for a project at the Spelderholt to adapt the battery cage to chicken welfare requirements. This time, they want to keep the project free from government interference.
20 In 1984 an alternative system was in principle already available for the Dutch market because in 1981 LACO BV, a Dutch producer of laying batteries, had started producing aviary systems for the Swiss market. LACO BV also sold these aviaries in Belgium, Great Britain and the Netherlands. Obviously, the politicians did not see the existence of this system as proof that a feasible alternative was available. And apparently the politicians looked to the Spelderholt for the answer to the question: "Can a feasible alternative be developed?"
21 About 20% of the Swiss laying hens are kept on farms with about 2,000 hens and 10% of the chickens are kept at farms with about 500 hens. By comparison, in the Netherlands, in 1984, about 10 % of the hens were kept at farms of 5,000 hens or less (C).
22 Information from the Dutch Product Board for Livestock, Meat and Eggs. Swiss tariffs are possible because Switzerland is not an EU member.
23 Runs on banks and stock-exchange panics are good examples (and also classic illustrations of the irrational unintended effects of rational actions by individuals).
24 Boudon (1986: 121-153), who argues that the search for a prime mover of social change is futile.
25 It could be argued that in all cases where the legitimacy and value of different goals and means to reach these goals is at stake, or where the legitimacy of various stakeholders is contested, political coordination plays a role. In principle, this applies to all technologies. Moreover, in modern societies governments can always mobilize a mandate to intervene in the design of a technology. Hence, even non-intervention is a political act. Becker (1982: 185), who states that government always plays a role in art worlds because it "might intervene to prevent the production or distribution of art works. Even if it seldom or never does, failure to act is a crucial form of cooperation in artistic activities" (emphasis by Becker). The same can be said of design activities.
26 See, for example (M: 4; BB: 40).
27 He names the classical ethologists Lorentz and Tinbergen. Typically Lorentz' model has been used both to prove and to refute the assertion that animals suffer in battery cages (CC).
28 This line of thought is visible in the work of Marion Stamp Dawkins (H) and in the work of the Dutch biologist Wemelsfelder (GG).
29 There are also other, non-scientific, reasons why it is doubtful whether the emergence of a new paradigm is a prerequisite for actual progress in the well-being of livestock animals. The public and government may, for example, be persuaded by popular arguments and other factors may determine the actual success of the pursuit

of animal well-being. Nevertheless, it could be argued that since the design of a technology always requires cognitive resources — albeit sometimes tacit ones — redirecting the design of a technology may require mobilization of new knowledge, heuristics, models, testing methods, and so on. Hence coordination by paradigms, technical models, prototypes, etc., always plays a role in technological development. The coordinating power of paradigms has in fact already been described by Kuhn (1962). The coordinating power of technical models is described by Disco et al. (1992). See also Van der Meulen (this volume). While Disco et al. (1992) stresses the epistemological roles of technical models and their roles in divisions of design labor, Disco (1990) has also shown that technical models are a means to reach "social closure" within technical communities. Technical models thus also fulfill political roles in defining social boundaries. (A role which Kuhn also ascribed to paradigms in relation to scientific communities.) The coordinating power of prototypes is described by Henderson (1995).

30 This combination of independence and interdependence can sometimes have ironic effects. For example, the technical fact that aviaries lead to higher ammoniac emissions pose a serious political (ideological) problem for environmental groups. On the one hand, such groups tend to support the animal rights groups in their protest against the battery cage. On the other hand, they favor lower ammonia emissions. They are now trying to solve this ideological dilemma — forced on them by the technical properties of the aviary — by proposing "integral" solutions.

31 The following description of design-regimes is based on Disco et al. (1992), Boudon (1986), and on Anselm Strauss' social worlds theory (see, for example Strauss (1978, 1982, 1984), Becker (1982), and Clarke (1991).

32 One can ponder at great length where to locate the boundaries of a design regime. Sometimes the chains of interactions connecting particular actors to a design regime are rather long. Think of the supermarket customer, buying biscuits produced by a commercial bakery which in turn buys the eggs for the biscuit from a farmer who keeps his chickens in a battery cage designed and produced by a battery cage producer. Clearly, the role of this end-user in the design regime is very marginal. Should we define this end-user as an outsider? I choose to restrict the regime to the designers, producers, direct users of (farmers) and researchers on battery cages. All those groups have, in principle, a rather direct impact on the design of housing for laying hens. In doing this, I follow the methodological rule for tracking social worlds as given in Becker (1982: 35). This rule can most simply be stated as "follow the activity" — in this case designing — until you reach actors that play only a marginal role on this activity or have only a marginal impact on the activity.

33 Note that in our case the product of the regime is eggs, because we defined the farmers as insiders to the regime.

34 For the same reasons the commercial production of scratching systems and aviaries by battery cage producers is not seen by farmers as illegitimate because these producers respond to "market demand."

Sources

A. *Agrarisch Dagblad,* November 8, 1990 (translation by the author).
B. Bell, D. (1993): "The egg industry of California and the USA in the 1990s: a Survey of Systems," *World's Poultry Science Journal*, 49: 58-59.
C. Bloem, L.J., W.A. Evers and P.I. Hautsen (eds.) (1986): *Handboek voor de Pluimveehouderij*. Beekbergen: Consulentschap in Algemene Dienst voor Pluimveehouderij.
D. Blokhuis, H.J. and J.H.M. Metz (1992): "Tussentijdse evaluatie van het programma. Ontwikkeling en praktijkbeproeving van volière- huisvestingssystemen voor leghennen," Nota P-92-59 van COVP-DLO en IMAG-DLO.
E. Blount, W.P. (1951): *Hen Batteries*. London: Baillière, Tindall and Cox.
F. Dekker, T. (1991):. *Scharrelprodukten: Richtlijnen, controle, omzet en promotie*. Den Haag: Nederlandse Vereniging tot Bescherming van dieren.
G. Dawkins, M.S. 1983. "Battery hens name their price: consumer demand theory and the measurement of ethological 'needs'," *Animal Behaviour*, 31, 4: 1195-1205.
H. Vines, G. (1994): "The emotional chicken," *New Scientist*, January 22: 28-31.
I. Elson, H.A. (1989): "A welfare update on laying cages," *Poultry,* April/May: 33-35.
J. Eurogroup for Animal Welfare (1989), *Poultry*. April/May: 33-34.
K. Eurogroup for Animal Welfare (1989): "Summary of legislation relative to animal welfare at the levels of the European Economic Community and the Council of Europe," Revised September.
L. Hanke, O.A., J.L. Skinner and J.H. Florea (eds.) (1974): *American poultry history 1823-1973*. Savoy, Ill: American Poultry Historical Society.
M. Harrison, R. (1964): *Animal machines; The new factory farming industry*. London: Vincent Stuart.
N. Harrison, R. (1991): "The myth of the barn egg," *New Scientist*, November 30: 40-43.
O. Harrison, R. (1993): "Case study: farm animals," in: R.J. Berry (ed.): *Environmental dilemmas: Ethics and decisions*. London, Chapman and Hall: 118-135.
P. Hartman, R. and D.F. King (1957): *Keeping chickens in cages*. Redlands, CA: Roland C. Hartman.
Q. Ketelaars, E.H. (1992): *Historie van de Nederlandse Pluimveehouderij*. Barneveld: BDU.
R. Kuit, A.R., D.A. Ehlhardt and H.J. Blokhuis (1989): *Alternative improved housing systems for poultry*. Beekbergen: Ministry of Agriculture and Fisheries of the Netherlands, Directorate of Agricultural Research.
S. *Lekker Dier* (1985): Information brochure on the bio-industry (translation by the author).
T. Meierhans, D. (1992): "Legehenenhaltung; Alternativen in der Schweiz (I)," *DGS*, 43: 1251-1257.
U. Ministry of Agriculture, Nature Management and Fisheries (eds.): *Dutch poultry farming world-wide*. The Hague.
V. *Misset World Poultry,* 7,4 (1991): 32-33.
W. *Misset World Poultry,* 7, 5 (1991): 32-35.
X. Nederlandse Vereniging tot bescherming van dieren. *Scharrelkrant* (translation IvdP)
Y. *Pluimveehouderij*, June 30, 1989: 6 (translation by the author).
Z. *Pluimveehouderij,* April 26, 1991: 4.

AA. *Publications Pluimveehouderij 1990:* 4 (translation by the author).
BB. Schenk, P. (1988): "Het nuttige dier," in: M.B.H. Visser and F.J. Grommers (eds.): *Dier of Ding, objectivering van dieren.* Wageningen, Pudoc: 31-50.
CC. Stern, G.H. (1980): "Familienkrach der Verhaltensforscher," *Der Spiegel,* 32: 50-58.
DD. Study Commission on Intensive Livestock Farming (1987): "Alternatives for the battery cage system for laying hens," *Report of the Study Commission on Intensive Livestock Farming, established by the Dutch Society for the Protection of Animals.*
EE. Ubbels, P. (1961): "Onderzoek als grondslag van moderne pluimveehouderij," *Landbouwkundig Tijdschrift,* 73, 20: 869-987.
FF. Wegner, R. (1990): "Poultry Welfare — problems and research to solve them," *World's Poultry Science Journal,* 46, 1: 19-30.
GG. *De Volkskrant,* June 19, 1993: 13.
HH. Wood-Gush, D.G.M. (1988): "History and development of poultry ethology," *Poultry,* April/May: 8-9.
II. Zonne-Ei-Farm BV (ed.): Advertizement Brochure.

Chapter 6
Aligning Gender and New Technology: The Case of Early Administrative Automation

Ellen van Oost

1 Introduction

In its early phases, the automation of administrative services in the Netherlands was characterized by rigid patterns of gender segregation and exclusion. In the process of shaping an automated office, not only were electronic computers brought into the organizations but new jobs were created as well. Jobs like computer programmer, systems analyst, computer operator and data-typist came into being. In the Netherlands this new occupational domain became highly gender-segregated from its very inception. Programmers, systems analysts and operators were predominantly men. Preparing the data for the computer — at that time done by key punching cards or tapes — was defined as work for young women.

The masculine gender typing of computer programming in administrative automation was by no means a foregone conclusion. Women had had an important role in operating and programming the early computers used for making scientific and military calculations. The best known female programmers were probably the "ENIAC-girls," who programmed the world's first electronic computer. The first Dutch computer, the ARRA, was also programmed by women. All things considered, therefore, one might at the least have expected a less uncompromisingly masculine gender typing.

Prevailing theories of gender segregation in the labor market are not able to explain the gender typing of *new* jobs, and are unable to comprehend the role of technology in such typing. Conceiving of administrative automation as a process of sociotechnical change offers a more promising avenue of approach. I will view the shaping of an automated office as a complex process of sociotechnical change in which various human and non-human actors are aligned in order to produce a functional computer system. The question is how this process is influenced and structured by prevailing gender symbols, structures and identities. The gender typing of the new computer jobs can then be understood as emerging from this gendered process of sociotechnical change.

2   Gender and Technology in the Making

The relation between gender and technology has drawn attention from scholars both in Women's Studies and in the new Technology Studies. At this intersection, technology as well as gender are conceived as socially constructed, as culturally and historically situated (Cockburn 1983; Cockburn and Ormrod 1993).

In Women's Studies, gender is conceptualized as a socio-cultural construction of femininities and masculinities. The concept gender emphasizes the fundamentally social character of a number of distinctions made on the basis of biological sex, thus denying biological determinism. As a rule, three levels are distinguished on which the gender dichotomy functions: the symbolic level, the structural level and the individual level. At the symbolic level, a cluster of dichotomies like dark/light, passive/active, emotional/rational are linked with femininity/masculinity. This gender symbolism is used to legitimize the different positions of women and men in social structures. At the level of individuals these symbolic and structural dichotomies are incorporated in the gender identity of concrete women and men. By emphasizing the relation between these three different levels of gender, one can better understand the myriad ways in which it structures our society (Harding 1986: 57).

But gender is not a neutral way of structuring our society. It is, rather, "an *asymmetrical* category of human thought, social organization, and individual identity and behavior" (Harding 1986: 55). The historian Scott (1986) has defined gender as an analytical concept signifying relations of power. The masculine side of the dichotomy is more highly valued than the feminine side. By ascribing masculine images to societal entities, the entity is strengthened, whereas an entity will be weakened if it is assigned feminine characteristics. Scott (1986) gives an example of the rhetoric used in a war situation: the enemy is weakened by ascribing feminine images while one's own army is empowered by masculine symbols.

Not only individuals, but also aggregated socio-cultural entities like organizations, can be ascribed meanings in terms of gender and can be analyzed in terms of gender. Acker (1990) defined "gendered organization" as follows:

6 Aligning Gender and New Technology   181

> To say that an organization, or any other analytic unit, is gendered means that advantage and disadvantage, exploitation and control, action and emotion, meaning and identity, are patterned through and in terms of a distinction between male and female, masculine and feminine. Gender is not an addition to ongoing processes, conceived as gender neutral. Rather, it is an integral part of those processes, which cannot be properly understood without an analysis of gender (Acker 1990: 146).

In that very same sense, processes of sociotechnical change can have a gendered character too. Gender, as a category by which persons give meaning to and shape their environment, is an integral element in the actions, interactions and policies of actor groups. Gender is an analytic concept that contributes to explaining specific inclusions and exclusions and the relative success of particular alignments over others.

How to analyze the shaping of early administrative automation as a gendered process? Automation was an issue at the national level as well as at organizational levels. Notwithstanding, the actual shaping of the automated office took place in concrete and inevitably idiosyncratic organizations, forcing us to consider the organization as a prime research site. In this chapter I have chosen to investigate the automation of the Dutch Postcheque and Clearing Service during the late fifties and early sixties. This was one of the first successful efforts in the Netherlands to automate a mass administration.

However, the diffusion of computer technology also stimulated public debate at a national level. In the course of this discussion, symbolic frames emerged which ascribed specific meanings to the new technology. These frames subsequently influenced the shaping of automation in concrete organizational settings. From the perspective of an interest in understanding the shaping of gendered technologies, this leads to a focus on, first, the nature of gender symbolism in the emerging frames of meaning for computer technology and, second, the gender structure of the collective actors involved and the alignments accomplished by them.

3 The Gendering of Computers in the Process of Their Socio-cultural Embedding

In the mid 1950s electronic computers emerged from their relatively cloistered service in scientific and military laboratories and began to assume a role in the administrative operations of large companies and government institutes. The implications of this development were a topic of public debate. Some saw office automation only as another method to further rationalize office work. Others argued that society was on the eve of a new revolution. At the same time, efforts were made to stimulate and coordinate automation in concrete organizations. Which social groups were involved in these public debates and in coordinating automation? What gendered meanings did these groups give to computer technology and how did this influence the way the computer became embedded in society?

A   Gendered Frames of Meaning

An essential element in embedding a new technology in society is the development of conceptual frames for giving meaning to the new phenomenon. In the 50s and 60s public debate about computers was fierce, even though their actual use was limited. In the press, which was an important arena of the debate, two frames prevailed: The "computer as electronic brain" and "the computer as business tool."

The "electronic brain" frame problematized the borderline between humans and machines, but did so in a Janus-faced manner. On the one hand there was a fear that computers could and would replace human beings. So it evoked feelings of potential danger. But on the other hand the frame gave vent to wonder. The computer was seen as "one of the most fascinating products of our technological potential," to cite one of the newspaper articles. It was considered nothing short of miraculous that a machine "could multiply two numbers of ten digits in less then 1/10 of a second." The meaning of the computer as an electronic brain thus incorporated values like technological virtuosity and control, technological modernism, and the excitement of unlimited progress.

The other prevalent frame painted the computer as basically a new and unusually powerful "business tool." In this frame the computer was represented as essential for increasing efficiency, for realizing economic growth, for moderating the pressure of work, for improving performances ("the computer makes no mistakes"). This frame, too, is a two-sided coin. On the one side, it was argued by managers and economists that the computer could be instrumental in achieving what they saw as a necessary increase in productivity. On the other hand, the threat of massive unemployment inevitably reared its head. This particular connotation of automation, however, failed to resonate in the Netherlands at that time in view of the severe post-war labor shortage. So in general, the meaning of the computer as business tool expressed economic values of growth, development, and rationalization.

The two frames implied a different gendering of the computer. The computer as electronic brain assigned the computer and its operation to the masculine realm, and thus excluded feminine connotations. The two faces of the electronic brain, the exciting and the frightening one, both call on the masculine identity of being in control of sophisticated technology (Wajcman 1991). A 1963 newspaper article emphasizes the virtuosity of computer programming in terms of a brain metaphor. The work of a young male programmer is described as follows:

[...] channels the irresistible development of this electronic annex to the human brain with its unforeseeable potential while at the same time providing benefits to others, who gaze on these labors in wide-eyed wonder (C: 11).

This quotation illustrates how in the frame of the brain-metaphor, computers became associated with control, power, mystery and progress. This symbolism of masculine virtuosity resonated with the identity and aspirations of a specific social group: namely young, middle-class men (Pacey 1983).

The physical settings in which computers were placed also reinforced this masculine-oriented web. The classic configuration was to display the computer in a white, modern, hall with glass walls, and surround it by white-coated men. Sometimes the new computers were even visible from outside the building — like IBM's first electronic calculator in 1948 (B). This paradoxical construction — the mysterious black box in the glass cage — reinforced rather than unraveled the mystery.

The frame of the computer as business tool evoked a less rigidly gendered meaning. In this frame, programming a computer was not so heavily burdened with values of masculine virtuosity. On the contrary, the qualifications of accuracy and conscientiousness which were used to describe computer programming, even suggested possibilities for feminine labeling. Within this frame the computer was not so much mystified as an impenetrable black-box, but rather discursively deconstructed as a tool with comprehensible components and functions. Its aggregate behavior was explained in terms of the basic binary concepts underlying the principle of electronic computers. It was represented as inherently dumb, unable to do anything without its human instructors and operators.

The business tool frame imposed two kinds of interpretations on computer-related jobs. On the one hand, it deconstructed the image of the mysterious wizard-programmer. Computer programming was described not so much as an evocative leap of imagination but as a tedious process of writing precise instructions. This frame sometimes linked programming with specific images of feminine competence, as illustrated in the next quotation from a newspaper article:

It is said that the best programmers are intelligent girls who are good in embroidering because they have enough patience and neatness for this kind of job (K: 7).

On the other hand, potential new computer jobs were highly rated in order to suggest that they would more than compensate for the loss of current jobs due to automation. Indeed, as I noted, the other side of the business tool coin was the threat of unemployment. Although at that time unemployment was *de facto* non-existent in The Netherlands, public debate was colored by news from the United States where automation-related unemployment was definitely an issue. In general, proponents of automation did not deny that the computer would in fact replace human workers. However, they argued that the computer would also generate new highly qualified jobs, like computer operator, programmer and systems analyst. Along similar lines the public was given to believe that the computer would take over chiefly dull, monotonous, labor, thus suggesting that automation would improve the overall quality of labor.

Efficiency discourse linked electronic computers less to masculine values than did the brain metaphor. The image of the tedious programmer linked the computer with traditional female qualities. By contrast the "prospective jobs" suggested a bias toward male groups, since the emphasis on career possibilities was — at that time — considered to be more relevant for men than for women.

Both frames emerged out of the public debate on automation. However, the brain-metaphor was the dominant one in the early period. Journalists — most of them men — were more interested in writing about intriguing machines with a potential to change the existing social order, than about dull machines that would rationalize organizations. As a result, in the public mind the electronic computer came to be linked more with values of masculine virtuosity, than with feminine virtues like neatness and accuracy.

B   Aligning Computers with Male-Dominated Actor Groups

During the reconstruction of Dutch economy and society after World War II, efficiency became a central aspect of organizational thinking in both governmental organizations and private companies.[1] The striving for efficiency in the Netherlands was stimulated by the tight labor market in the fifties and sixties. The shortage of labor supply was reinforced by the low participation of married women in the work force. These three aspects — efficiency, the tight labor market and the domesticity of married women — were closely intertwined. The tightness of the labor market was not only a factor in stimulating efficiency but also a *result* of the striving for efficiency. Efficiency measures stimulated labor productivity and as a result unions were able to negotiate relatively high wages. This meant that the prevailing ideology of the family: A male breadwinner and a domestic, stay-at-home, wife could be realized in practice.[2] In turn, the domesticity of married women further limited the supply of labor. This complex of interrelated factors was — and to a certain extent still is — a structural feature of the Dutch labor market.

The economic climate of the late fifties favored labor process rationalization. For this reason, the electronic computer aroused considerable interest among the higher management echelons of large companies and governmental organizations. Quite a number of Dutch managers traveled to various locations in the United States to study how computers were being deployed for administrative purposes. Nonetheless, throughout the fifties, actual cases of implementation of electronic computers in Dutch administrative work were few and far between. In 1957, the Heidemaatschappij, a civil engineering consultancy firm, installed the first American electronic computer, an IBM 650, for calculating the weekly payroll. In the following years, the number of computers grew slowly. The introduction of the transistor-based computer in 1958 stimulated additional interest in automation. In 1963 it was estimated that there were one hundred computers in the Netherlands; three years later estimates varied from 300–450 computers (I).

But electronic computers were far from a self-evident solution for rationalizing business administrations. The early electronic computers were extremely expensive. For example, the IBM 1401, a medium-sized second generation computer (4 K memory) cost 1.5 million Dutch guilders in the early sixties. This meant that automation was simply beyond the reach of most organizations. Less expensive alternatives, like the tried and tested mechanical keypunch machines, were used more often.

Nevertheless, by the late fifties several large Dutch companies and governmental organizations were familiarizing themselves with the ins and outs of automation. The managers and corporate economists involved felt a growing need for coordination of their often isolated activities since it was clear that each organization was re-inventing the wheel on its own account. In Amsterdam in February, 1958, on the occasion of a series of lectures by John Diebold, a number of Dutch managers and economists interested in automation gathered to discuss the problem of duplication of effort and to take initial steps in founding a national association for stimulating and coordinating knowledge of administrative automation.

This association, the Netherlands Automatic Data Processing Research Centre, was officially established in July, 1958, and affiliated with the Department of Economics of the University of Amsterdam. The Research Centre's official aim was "to engage in research on and diffuse knowledge of administrative automation (defined as 'the gathering and processing of data in the most general sense') within companies, government and other economic units." The Research Centre was set up on a strictly non-profit basis. Its ideology stemmed from academic values: independence, objectivity and the pursuit of scientifically justifiable knowledge.

Within the year, the initiators of the Research Centre managed to create a powerful network which bound together all actors that were seen as relevant to administrative automation in The Netherlands. Actors involved in the Centre were leading persons in the fields of accountancy, large industry, government agencies, and scientific domains like economics, engineering and mathematics. The Research Centre became the central organizational locus of administrative automation in the Netherlands in the 1960s.

The center unquestionably labored under a meritocratic bias. It was their strong belief that the question of the social acceptability of automation could be best addressed by a small elite of experts. This kind of paternalism by a male elite was characteristic of Dutch socio-political culture at that time. This culture was also reflected in a strong national ideology of the nuclear family, headed by a *pater familias*.

The Research Centre played a crucial role in articulating demand for the new computer technology. One of their central aims was to provide Dutch companies and government organizations with "objective" information about the possibilities of administrative automation. Informing higher management about relevant possibilities and developments was the keystone of the Research Centre's approach to stimulating administrative automation, since they strongly believed that implementation was a top-down process. In their view, a commitment by higher management — at that time composed almost solely of men — was the first crucial step in any successful automation trajectory. Twice a year the Centre organized special conferences for ranking executives in luxurious settings, thus reinforcing old-boy networks. Participants were informed about what computers could and could not do; about how automation feasibility studies should be done; and about the organizational and psychological impacts of automation.

A second core activity pursued by the Research Centre was developing and providing training courses for future computer programmers. As a rule, computer firms offered training for client personnel to assist them in programming the purchased (or leased) computer. The Research Centre, however, stressed the need to go beyond these practical courses. Accordingly, they developed introductory courses outlining the general principles of computer programming. The team that developed these programming courses was composed chiefly of electronic engineers and mathematicians. Their implicit goal was to develop an appreciation of the scientific fundamentals of computer programming. Consequently, their programming courses were steeped in the kinds of mathematical formalism that underlay their own disciplinary backgrounds. In Western culture, mathematical formalism is strongly associated with male capabilities (Mahony and Van Toen 1990).

Thus, in giving form to its programming training, the Research Centre aligned the programmers' qualifications with a putatively masculine domain.

In sum, during the early phase of societal embedding, the computer became aligned with specific male–dominated social groups by gender symbolism as well as by the prevailing gender structure. In the process of cultural embedding, computer technology became linked to a set of values perceived in our society as embodying masculine virtuosity, hence allowing men to incorporate computer technology as part of a masculine identity rooted in male virtuosity.

In developing national coordination of administrative automation, the Dutch Research Centre facilitated the societal embedding of computer technology among specific male–dominated actor groups: managers, business economists and mathematicians/electronic engineers. By reason of its structure and *modus operandi*, the Research Centre linked the computer to a distinctly male elite group and to a set of masculine values shared by the members of these groups.

Now that we have seen how, through nationwide processes of cultural and structural embedding, the new electronic computer technology was gendered into the male/masculine domain, we will shift our attention to the level of the concrete organization. Organizational processes of sociotechnical change are another important locus where new technology gets embedded into a previously gendered structure. Local processes of change do not of course occur in a socio–cultural vacuum. On the contrary, developments at national and local levels continuously interact with and influence each other. Concrete organizations act within and are influenced by the broader existing sociotechnical (and thus also gender) order. In turn, local constructions contribute to and co–construct the wider socio–technical order. The case of the Dutch Postcheque and Clearing Service provides a clear example of how the actual gendering of the new machines and the associated functions were influenced by the specific organizational context. The newly created computer-related jobs in this organization were strongly gender-segregated in such a way that this local outcome strengthened the evolving masculinity of computer jobs at a national level. The Postcheque- and Clearing Service was one of the first Dutch organizations to succeed in automating a mass administration and it attracted much attention, both from professional organizations and the mass media.

4 The Automation of the Dutch Postcheque- and Clearing Service

The Postcheque and Clearing Service was a semi-governmental organization that administered the giro-accounts of a million Dutch citizens and thousands of organizations. The organization was established in 1918 as a subdivision of the Dutch Post, Telephone and Telegraph Service (PTT). In the 1920s, an overly ambitious effort to mechanize the clearing of cheques had turned out a complete fiasco, leaving a legacy of deep mistrust of all kinds of mechanical processing.[3] As a consequence, machines were more or less banned from the organization and the daily account bookings were predominantly processed by hand.

In the post-war period, the shortage of labor was endemic. The labor-intensive character of the Postcheque and Clearing Service made it especially vulnerable to the tight labor market. The postwar economic revival led to a rapid growth in the number of accounts and transactions, thus increasing the average work load for personnel. Despite severe shortages of personnel, the Postcheque and Clearing Service persevered in offering their one-day service for account holders. This implied that all incoming transactions had to be booked that same day, causing a very high work pressure. It is not surprising that the Postcheque and Clearing Service was considered a poor employer and had problems attracting qualified personnel. This degenerative spiral convinced the responsible managers that only structural changes could save the Postcheque and Clearing Service from total collapse.

In the mid-fifties, two different strategies for solving these problems were developed: organizational changes versus automation. On the one hand we see the executive board of the Postcheque and Clearing Service stimulating organizational changes while introducing a few mechanical bookkeeping machines. Anything approaching automation was, at that time, still well beyond their scope. If the cheque clearing process was to be automated, therefore, initiatives would have to come from outside the Postcheque and Clearing Service. In 1953, H. Reinoud, the PTT director responsible for the Postcheque and Clearing Service, in fact formed a small study group to investigate the possibilities for mechanical processing in mass administrations. Reinoud, together with two eminent engineers employed at the Dr. Neher Laboratory (the PTT's research institute) formed the core of this study group. Reinoud's strategic interest in computer technology derived from the expectation that successful automation would strengthen his position within the PTT board of directors (I).

The study group visited the United States on several occasions to acquaint themselves with the latest technology and to study actual administrative automation projects. Although the group applied itself to the study of automating mass administrations in general, the Postcheque and Clearing Service was considered to be the most urgent case. In their first report, published in 1955, they stated that "in order to avoid a total breakdown, the automation of the Postcheque and Clearing Service was inevitable," and subsequently they presented the global outlines of an automated giro account booking system. Up to then, the board of directors had not been involved in developing ideas about automation. As already noted, the focus for change within the Postcheque and Clearing Service was less on technology than on the organizational structure and culture. In 1956, for example, a second office was opened in a region of the Netherlands that had a relatively plentiful supply of labor. That same year witnessed a report on labor relations at the Postcheque and Clearing Service, written by an outside academic researcher, H.A. Hutte (G). Hutte criticized the Postcheque and Clearing Service for being too bureaucratic and hierarchical. One of the recommendations was to replace the promotion system based on seniority by a more meritocratic one based on individual capacities. The implementation of Hutte's recommendations caused drastic changes in personnel policy. The new policy was based on the slogan "the best man in the right place" and, among other things, had a serious impact on gender relations in the organization (F).

In the 1950s, the Postcheque and Clearing Service exhibited an unusual gender structure: almost 50% of middle management was composed of women, whereas the overall proportion of female employees did not exceed 40%. As a rule the gender structure in an organizational hierarchy follows a pyramid shape; the higher the position the smaller the proportion of women. The deviant gender structure can be explained by the fact that in the 1930s employees of the Postcheque and Clearing Service had been predominantly female (70%). The director at that time was convinced that women could process the transaction orders more efficiently than men. Most of these women quit their jobs at marriage, as was customary at that time in the Netherlands. But a number remained single, continued to work, and by dint of more or less automatic promotion based on seniority achieved middle management positions in the fifties, since the promotion was based on a seniority system. Hutte found that especially younger male employees had problems with these middle-aged women managers, but he tended to blame the women. In the report he stated:

Due to the fact that in earlier times there was more female personnel than nowadays, there is a number of unmarried women that have acquired responsible positions, probably not so much because of their managerial capacities as because of the diligence and devotion they have developed in order to compensate for missed life chances (G: 29).

These women managers became more or less the embodiment of the traditional bureaucratic organization, thus complicating their placement in the "new" organization.

Although the first report on automating the Postcheque and Clearing Service was presented in 1955, it took until 1961 before the decision was made to automate. In that period, automation and reorganization remained two rather distinct trajectories. Reorganization involved the bulk of employees of the Postcheque and Clearing Service, whereas preparing for automation was relegated to specialized working groups and committees. When, in the early sixties, conversion to an automated booking system was launched, the changes already achieved in the organizational structure and culture exerted considerable influence on the subsequent shaping of the (gendered) automated organization.

The study group's 1955 report stated that the technology to realize an automated Postcheque and Clearing Service did not yet exist. Two problems were mentioned. The first one pertained to the data preparation for computer processing and the second to the actual processing itself. The problem of data preparation was in fact generic to mass administrations. It was not the nature of the data that presented problems but rather its massive character. At the Postcheque and Clearing Service, more than half a million transaction orders had to be processed daily. These orders, most of them hand-written paper forms submitted directly by account holders, were unsuitable for machine reading. The second problem pertained to the method of computer processing. The study group strongly preferred a random access method that allowed instant debiting and crediting of a given transaction order. To realize this, all the millions of accounts had to be accessible and suitably large and fast random access memory systems were not yet available.

Developing and negotiating solutions for these two problems dominated the period prior to a final forced decision in the early sixties. This preparatory period was, as we shall see, characterized by a continuous interplay among the wishes, promises and judgments of the different actor groups involved. Not only was new technology at stake here, but also the construction of new social, organizational and gender relations.

A   Punch Cards and the Construction of the Data Typist

In theory there was a solution for the data entry problem: the information on the hand-written forms could be manually transferred to machine-readable punch cards. The study group considered this solution and they estimated that its implementation would require more than two-thousand key punch operators. This was perceived as both impossible and undesirable. The tight labor market could not in any case provide that many key punch operators, but perhaps of greater salience to the members of the study group was that key punch technology was seen as "old fashioned."

Punch cards and key punch operators had both long been an integral part of mechanical data processing. At the end of the nineteenth century Herman Hollerith had developed a mechanical way of processing large quantities of data. He developed a punch card as the information carrier and machines that could perform tasks with these cards: sorting, collating and tabulating. On the punch cards each character was represented as a punched hole (or combination of holes) at unique positions. Processing these cards was based on diverse electromechanical mechanisms: reading a hole caused an electric signal that implied a specific mechanical action, e.g., opening a specific sorting-compartment or adding one to a mechanical counter. The punch card also turned out to be an adequate input medium for electronic computer processing.

Punching data onto the cards was labor-intensive and the work was regarded as repetitive and monotonous. Employers hired chiefly young women to perform these tasks, since they were presumed to work faster and more conscientiously and — probably more important — could be had at minimal wages. Key punching was not a popular job among young girls and organizations using mechanical data processing in the fifties experienced great difficulties in attracting enough key punch operators (Van Oost 1994).

But there was more at stake. An ideological mainstay of the era of the electronic computer was that dull and routine jobs would be taken over by the computer, to be replaced by new and interesting computer-related jobs. The aura of the computer conformed to a modernist ideal: clean, silent and high quality work. The image of thousands of young girls engaged in drudgery on noisy mechanical machines did not fit in at all (J).

Clearly, it was not the technology itself that caused the data-entry problems — punch card technology was well established and reliable — but rather the organizational and ideological impact of applying this technology to mass administrations. Solutions were sought in new technologies. The desire for an automatic reading system congenial with the modern, electronic, era, became a powerful frame for coordinating action among the technically oriented actors involved.

In 1955, the study group began to discuss the problem of automatic reading intensively with IBM. IBM's top management showed interest in developing a technology that could automatically read information originating from the general public. IBM and the Neher Laboratory of the Dutch PTT launched a joint project to develop an automatic reading system, for which the Postcheque and Clearing Service would serve as a pilot. The Neher Laboratory had had a brief flirtation with developing electronic computer technologies. An employee of the Laboratory, W.L. van der Poel, had developed an electronic computer in the late forties and early fifties. However, in the wake of this project the Neher Laboratory abandoned computer development since they judged it impossible to keep up with the fevered pace of developments in subsequent years. Now that Neher had became involved in the automation of the Postcheque and Clearing Service, the old engineers' dreams came to life again. The Laboratory clearly took pride in playing an active role in the development of path breaking technologies. But the dream was short-lived. Shortly after the start of the project, IBM reneged on the agreement and pulled out, leaving behind a disappointed and angry PTT/Neher Laboratory. IBM's rationale was that the American banks had opted for a different system with readable magnetic ink.

The Neher Laboratory, however, decided to continue developing an optical reading machine on its own. The essence of the optical reading problem was that the data on the hand-written forms deviated widely from the uniformity requirements posed by the computer. Two possible solutions presented themselves: either teach the computer to recognize hand-written characters or teach the account holders to fill out their forms in such a way that the information could be read automatically. The Neher Laboratory opted for the latter course. Their concept of an automatic reading system was based on the use of so-called marking grids. Marking a specific part of the grid corresponded with a specific digit. This technology had already been used in the United Kingdom for the automatic reading of Toto-forms. The marking grid (see Figure 1) is a compromise between the way humans read information and the way machines read information, and so it aims to align account holders with the optical reading machines.

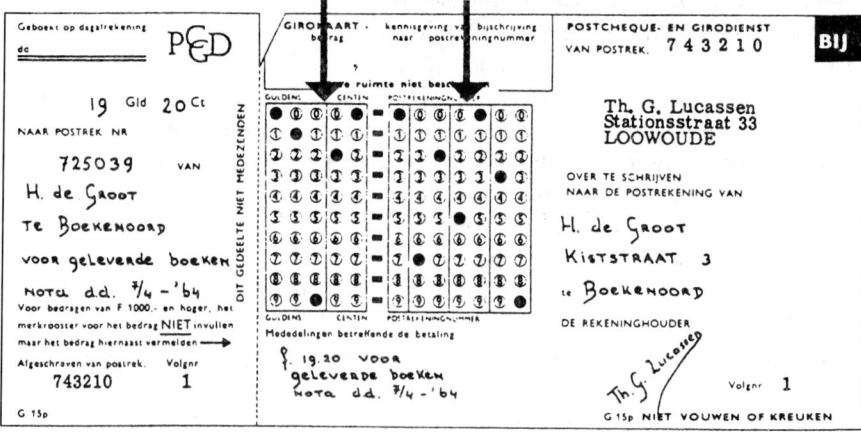

Figure 1: Giro-form Showing Marking Grid

In the period 1957–1960 the Neher Laboratory undertook several experiments with differently configured grids, but in all cases the account holders experienced persistent difficulties in filling out these types of forms.[4] The engineers nonetheless remained optimistic and maintained that the problems would be resolved within a short time. But in fact the Neher Laboratory could do no more than offer *promises* and was hardly in a position to offer guarantees.

Meanwhile, the Postcheque and Clearing Service was experimenting with the use of punch cards as a remittance form. The advantage of a punch card above a flimsy form was that punch cards could hold previously digitized information, for instance the client's account number. In this sense it was a dual-purpose card: it could contain both handwritten information provided by humans and encoded information directly processable by machines. The use of punch cards as remittance forms would not make key punch operators redundant, but it was expected to reduce, by approximately half, the key punching capacity that would have been required for transferring *all* the information from the flimsy form to a punch card. Another aspect of the dual purpose punch card that diminished the number of key punch operators needed, was the possibility for "external integration." This implied that organizational clients using punch cards and punch card machinery could employ the card in their own administrative procedures. These clients (for example, housing agencies or public utilities) could use pre-punched cards supplied by the Postcheque and Clearing Service in their own administrations and, after remittance by their customers, the same cards could be used for processing the transactions at the Postcheque and Clearing Service.

The management increasingly favored the use of punch cards, since their processing was reliable and secure. But they also envisioned problems in the use of punch cards as remittance forms. Up to then, punch cards had been used predominantly *inside* organizations and been handled by trained personnel only. It was expected that in the hands of lay people the cards could easily be damaged so as to make mechanical processing impossible or unreliable. However, in preliminary trials which included instructions on proper handling, it turned out that by and large the account holders took good care of their cards. The few cards that came in damaged could be processed in the "Carditioner," a new invention that could flatten crumpled cards successfully.

By the end of 1960, the burden of work at the Postcheque and Clearing Service had again become unacceptably high. In early 1961, the organization had to freeze the opening of new accounts. This undesirable measure was the straw that broke the camel's back. The management forced a decision, against the better judgment of the Neher Laboratory, to automate on the basis of pre-coded punchcards. The Postcheque and Clearing Service found an ally in IBM who affirmed that punch card technology was the only reliable input technology then available. Reliability was a core criterion because, in light of the collective trauma of the mechanization fiasco of the twenties, another debacle could not be afforded.

This decision required creating about 700 new key punching jobs all to be filled by young women. Management saw these new jobs as a necessary, but fortunately only temporary, evil. At the time, both managers and engineers believed that it was only a matter of time (the estimate was five years) before a properly functioning automatic reading system would be available. The Neher Laboratory was in fact pursuing the development of such technology with unabated energy. A passage from an internal report clearly illustrates the construction of the key punch operator as a temporary evil:

> She (the key punch operator, EvO) is a phenomenon that exists only during the rise of the automated office, and is not a worker that belongs there. She is indeed a purely temporary phenomenon, corresponding to the phase of rationalization that office work is going through at the moment (H: 12).

The report concluded that this job profile was best suited to young girls. These girls were expected to leave the organization voluntarily within the five-year period in order to get married and the Postcheque and Clearing Service defined the key punching job as an "opportunity for these girls to have a full-fledged job until their marriage" (H: 87). Here we see a co-construction of technological promises and job characteristics on the one hand, and the gendered profile of the ideal key punch operator on the other.

B Shaping a Computer System and Computer Jobs

A second item that was discussed intensively during the preparatory phase was the choice of computer system. The proposal made by the study group in 1955 recommended random access processing above batch processing. Batch processing was based on performing one operation (e.g., debiting) on a fixed sequence of account numbers. This way of working was quite similar to manual processing. In random access processing, the debiting and crediting of a given order could be performed simultaneously. The members of the study group saw the latter method as the ideal automated transaction posting system. The random access method, however, placed inordinate demands upon the size and accessibility of the computer's memory system. At that time suitable magnetic disk units were under development but not yet commercially available at an affordable price.

In 1956, the study group conducted a practical trial of a first generation tube-computer, the IBM 650, in cooperation with IBM. The trial showed that about thirty to forty of these computers would be needed for the automation of all giro accounts. The study group did not see this as a realistic solution, since "the Postcheque and Clearing Service would become too much like a large factory"(E).[5] Tube-computers were large and required corresponding cooling systems.

In the meantime, organizational measures like the opening of a second office had helped alleviate the work pressure. So the final decision about automation could be put off. The engineers at the Neher Laboratory strongly advised waiting for the transistorized computer then being developed. They expected it to be much more reliable and faster than the tube computers. So, the pursuit of automation was put on the back burner for a while.

In 1959, when transistorized computers appeared on the market, a new technically oriented working group was formed with the participation of the Neher Laboratory, IBM and Philips. The group's task was to formulate a set of computer requirements for the Postcheque and Clearing Service. In its final report, this group also strongly advocated the most advanced random access method.

In the event, the other method provided the final solution. The choice fell not on the most advanced technology, but on the technology that was most suitable from an organizational point of view. In 1960, work pressure increased again to such an extent that the board of directors forced a decision, not only with respect to the method of data entry, but also with regard to the type of computer. Again, reliability was the crucial criterion for the top echelon.

In cooperation with IBM, management proposed a trial with an IBM 1401 computer. The trial was successful and it was decided to continue with the IBM 1401. This computer, relatively small but fast and reliable, was suitable only for batch processing. While this choice was surely not the most advanced from a technical point of view, it did offer some crucial advantages from a management perspective.

The most important of these was that with the 1401 the Postcheque and Clearing Service could convert the transactions posting process gradually. One IBM 1401 computer could handle only a limited number of accounts and a total conversion would require approximately sixteen computers. Gradual conversion made it possible for management to respect their promise that automation would not lead to forced layoffs. This promise had been made to ensure the consent and cooperation of the employees. Since management foresaw an ultimate decrease of more than a thousand employees, a gradual conversion — paced to the normal rate of employee turnover — was seen as the most suitable solution.⁶ The aim was to complete the conversion over a period of five years.

A second argument for gradual conversion put forth by management was that this allowed for better control of the process. This argument should be interpreted in the light of the disastrous mechanization project of the twenties in which conversion to the new system was scheduled for a single day. The management very much feared a repetition of this fiasco. Nonetheless, gradual conversion did have one major drawback: It necessitated working simultaneously with automated and manual processing during the conversion period. During the period of conversion, jargon signified this by referring to the "new" and the "old" organization.

How did this "new" organization evolve out of the old one, and how did the gender patterns in the old organization influence the gender structures in the new one? I will elaborate both on how the organizational (gender) structure and the new computer jobs were defined, as well as on the gendered selection of incumbents of these new jobs. Both related processes contributed to the male gendering of the computer jobs.

The management decided to deploy the computers in the eight existing sub-departments of Current Accounts (CA), instead of in one main computer center, in order to circumvent drastic organizational changes. The Current Accounts Department performed the actual posting of transactions, i.e. the debiting and crediting of accounts. As such, it was the heart of operations at the Postcheque and Clearing Service, employing 1,800 of the 4,000 employees. CA employees had a high status within the Postcheque and Clearing Service, a ranking reflected in the educational requirements needed to work there (secondary school diploma at "MULO" level). In the automated, "new" organization, each of the CA sub-departments were divided into two sections called, respectively, Order Control and Electronic Processing. In 1965, after the conversion was completed, each of the eight sub-departments had a section for Electronic Processing equipped with two computers and four teams (two computers, each with separate teams for day and night shifts). All told, there were 16 computers, 131 operators (4 female) and 32 chief-operators (5 female) employed in the new electronic processing sections.

The decentralized organizational embedding of the computer processing conformed closely to the old organizational structure. However, the separation between the Order Control and Electronic Processing sections generated a new status distinction. Electronic Processing had higher status and the career possibilities were better than at Order Control. The computers were placed in special glass-walled rooms and only authorized persons were allowed access. The operators were required to wear white coats. So here we can see a faithful mimicry of the images created in the public media. It is not coincidental that the proportion of female employees in Electronic Processing was significantly lower than in Order Control.

In contrast with the computerized (and manual) processing of transaction postings, which was distributed throughout the organization, programming and systems analysis were centralized activities. A Staff Department for Automation had been established as of 1956 to oversee the first trial with an IBM 650 computer. This new department was staffed by five internally recruited and trained employees from the Postcheque and Clearing Service, all young men. By 1966 this department had grown to employ about 40 programmers and analysts — still all men. This growth of course reflected the increasing use of computers, but it was also part of a general post-war trend of strong growth in staff-departments as compared to the line-departments. Most staff department jobs required more educational credentials than those in the line departments, like the Current Accounts Department. Transfer from Current Accounts to a staff department was regarded as a promotion.

The management recruited personnel for the new computer jobs from within the organization. This was a basic promise made by management to the employees prior to commencing automation. Like the promise to avoid layoffs, this promise also functioned as a social shock absorber (Steenbergen 1960). Another rationale for internal recruitment was the fact that a labor market for these new types of jobs had not yet come into being. Hence, all the programmers and operators were recruited from among the employees, as it turned out, most of them from the Automation Department. The personnel department had developed a selection and training procedure for filling these new jobs. Judged by outcomes, this procedure was not devoid of gender bias. In all the new computer-related jobs, the number of women was significantly lower then one would expect based on the gender structure of the group from which the incumbents of these new jobs were recruited. There were no female programmers or systems analysts, and only 16% of the chief-operators and 3% of the operators were female. In 1960 the CA Department had about 25% female employees, of whom a relatively high proportion (50%) occupied middle management positions as I have described above.

To understand the gendered character of the selection process, I will focus on the *mutual* shaping of both the jobs and job qualifications as well as the job incumbents, taking programming as an example. Management had defined the job of programmer in such a way that it entailed a significant career advance for those applying from other positions in the organization. The high quality of these jobs (as replacements for the volume of labor about to be made redundant by automated processing) was part of the management strategy of creating a social shock absorber. In spite of binding civil service salary and promotion policies, the Postcheque and Clearing

Service succeeded in establishing an accelerated career path for programmers on the basis of a rather severe selection procedure. CA Department employees were thus more or less "seduced" to favor automation by offering them the possibility of an interesting new job.

The combination of shaping new, career-oriented jobs with a severe selection gantlet was also not gender neutral with respect to the prevailing gender ideology. The focus on work and career was a severely gendered one. At the time, marriage and motherhood were considered to be the most important life goals for women. Employment and career perspectives were considered to be more important for men than for women. Hutte's research clearly showed that both male employees and management strongly adhered to this opinion.

An employee seeking to become a programmer had to pass three hurdles in a closely managed selection process. First, potential candidates for a programmers' position were selected by their department heads. Second, the candidates were given a programmer's aptitude test. Only those who scored good or very good on the test were admitted to a special internal training program. Third, they had to complete the one-year training program. This "meritocratic" selection procedure defused potential resistance within the organization. Those who had passed the various hurdles in the selection process were satisfied with their new career oriented position. To those who failed, management could simply argue insufficient competence.

In the actual practice of selecting future programmers the Postcheque and Clearing Service clearly discriminated against female employees. Eleven women were nominated by their chiefs (of 200 in total), but only one was given the programmer's aptitude test — in contrast to the male candidates, all of whom were allowed to have a go. In 1961, the board of directors deliberated whether female employees could be selected for the programmers' course. Although the official answer was that "female candidates could not be excluded in advance," the discussion revealed an implicit conviction that programming functions within the Postcheque and Clearing Service were actually male gender-typed jobs. Some of the women who had been nominated later achieved the position of chief operators, explaining the relatively high percentage of women in this function. The definition of the chief-operator as a managerial (rather than technical) job, facilitated the relatively "high" proportion of women (Van Oost 1994).

It took until 1968 before the Postcheque and Clearing Service recruited the first female junior programmer. By that time the Postcheque and Clearing Service was able to welcome her with open arms, since she could no longer threaten the already stabilized male gender typing of programming functions. On the contrary, she implicitly reinforced the rule by becoming the well-known token exception. This woman had applied in response to the first publicly advertised job openings for junior programmers. The Postcheque and Clearing Service required a high school diploma with emphasis on mathematics, which exceeded the educational level of most of the employees in similar functions. Personnel selected nine young candidates — the only female among them included — out of eighty applicants. The gender-biased composition of the group of applicants illustrates that male gender typing of the computer and related functions was already stabilized in Dutch society at that time.[7]

Why did the programmers' job within the Postcheque and Clearing Service inevitably acquire a male gender typing? In this specific setting, male gender typing evolved from a combination of a negative image of the group of middle-aged female employees and a strong pressure to avoid a repetition of the earlier mechanization debacle. One thing was clear to those who initiated and coordinated the automation process: failure was unthinkable. This implied that the reliability of the workers charged with carrying out the automation was crucial. While in the main computers were believed to be and were touted as conscientious and reliable actors, they were also viewed with suspicion because they might break down at the most unexpected times. To compensate for the perceived capriciousness of the apparatus, management stressed the importance of the reliability of the computers' attendants. The ideal reliable worker was not a female employee.

Another specific feature of this case was that the automation of the Postcheque and Clearing Service was supposed to be the jewel in the crown of the modernization process that had been started in the mid-fifties. The female middle managers implicitly symbolized the old, bureaucratic organization. Automation was meant to raise the status and prestige both of individual employees and of the organization (A). Women with "missed life chances" — as Hutte characterized the female middle managers — could not be expected to contribute to this envisioned prestige.

I do not want to suggest that all this was a plot against women. On the contrary, gender-typing of machines and functions and the relative exclusion of women just happened.[8] None of the actors involved would have thought explicitly in gender terms, but as an analyst one can still isolate reasons for this specific course of events.

5 Conclusions

In the pioneering phase of the embedding of new computer technology, the simultaneous use of two frames of meaning was crucial for embedding, that is, for the successful construction of a new socio-technical order. The virtuosity frame was crucial in stimulating interest among a strategic social group, top managers. Engineers presented their new product, the computer, as the fulfillment of a great intellectual challenge. Now, it was the turn of top managers to take up the challenge of automating their organizations. This mechanism was clearly visible in the case of the Postcheque and Clearing Service and it was also the keystone of the strategy of diffusion pursued by the national study center. As a result, in the initial phase of the societal embedding of computer technology, alignments were created with a strongly male-biased set of actors.

Gendered images of the computer (and other artifacts) are intentionally or unintentionally used as a *means or a strategy* to influence successful embedding in society. For a successful social embedding of a new artifact, links must be made between the new artifact and elements of the social structure. Gendered representations are intentionally — or perhaps more often unintentionally — used to establish successful links with an intended relevant social group, like top managers. In its infancy, the computer was a very expensive piece of apparatus. Consequently decision-making on automation was a matter for top management, hence, for men. The financial and organizational consequences were often enormous. Why would management make such a hazardous move, the more so in view of the alternatives available in the field of mechanical data processing. By surrounding the computer with masculine virtuosity values, early champions of automation could mobilize key managerial groups. The computer became associated with challenge, with expanding horizons, taking risks, status and prestige for managers. Managers, too, became fascinated, in thrall to the electronic brain. In its turn management was able to use the masculine, status-laden representation of the computer to propagate the desirability of automation among an important part of the male staff. This is why the business tool frame is continuously mixed with elements of the virtuosity frame.

In the case of the Postcheque and Clearing Service, both frames of meaning associated with computer technology were simultaneously present and mobilized. Arguments from the business-tool frame officially played the central role for most actors. But we have seen how engineers at the Dr. Neher Laboratory and the pioneering PTT director, Reinoud, very much favored the latest and most advanced technology. They linked their professional identity to the virtuosity of computer technology and they saw automation as an opportunity to enhance their status and position (DeWit 1994).

The male virtuosity image of computers and computer jobs also played a crucial role in manufacturing consent among the employees. The dominant public image of the Faustian magician in control of a mysterious and powerful apparatus was imbued with masculine symbols. Men and boys could easily incorporate this image in their masculine identity. This made the group of male employees especially eager to occupy those new jobs, and consequently from a management perspective a more dangerous source of resistance in case of perceived exclusion.

The tension between the masculine virtuosity frame and the more gender neutral business-tool frame is of wider relevance than computers alone. Many processes of sociotechnical change — that is, the construction and societal embedding of new technology — will be influenced by meanings and evaluations implicit in these two frames. In cases like the computer, where the virtuosity frame dominates, a masculine image of the new technology will be created. An explicit masculine image of the new technology will most probably lead to a reinforcing of existing gender inequalities in the changing sociotechnical structure. On the level of organizations, one can predict that equal-opportunity policies aiming at neutralizing gender inequalities in the organizational structure or culture, will not suffice for avoiding the development of new gender inequalities. Sally Hacker (1979) discovered this in her AT&T study: the effects of many years of positive action were completely undone in a brief period when new technology was introduced into the organization. If we want current sociotechnical change to improve gender equality in this time of exuberantly virtuoso technology, we can only succeed by seriously and incessantly trying to deconstruct the triple alliance among technology, virtuosity and masculinity.

Notes

1 Financial aid in the framework of the Marshall Plan (started in 1948) intended for reconstructing European economies and society in the wake of World War II, emphasized the importance of efficiency. In the Dutch case this was effected in part by the installation of a Marshall-related committee for the increase of productivity, the so-called COP (Commissie Opvoering Produktiviteit). The COP offered advice to firms on increasing efficiency from the well-known Human Relations perspective.
2 In other countries with the same ideology, but lower productivity, married women participated to a greater extent in the labor market (Plantenga 1993).
3 The Dutch historian De Wit refers to this episode as a "historical specter" (De Wit 1994).
4 Several tests were done with different grids to find out what would be the best design. But in all cases too large a proportion of the account holders made mistakes filling them out.
5 Hutte's research had indicated that employees placed a high value on the specific office ambiance. From their point of view, a few bookkeeping machines were enough to ruin this ambiance (G).
6 At the start of the automation trajectory a conversion period of five years was deemed necessary (1962-1967). However, in the course of the project, management

decided to accelerate the process so that by 1965 automation was completed (De Wit 1994).
7 The mathematics requirement in and of itself cannot explain the extent of the gender-biased composition, since fully 25% of female secondary school students took math.
8 The Dutch anthropologist Don Kalb draws a similar conclusion with respect to the personnel management of Philips, which he characterized as systematic exploitation of child-rich families (D).

Sources

A. Bijleveld, H. (1968): "Sociale Aspecten in verband met de voorbereiding en invoering van de automatisering," in: H. Reinoud (ed.): *Een halve eeuw Postcheque- en Girodienst*. Utrecht, Marka: 313-328.
B. Bowden, B.V. (ed.) (1953): *Faster than thought. A symposium on digital computing machines*. London: Pitman and Sons.
C. *De Telegraaf*, October 6, 1963.
D. *De Volkskrant*, June 10, 1995.
E. Groen, G.J.F.A. (1966): "De Automatisering van de Postcheque- en Girodienst," in: H. Reinoud et al. (eds.): 205-223.
F. Groen, G.F.J.A. (1966): "Sociale Aspecten van de automatisering van de Postcheque- en Girodienst," in: H. Reinoud et al. (eds.): 315-350.
G. Hutte, H.A. (1956): *De Arbeidsverhoudingen bij de Postcheque- en Girodienst*. 's-Gravenhage: PTT.
H. PTT Personnel Department (1963): *The keypunch typist, her work and working conditions*. PTT Internal Report: 12.
I. Reinoud, H., G.J.Hennephof and F.J.E. Hodewind (eds.) (1966): *Automatisering en Overheid. Computertoepassingen in de Rijksdienst, bij PTT, Spoorwegen, en Staatsmijnen*. Utrecht: Marka Pocket.
J. Ulbricht, D. (1964): "Mogelijkheden en beperkingen van automatisering," *Kantoor en Efficiency*: 1098-1103.
K. *Vrije Volk*, December 17, 1966.

Chapter 7
Expectations in Technological Developments: An Example of Prospective Structures to be Filled in by Agency*

Harro van Lente and *Arie Rip*

1 Introduction

This chapter analyzes the dynamics of expectations in technological developments as a mode of coordination. Technology is an interesting research site, since here novelty is created: not just new products, but also new linkages and emerging technical–scientific fields. On the basis of case studies we will discuss how expectations structure activities differently than structures normally do: expectations help to interlock and to coordinate activities and to build up agendas. Because of their "script," expectations allocate roles for selves, others and (future) artifacts. When these roles are adopted, a new social order emerges on the basis of collective projections of the future.

In sociology, several theoretical attempts have been made to fill the gap between a functionalism which forgets creative actors, and an interactionism which forgets about constraining structure. Such attempts (e.g., Giddens, Berger and Luckmann, Boudon, Strauss) seek to overcome the dualism of structure versus agency. The key issues are how actions lead to structures, and how these structures enable and constrain action.

Technology dynamics teaches us that structure emerges and action is shaped in a way in which content matters as much as traditional sociological categories of explanation. It is the content of orientations and scripts which pulls actors together and which, when they enact appropriate strategies, generates a structure which shapes further action. The key point here is that structures can be "merely" prospective, and still be influential. To emphasize this, we introduce the (paradoxical) term *prospective structure*. A prospective structure is made up of links which can appear in texts. In the subsequent actions and reactions, the structure is filled in, modified, reshuffled; and becomes social structure (in its various forms).

2 Technology and the Problem of Structure–Agency in Sociology

Technology is an interesting research site on a number of grounds. For one thing, several key modes of coordination are in evidence. In this chapter we will focus on one in particular: the way actions become coordinated through the prospect of a new technology and its functions while this emerging configuration is simultaneously shaping the technology-to-be. The coordination is actively sought by actors, but the patterns that may eventually emerge cannot be attributed to any one particular actor. This is clearly an example of the well-known phenomenon of unintended collective effects of the actions of individual actors.

Phrased like this, we can see a clear relation with the problematic duality of structure versus agency.[1] Sociologists have taken (sometimes extreme) positions on the relative weight of structural arrangements versus the creative power of voluntaristic agents, and on the way structure influences or even determines agency, and *vice-versa*. Such questions can be rephrased as issues of coordination: who or what is the source of order? Curiously, sociology (with the exception of some strands of game theory) has paid little attention to processes and outcomes of coordination.[2] This is partly the result of the traditional split between structure- and actor-oriented sociologies. But it also results — and this is our contention in this paper — from a neglect of prospective stories and diffuse scenarios as key phenomena of social life. Technology dynamics, with their interest in change and the introduction of novelty, have something to offer here.

Giddens has suggested that the discipline of sociology has a specific division of labor (Giddens 1981: 167). In his opinion, the sociological work at hand has been divided into two streams, especially since World War II. Roughly speaking, functionalism has dealt with structures, while phenomenology and symbolic interactionism have dealt with agency. As a result they condemned themselves and each other to *halbierte* theory.

To elaborate this point briefly: Functionalism is limited in the sense that it tends to treat structure as some external factor working behind the backs of actors, as it were, thereby ignoring interaction processes, and the social shaping of reality by agents. In other words, "agents in situations are not seen as doing very much; and structure is seen to march along disembodied from the actors who are involved in its reproduction or transformation" (Turner 1986: 459).

The other party in the division of labor, symbolic interactionism (and phenomenology) is limited in the sense that it tends to ignore structural constraints on social processes. It focuses on face-to-face (symbolic) interaction, e.g., on role-taking (Mead) or on dramaturgical presentation of self

(Goffman), while little effort is expended in explaining how aggregated outcomes may result from ongoing interactions. In interactionists' accounts, social structure evaporates, so to speak, due to the focus on here-and-now, face-to-face, encounters, and due to the interest in the irreducibly creative and personal side of social life. Structure enters only through the perception of constraints by actors. Other effects cannot be discussed; effects like some courses of action being routinely more successful than others given constraints visible to an analyst (i.e. structure).

Several attempts have been made to bridge the gap between a functionalism which forgets creative actors, and an interactionism which forgets constraining structure. For example, Berger and Luckmann (social construction of reality), Giddens (structuration theory), Burns and Flam (social rule system theory), Shibutani (social processes), Strauss (social world), Boudon (transformation processes). These attempts share a belief in the need to overcome the dualisms and dichotomies resulting from the division of labor in sociology: structure versus agency; determinism versus voluntarism; objectivism versus subjectivism. The key issues then become how actions lead to structures, and how these structures enable and constrain action.

While these attempts are laudable, they are, in a sense, still prisoners of the duality of structure and agency, because they look for solutions in terms of complementarities between the two. This is admittedly too coarse a treatment of complex and sophisticated theories, but it does allow us to make the point that one could, and should, consider a third source of change: *expectations* about possible developments, especially as these are put forward and taken up in statements, brief stories or scenarios. These lead to action, not because there is a structure behind the backs of the actors, but because now actors are creating one before them.

We shall come back to these general points in the concluding section. Here, it is important to note that similar dichotomies are endemic in technology studies, where the structure of a technology has been seen as determining, or, alternatively, where actors have been seen as freely constructing and modifying technological configurations. Again, the contrast between artifacts or actors as sources of social agency can be transcended by introducing a third possibility, which we have called "agenda" (Van Lente 1993: Chapter 1). The emergence and stabilization of agendas can usefully be traced in terms of interlocking stories. The study of texts, in linguistics, new rhetorics and cultural studies, then becomes relevant and, in this sense, one can think of technology as a generalized text. This is important for our overall topic of coordination, because texts are nothing if not internally coordinated, among other things because of their story-line. Here, we have the basics of a mechanism of coordination, which we shall develop in the body of the paper.

With our focus on expectations, and texts about the future world in general, we are in a good position to see how story-lines may structure action before the fact. It is because of their content that these stories help to create new patterns and institutions. The basic mechanism is that actors position themselves and others (and future technology) in a story or plot, and so make others into characters in the story. In reaction, others will become enunciators of stories of their own, which are inevitably linked to the original one. Since implications for action are drawn, the stories become assembled into a repertoire used by actors to define possibilities and strategies, as well as to evaluate the actions of others. Thus, *prospective structures* emerge, i.e. arrangements that do not yet exist, but are nonetheless forceful due to the perceived implications of the projected future.

These are the two points we want to illustrate and elaborate with the help of three case studies, first in relation to technology dynamics, and then more generally.

3 Some Examples of Emerging Patterns in Technology[3]

A Example 1: Moore's Law as a Self-Fulfilling Prophecy

The first example is about the development of memory chips, and involves all the large microelectronic firms in Japan, Europe and the United States. In 1964, Gordon E. Moore, research manager of Fairchild Semiconductor, observed a regular periodic doubling of the number of "gates" (a measure of complexity), and claimed, by extrapolation, that this would continue. This prediction has come true so beautifully, that nowadays we speak of "Moore's Law," as if it were a law of Nature. The validity of this law cannot be understood from the technical procedures by which the chips are made. The fact that the law holds so well is an effect of the way actors (in industry, in science and in government) judge their own and each others' accomplishments with respect to what Moore's Law predicts. They direct their efforts towards achieving the predicted values. Laboratories evaluate and plan their efforts in terms of Moore's Law; when there is danger of specifications falling short at the predicted moment, extra effort is expended. Firms use the law to guide investment decisions in specific technologies; for example whether or not to develop products that need chips with the predicted capacity — such as calculators or compact disc players. Governments are willing to provide subsidies in order to help firms avert the danger of not meeting the predicted value.[4] All actors exert themselves to measure up to the predicted competition and to stay in the race. Moore's Law is the yardstick for the behavior of chip producers and governments in Japan, the United States and Europe, and it shapes their mutual dependency in the strategic game they play with one another.

Because it holds so well, other firms use a modified version of Moore's Law that claims a reduction in the price per gate of 30% annually. This version allows one to predict the commercial opportunities for products like pocket calculators, compact disc players and videotext systems. Assuming, for instance, that chips more expensive than $10 to $20 are a barrier to widespread use in consumer electronics, one could predict a market for pocket calculators, with their 1 K (= 1024 gates) chip, only after 1972, when the price per gate had fallen to about $0.01. By the same token, firms saw opportunities for producing a compact disc player (with its 70 K chip) in 1985, when the price per gate had dropped to $0.00015.

We may speak of a self-fulfilling prophecy, but the fulfilling did not occur because it was a prophecy, but because actors took up the prophecy and acted accordingly. This was a basis for other actors to accept the expectations and act accordingly, etcetera. The promise has now become part of a *prisoner's dilemma*: firms and governments in Europe, Japan (and now also other countries in South-East Asia) and the United States stay in the race for superior chips, even if this requires huge investments, because they do not want to run the risk of falling behind the other parties in the triad. And while not absolutely certain, they strongly suspect the others will continue — simply because chips are a promising technology.

Prisoner's dilemma situations (and similar phenomena analyzed in game theory) are widespread and explain why actors remain involved. However, the model does not explain the nature of decisions and mutual coordination through the content of the technology. The vocabulary of generations, and the regularity of generations following each other according to Moore's Law, allow decisions to be made in terms of: which generation are we working on? Can we leap toward the next generation?[5] The antagonistic coordination of the chips race is made possible by "reading" the opportunities in and requirements for the coming generations of chips. It is only because of this reflexive agency that the division of labor and the strategies allow one to speak of a "self-fulfilling prophecy."

The lesson of this example is that actors start to take mutual account of each other because of the opportunities they perceive in the future technology. Initially, the participating actors belong to different organizations and different sectors, but by commonly anticipating a future technology, they become interconnected. These interconnections are not like producer–client relationships or hierarchical relations: actors do not exchange products, but ideas about technology and technical opportunities.

Note that in this first example, coordination is clearly articulated. Actors (firms, governments and research institutes) know the other actors, they know the stakes and the rules of the game. And they act accordingly. In this case, coordination is established through expectations about the next generation of memory chips, and through the more or less stabilized strategic game.

Coordination through games is a well-recognized phenomenon. What we have added here is the role of a promising technology in assembling the actors and setting up the rules of the game, as for example in the notion of a "next generation" of chips.

B Example 2: The Emerging World of Membrane Technology in the Netherlands

In the next example, we present a more diffuse situation: the emergence of a promising technology and all the activities and relationships that go with it. Membrane technology, a technology which is now high on the agenda of technology policy makers, researchers and industrialists, did not exist as such before the late 1970s. There were a variety of separation techniques, and their names indicated the way the specific function was performed: membrane filtration, ultrafiltration, reverse osmosis, etc. Partly for internal scientific and technological reasons, partly because of the external interest in strategic research and new technologies, a new construct emerged: membrane technology as such. At first it was only a programmatic entity, but gradually the construct got filled in: cognitively, technically, industrially, and socially. We shall trace this evolution in the Netherlands (that country being a leader in the new field of membrane technology) and in some detail because we shall return to this case in section 4 where we discuss mechanisms.

Our story starts in 1981, when the Dutch Ministry of Economic Affairs and the Office of Science Policy initiated a range of innovation stimulation programs in specific areas, the so-called Innovation-Oriented Research Programs (IOP). The first was on biotechnology (Rip and Nederhof 1986), followed in 1984 by membrane technology. Previously, Professor Kees Smolders of the University of Twente had been a central figure in the development of membrane R&D in the Netherlands. He had investigated membranes in the early 1970s, among other things as part of a research program in macromolecular chemistry and polymer studies. Linkages with other actors followed on collaborations with a variety of firms (Wafilin, AKZO, Shell)[6] and research institutes (TNO, NIZO, KIWA).[7] At the same time he tried to promote membrane filtration by organizing meetings and publishing in chemical engineering journals on the importance of membrane filtration. In the journal *Procestechniek* (Process Technology), he wrote editorials and articles with titles like "Why Membrane Filtration?" "Membrane Filtration as Technological Innovation" and "Membrane Filtration: a Steady Growth" (our translations). Smolders emphasized that the governments of Western industrial countries had become interested in innovation–induced economic growth; he argued that membrane filtration was an immature, yet promising, technology; and he accordingly recommended the creation of a government–funded research program on membrane filtration. We note that his arguments typically moved from a specification of industrial (or societal) problems to solutions based on special separation techniques, and never failed to include the additional diagnosis that more research was needed to develop the full potential of these techniques.

While he did indeed acquire funds for several research projects, his major success in terms of the present argument was the leadership of an investigative mission to Japan in the Spring of 1982. The mission included members of organizations with which Smolders already had contacts: TNO, Wafilin, AKZO, KIWA; i.e. the traditional membrane industries and research institutes. The investigators, not surprisingly, concluded that Japan was investing a lot of money in membranes and that a research program on membranes was badly needed in the Netherlands. In addition — and this is an important shift — they introduced the umbrella term "membrane technology."[8] Their official mission was to report on "membrane filtration technology" in Japan, but upon returning they flashed the new concept "membrane technology" in the title and text of their report (D). The introduction of the umbrella term was not just a rhetorical re–description of what was already going on, but also a key step in the "reversal" of membrane technology, i.e. from a dependent entity appearing at the end of a long argument, to an independent entity with a life of its own — an entity, moreover, which had to be taken into account by policy makers (who had commissioned the investigative committee) researchers and industrialists. In other words, it was a key step in the establishment of membrane technology as a macro–actor.[9]

One could view this as a prosaic example of mobilizing rhetorical and other resources to further one's interests — but the interests were defined in terms of the promise of a new technology in which other actors could and should participate. And indeed, there was considerable coincidence of interests among the heterogeneous actors involved. A policy maker who was responsible for the programs at the Ministry of Economic Affairs later recalled how Smolders "with his enthusiasm, expertise, technical knowledge and his vision of the future, succeeded in selling this [membrane technology]." He himself thought it was an "interesting opportunity for the Ministry to try out its innovation stimulation measures," but it was not easy to sell it to others within the Ministry (which had a tradition of fiscal measures and links with actual industries, and had only recently started up a technology policy). Membrane technology lacked a link with existing industrial activities and "therefore, within the Ministry, it was not such a favored topic" (K: 8). Nevertheless, the Ministry of Economic Affairs and the Office of Science Policy established an eight-year Program on Membrane Technology with a total budget of f25.8 million, as had been strongly advised by the investigative committee. Universities and research institutes were invited to propose research projects; each project was to have a supervisory group, preferably including a representative of industry. Smolders was appointed chairman of the Program Committee of the IOP (E).

It was not difficult to show that Smolders' interests were directly served by some of the outcomes: e.g., the main recipient of IOP funding was Smolders' own research group (about a third of all projects) and by the end of the 1980s his research group had become the largest university group on membrane research in the world. However, the growth of his research group and his reputation were not the only outcomes of the IOP. The Program Committee also actively drew attention to the promise of membrane technology in those firms they regarded as relevant. In this connection, the Committee wrote two booklets about membrane technology in the Netherlands (I), in which it argued that such a thing as membrane technology existed, that it was increasingly used by industry and that it had a brilliant future. So, membrane technology itself, defined as a promising technology, was coordinating action — at least rhetorically in texts.

That it also coordinated more palpable sorts of action becomes clear when we look at the emerging membrane *world*. While activities oriented by the promise of membrane technology were being pursued worldwide, and as Dutch actors positioned themselves in relation to those activities, we will focus on the Dutch IOP Committee again to show some of the dynamics involved. The committee tried to involve industry by organizing annual poster sessions and public lectures at the so-called IOP days. These sessions eventually appeared in the journal *Membraantechnologie*, published and circulated by the Committee. All firms presumed to be "potentially interested" were invited to the IOP days. Table 1 shows how this worked out.

Table 1: Backgrounds of Participants of the IOP Day, 1989

Background	# of Posters	# of Persons	Ratio
University	26	72	0.36
Research Institute	19	40	0.48
Firm	6	82	0.07

IOP research results do not, or do not necessarily, have any direct instrumental value for the firms. The ratio poster/persons in Table 1 suggests that at an IOP day researchers were presenting themselves to firms, hoping to get support for their projects. The firms, on the other hand, complained that the posters were "too scientific." While they did not question the scientific value of the presentations, they felt that their economic relevance — purportedly the reason for supporting the research — was not self-evident. They went so far as to suggest to the Program Committee that for each poster a statement of technical and economic relevance be made obligatory. As one participant from industry formulated it, "now they are just organizing a scientific meeting, paid for by IOP money."[10]

Nevertheless, firms do come to these events, "because you hear a lot about membranes nowadays, and you want to find out what's going on." Another indication that these events are seen as important is that even actors not working on IOP projects are willing to present their activities; for instance, new firms entering into the production of membranes or separation facilities. They have no particular obligation towards a funding agency to attend. However, they do feel obliged to present their activities because a relevant audience is being constructed, and they want to be in the picture. In so doing, they further reinforce interest in membrane technology, giving it a life of its own. What was only a label at the outset, is now becoming an entity that exerts force on others.

So the IOP Membranes was not just an innovation program: it was a node in the emerging network of actors, activities and texts that helped shape the new world of membranes. The IOP built on, and then reinforced, a transition to an interest in a particular technology as such and tried to construct a new sector or field around it. And the technology as such was supported, as we indicated, by a cluster of actors: innovation policy makers, intermediary actors in the R&D system, scientists from universities and research institutes, technologists and firms. The audience has now stabilized, and while participants may shift, membrane technology will survive because actors feel compelled to be interested in it.

There are structural changes, like the increase of the number of membrane researchers at universities, the entrance of new firms into the membrane business and the various small-scale collaborations. These are paralleled by cultural changes: expectations and search behaviors have changed in universities and research institutes, but also in industry. Developments under the umbrella of membrane technology lead marketing people and engineers in firms to ask: "Shouldn't we do something with membranes? There's so much talk about them that we have to try them out."[11] Heuristics in the design and development of separation devices now contain the prescription to check on the possibilities of membrane technology. In 1990, at a conference about membranes and biotechnology, for instance, several participants from industry said that they did not have any experience with membranes but "you hear a lot about them, so we have to find out what's going on. Besides, our customers also keep asking about it." People are now better informed about where membrane research is carried out, which firms are important and which developments it is important to keep track of. And there is articulation of relevance: what is membrane technology about? Which actors are important? What should be taken into account? What is on the agenda?

Note that the activities interlock on the basis of what actors have to do and on what they expect others will do in the future. It is not necessary that actors have shared views about the world; they may have different opinions about the common cause. What counts is not "saying the same," but "knowing what to expect." For instance, a firm may say that membrane technology may be helpful in cleaning waste water, a policy maker will say that the Ministry of Economic Affairs needs strategic programs in its portfolio, and the scientists will say that the mechanism of transportation across membranes is an unsolved problem. They do not say the same, but all of them have good reasons to support a research program on membrane technology. The projected future of membrane technology as an innovative research area generates a division of tasks.

The emerging world of membrane technology is an example of *diffuse coordination*. The main structuring processes in this example are (i) a gradual agenda-building and (ii) a reversal of membrane technology from a tentative possibility to an independent and compelling macro-actor. We will discuss the core mechanisms in these processes in a later section.

The additional point to be made is the fact that the agendas, which get articulated in due course, lead to specifications for further work, as a kind of socio-cognitive version of the way that requirements for a new product or artifact lead to design specifications. For a new technology, it is necessarily a diffuse process, which nonetheless tends to gel after some time.[12] This latter dynamic will be much more prominent when a new technical system with clearly projected functions is the focus of promise. This is what we see in the next, and last, example.

C   Example 3: HDTV, a Self-Justifying Technology

Before 1985, High Definition Television (HDTV) was held to be a promising idea among technologists, but it was not widely promoted, certainly not as any kind of national cause (Van Lente 1993; Rice 1990; OTA 1990; J), except perhaps in Japan, where NHK, the Japanese national public broadcaster, succeeded in coordinating a lot of research on HDTV from 1968 onwards. The shift to prominence took place because HDTV became the subject of a strategic game. The CCIR (International Radio Consultative Committee) meeting in Dubrovnik in May, 1986, marks the transition. At the meetings of the CCIR, held every four years, global standards for television production and broadcasting are discussed. In May, 1986, Japanese firms, under the direction of NHK, proposed their "Hi-Vision" system as a global standard. This was supported by some American firms. The European participants feared they would be relegated to a second-rate position in this future technology; lobbyists created a united European block against the proposed standard, asking for some studies and for alternative proposals,[13] and thus succeeded in delaying the CCIR decision. The proposals were to be presented at the next meeting, i.e. four years later. A month later, in June, 1986, the Eureka EU95 program was launched as one of the first research programs within the Eureka framework. The aim was to work on a European proposal for a global standard for HDTV; the budget was about $350 million for the first phase, and $500 million for the second phase (after 1990).

From that period on, HDTV in Europe became "Highly Debated Television," as it is sometimes called in the press, and it appeared on the agendas of European governments,[14] side by side with issues like the common European market and common currency, or the war in (former) Yugoslavia (H). In the resulting tug of war around HDTV, more is at issue than interest politics around standards and markets. It is obvious that interest politics will occur: the economic and political interests associated with HDTV are immense. The potential world market is estimated in terms of hundreds of billions of dollars and the size of one's share of this market will have important consequences for other industrial activities, for the labor market, and for the relative strength of national economies. So it is not surprising when parties interested in the development try to gather support or try to promote this (future) commodity. Notable, however, is how interests are defined here. It is only because HDTV is taken as promising that there are interests at stake, and that interest politics can come into play. If HDTV were not seen as promising, there would be no interest in the technology, and thus no actor's interests to be staked out.

The promise of HDTV is not a black-boxed promise: certain functions to be realized by the new TV system are emphasized, and these have rapidly turned into an agenda of requirements to be met by the new system. The dynamics of this process show a curious duality, which can be recognized in other cases as well, and may be characteristic for how our technological culture works.

The official legitimization of the proposed efforts uses an instrumental register: "novelty is necessary because the existing state of affairs is not good." In the instrumental mentality the new technology must be seen as the solution for a problem — but is there a problem? Present television is not criticized in terms of complaints — such as, for example, "after two hours of watching TV one gets a headache" — but in terms of what is missing compared with what is projected as technologically feasible. If the quality of the sound in our television is less than that of the best sound system available at the moment (the compact disc) this tends to get defined as a limitation. A report that investigates the potentials of HDTV notes:

The sound quality of the present systems is not very good, especially compared with the sound quality of compact discs (CD). Also the number of channels of the present systems is limited.[15]

The solution for such a problem is obvious: HDTV must (and will) deliver sound quality at the level of the compact disc. The same story holds for the size of the screen, the number of sound channels, the number of lines per image, the frequency of the scanning lines. All are less than what is possible, or, better, what should be possible, since the actual achievement of these specifications still requires much effort. Thus, the main shortcoming of the present system is that it is not HDTV. The present system is deficient exactly to the degree that it does not live up to the projected characteristics of HDTV. The present is measured by the yardstick of the technological promise, and found wanting.[16]

It is not without reason that the metaphor of a "new generation" superseding the "old" one is widespread. The notion of "generation" suggests that it is quite natural to trade in the old for the new. The argument proceeds by taking technological development itself as a proof that the old generation is outdated: black & white television is outdated by color television, which in its turn will be rendered obsolete by the introduction of HDTV; the typewriter has surrendered its place to word processors and the horse-cart is succeeded by the motor car. The tautology of self-justification is clear. But in the world of action, as opposed to the world of logic, even tautologies can be important sources of coherence.

While this point can be amplified into a general diagnosis of our technological culture, our present concern is the dynamics of action. Once something is defined as a promise, action is demanded. And one sees actors act, but curiously enough, with a sense that they cannot help it. For technologists, industrialists, and other participants, self-justifying technologies such as HDTV appear as a two-headed monster.[17] On the one hand (head), the development seems inescapable. HDTV is coming, whether we want it or not. Or as chairman Bögels of the Dutch Platform for HDTV (a body devoted to paving the way for HDTV) put it: "This train is in motion. This train can no longer be stopped" (F). So, ignoring the development is impossible; you may get off, but the train will continue without you.

On the other hand, leaning back leisurely and looking out of the window, knowing that the train cannot be stopped, is also impossible. On the contrary, hard work is necessary to keep things moving. The infrastructure has to be adequate; the knowledge level has to be kept up;[18] cable proprietors, program writers and broadcasting companies must adapt to the demands of HDTV; new services that can make the system profitable must be designed and tried out; legal standards for the new system must be formulated. If one of these fails — and the list is certainly not complete — HDTV is delayed, and this is considered intolerable and a spur to extra efforts. Bögels, for example, only a year after his comparison of HDTV to an unstoppable train, was worrying about political hesitation in setting a world standard for the new system. He began to press for clarity on this point, for the decision had already been postponed once, and "Another postponement would be quite bothersome, *because* this might delay the introduction of HDTV" (G, emphasis added). In other words, as soon as the massive inevitability of the technology starts to fracture, it must be mended immediately.

So while agency is presented as a foregone thing (we just have to ride with the train), at the same time, a lot of agency is necessary to actually keep this train going. The "train" is a metaphor for a future, or prospective, structure here. The two-headed monster of the self-justifying technology is reflected in a two-headed sense of agency: being (co-) driver of the train, as well as being driven. While actors may attribute this dual agency to the nature of (promising) technologies, we would argue that the duality is in the structure-agency tension itself, where agents will represent the structure as determining their action in order to protect their voluntarism against criticism. This is what confers a quasi-autonomous character on technological developments. In the last section of this paper we will conclude that studies of technological development will therefore shed light on basic features of the problematic duality of structure and agency. But, first, we shall identify a key feature of technological development which is apparent in all three of our examples.

4 From Promise to Requirement

The dynamics of the "unstoppable train," and of self-justifying technology in general, is a conversion of promises into requirements at the societal level. We will argue this by illustrating it for HDTV, but the point is applicable to all technological developments. Once technical promises are shared, they demand action, and it appears necessary for technologists to develop them, and for others to support them. At the same time, the options which are considered feasible and promising are translated into requirements, guidelines and specifications. This shapes concrete design and development trajectories, now at the meso- and micro-level, but also through conversions of expectations and promises.

In the case of HDTV, the specificities of the promise serve as goals for the R&D activities, such as signal processing, chips design, display and recording techniques. Passages from the OTA report (OTA 1990) prepared for the American Congress are indicative (emphasis added):

HDTVs *must* process huge quantities of information at speeds approaching those of today's supercomputers in order to display a real-time, full-color, high-definition video signal. HDTVs are able to do this at relatively low cost through the use of circuitry dedicated to specialized tasks (OTA 1990: 1).

The much higher quality pictures of HDTV, however, *will require* the transmission of substantially more information than current TV systems ([...] requiring) the equivalent of five TV channels. To reduce this a number of tricks are used (OTA 1990: 52).

HDTVs similarly place heavy requirements on memory technology. Access times needed for HDTV memory chips *must* be roughly 20 nanoseconds (ns) — 20 billionths of a second. Today's fastest DRAMs have typical access times of 60 to 80 ns (OTA1990: 65).

To truly appreciate HDTV, much larger high-resolution displays *are needed* than are generally available today (OTA 1990: 68).

However different these activities are, the technical promise sets their goals and coordinates the division of labor in terms of the overall objective of realizing the promising technology HDTV.

Technological promises function as a yardstick for the present and as a signpost for the future. The implication for the dynamics of concrete developments is that what starts as an "option" can be labeled a technical "promise," and may subsequently function as a "requirement" to be achieved, and a "necessity" for technologists to work on, and for others to support. Technological determinism is not given, but actively constructed. The option–promise–requirement–necessity sequence does not imply anything like an autonomous socio-technical process. The transitions do not occur automatically, but are the result of the actions and interactions of technologists, firms and governments. The transitions are a consequence of actors assessing what is "feasible," what is "obsolete," and what is "necessary," and of the efforts that follow these assessments. Moreover, the transitions are in principle reversible and can be undone — although entailing increasing costs and work. The pressure to at least recover and preferably to cash in on sunk investments also increases. As more and more is invested in a promising technology, any detour or even delay will encounter increasing resistance.[19]

When some overall promise has been articulated and become forceful, the functions that were seen as necessary (like more transmission and higher resolution displays in the case of HDTV) become specifications for design and development work. Promise–requirement cycles start up again, embedded in the overall framework (which itself continues to evolve: HDTV has now become Digital TV). This is the dynamic version of Clark's concept of design hierarchy (Clark 1985).

Behind the promise–requirement cycles lies the dynamic of expectations: as soon as expectations are shared they assume a life of their own. The basic point is that when expectations are shared they create a pattern into which the actors themselves may be locked. We will consider the mechanisms below, in Section 4, but here we note the link with the general issue of structure and agency.

Expectations, and stories about the future in general, reduce essential contingency in a non–deterministic sense, by providing blueprints that can be used in action. Such stories are thereby transformed into reality. Starting with contingency, a scenario is told, presented, read, and filled in by the actions of self and others. This creates a structure, which becomes the background for further stories. Afterwards, some actors or factors may be identified as driving forces, but this is attribution, and part of the process (as the attribution to HDTV of being an "unstoppable train" is part of the process). The process of filling–in the scenario creates the substance from which structure as well as the agency, i.e. attributed actors, are derived.

This implies that one can work with a minimalistic concept of structure: something not dependent on, or amenable to, voluntaristic action, and in that sense constraining. But its existence, in an operational sense, is in the claims about structure and structural determinants. This holds for the claims of actors (about "unstoppable trains," about competition with Japan, about economic forces), as well as for claims of analysts. When an economist or a sociologist makes a claim about structure, s/he is required to provide an account with more and different data and arguments than would be required of actors themselves. But "structure" still exists at the narrative level, i.e. there is no category shift as in the traditional debate (for example, when symbolic interactionism recognizes how actors introduce perceived structure in their situation, but forbids analysts to use "structure" — specifically as something "out there" — as an explanation of what happens in the situation).[20]

5   Mechanisms: Mutual Positioning and Agenda-Building

The three examples showed different degrees of coordination, from clearly articulated (Moore's Law and HDTV) to diffuse (membrane technology). In all three we saw a co-evolution of structure and agency. What are the core mechanisms behind this co-evolution? How can we identify them? Some suggestions have been made already. The workings of expectations and mutual positioning are at the core of this process, and we shall summarize here what we think is the general pattern.

Our starting point is the insight that expectation statements contain a "script," indicating promising lines of research and technical development to be undertaken by the enunciator of the statement and/or by others.[21] Thus, they mobilize support in specific ways. And they can be assessed as to their script, e.g., in discussions about priorities and strategic orientations in general.

There is a certain hierarchy in expectation statements and in scripts over time. When scientists and technologists ventilate expectations about the potential of membranes, and industrialists and government officials accept such statements and invest in their follow-up, a basic mold for activities in the emerging social world of membrane technology is created. On this basis, further opportunities for resource mobilization with a new round of more specific expectation statements present themselves. The groundwork has been laid, and one can go on immediately to discuss, say, the relative merits of different kinds of membranes.

Concrete expectation work is done by specific actors, but the outcomes are not just a matter of their adroitness in enrolling others. Even in bilateral negotiations, success depends on the extent to which the membrane world has become accepted. And actors invest in building up such a world, as we have shown in the way they mobilize goodwill through publications in magazines, through reports and public statements, and occasionally, through getting industrialists to express interest in the further development of membranes. Smolders, for example, who has been particularly active in this respect in the Netherlands and internationally, thus truly deserves the label "Mr. Membrane Technology" that is accorded to him. On the other hand, the fact that he is widely seen as the central actor and spokesman is an indicator of the stabilization of the membrane technology world, rather than just a reflection of Smolders' entrepreneurial capabilities.

Because expectation statements contain a script of the future world, they position the relevant actors, explicitly or implicitly, exactly as characters in a story are positioned. Since the expectations are public or semi-public statements that can be drawn upon by others, expectation statements require some response from the actors being positioned. An actor who rejects the role allocated by the script must react, either by protesting against the role, or by contesting the nature of the expectations. For novel, emerging, technologies and for possible priorities in science and technology policy, such contested expectations are a recognized phenomenon.[22]

The next step, after a priority is recognized, is that the (often agonistic) mutual positioning creates an agenda for activities in the membrane world. The agenda can focus on the specifications of the kind of membrane that is needed, or on options. An example is the competition between polymeric and ceramic membranes. We shall cite representatives of Dutch firms which work in very different parts of the membrane world, but which still take each other into account. This illustrates the encompassing nature of the (emerging) world of a promising technology, which connects actors that could not be related in terms of traditional economic or sociological theory.

The firms involved are X–Flow, a small company that originated from IOP-related research, and Hoogovens Industrial Ceramics (HIC), a subsidiary company of the large iron and steel firm Hoogovens. The former produces polymeric hollow fiber membranes; the latter tubular ceramic membranes. They differ in size: X–Flow is a small firm and has a modest R&D budget; HIC is a subsidiary of a big firm and its R&D budget is ten times that of X–Flow. Because they are big, HIC and Hoogovens feel obliged to innovate in that part of the promising area of membranes where smaller firms cannot compete with them: ceramic membranes, which need a lot of R&D and correspondingly ample financial resources. X–Flow feels obliged, because it is small, to stay close to the niche it evolved in, i.e. the IOP research and its derivative specializations. The two firms recognize that they are involved in mutual strategic positioning with respect to the new technology. The following quotes are indicative.

X–Flow about HIC:

The development of ceramic membranes is not an option for small firms. It is too expensive, you need a big firm behind you [...]. But I really doubt whether all this is worth the trouble: ceramic membranes will always be more expensive (M).

HIC about X–Flow:

For them it is more difficult, but at the same time also easier, than for us. They lack the support of a large firm, so they have few resources to start with. But on the other hand, that kind of small firm does not have to grow all that much. Having found a market niche, they can stay there. We are forced to grow in order to pay Hoogovens back their huge investments (L).

As inhabitants of the new membrane world, the firms position themselves in different ways. One further indication is the type of presentations they give at the IOP presentations and at membrane science conferences: X-Flow gives detailed scientific lectures,[23] whereas HIC only gives a global overview of their activities.[24] But both present themselves as being at the forefront of R&D, at least in the Netherlands. This positioning is accepted by the Program Committee which "uses" both firms as proof that the IOP formula works well. In a list of "successful results" published in an IOP brochure, the first two are accomplishments by X-Flow and HIC.

Why do X-Flow and HIC take each other into account when they seem to be located in very different parts of the market? Their technical and market expectations are drawn from an emerging agenda that is shared in the wider circle of the membrane world. In fact, both firms define their strategic goals and the internal agendas derived from them in terms of these shared expectations.

Both HIC and X-Flow expect that the importance of applications involving organic solvents (until now a difficult area for membranes) will grow. As a result, the research on membranes for such applications is high on the agenda in both firms, and both seek collaboration with firms already expert in this area. Given their self-positioning as leaders in membrane technology in the Netherlands, both HIC and X-Flow assess the polymer — ceramic distinction as one of the most important issues on the agenda of the membrane world in the 1990s. From a scientific or technical point of view alone it is not clear why ceramic versus polymer should be the issue of the next decade. It can only be understood if we consider the positioning of the firms (as leaders in the development of membrane technology) and the shared expectations (especially those with regard to the organic solvent applications). When the R&D manager of X-Flow talks about the research agenda, he frames specifications for the material he is looking for in terms of the polymer-ceramics distinction and the associated competition:

I am in charge of the R&D here. What I am interested in is the following: when will a polymer be available on the market in greater quantities that can resist higher temperatures, has a longer lifetime, which has such and such properties, and so on, and so on. [...]. Just to enhance the possibilities of applications of this [polymeric] material. Thus, to continue the competition with ceramics (L).

This last sentence, we note, was unsolicited. The effect of the strategies of X-Flow and HIC on the agenda of the world of membrane technology is unintended. X-Flow and HIC pursue their strategies in terms of their own interests, given their perception of their position.

The mechanisms involved in this agenda-building are such that the result is more than just an agenda. Mutual positioning is based on the interests of the actors, on their perceived interdependencies, and on their assessments of relevant environments. Thus, the outcome of the interactions is also a mutual accommodation of interests, and a specification of infrastructural possibilities. Coordination of action is created in the same movement as an agenda is built up through mutual positioning.

The mechanisms of mutual positioning and agenda-building are also visible in the two other examples given above. They occur at a lower level, under a better articulated "umbrella" of a dominant concept of what chips are and can do, or what HDTV should be. In the case of Moore's Law as well as for HDTV, there has already been further articulation of the positions and agendas, partly through the pressures of the strategic games being played. The patterns are more stabilized, the rules seem clear to the participants, and so the overall coordination is less tentative.

We have based our analysis of (mutual) positioning and agenda-building on a few case studies, and on cases of technological development at that. But our findings fit very well with two approaches in social science that tackle the structure and agency issue on a general level. One is the theory of positioning, as it has been developed by R. Harré, B. Davies, and L. van Langenhove (Harré and Van Langenhove 1992; Van Langenhove and Harré 1990; Davies and Harré 1990; Harré 1990; Harré 1975). Their central message is that people constitute themselves, as well as their practices and institutions, through conversations. By focusing on what is constituted, positioning theory aims to offer an alternative to the more static concept of role.[25] The work of Harré and Van Langenhove should be extended in three ways. First, positioning is not necessarily through conversation and immediate interaction, by also through statements and (public) acts. Second, positioning need not be restricted to the present, but may be located in the future. Third, in expectations statements about technology not only selves and others are located, but "things" (future artifacts, systems) as well.

A second congenial sociological tradition tries to understand social order as an achievement brought about and maintained by story-telling (Ricoeur, MacIntyre).[26] This tradition has a weak empirical and theoretical research program, with the exception of studies of life stories and identity. One can transpose the analysis of trajectories of personal lives and identities to analysis of the "life" of a technology. The scenario-like character of technological promises, which we have emphasized, supplements this tradition: The allocation of roles to actors in scenarios/stories, and the way these actors react (accepting their roles, contesting them, or exiting from the emerging world) contribute to the filling in of a shared scenario. Evolving social life can be studied in the same way.

6 In Conclusion

Instead of a static picture, portraying, as it were, a struggle between given structures and contesting agents, both "structure" and "agent" have now become fluid, evolving. Technology, injecting novelty, stimulates the co-evolution of agency and structure. But technology is not something outside the social fabric, merely impacting on it. It is part and parcel of perceptions, intentions, interactions, and interests.

One lesson from our analysis of co-evolution stands out, and needs to be discussed briefly before we conclude. It concerns the transformation of diffuse promises into specific requirements for further work in the context of a more or less powerful collective agenda arising out of agonistic interactions. The etiology of such agendas can be understood only when the role of scenarios, and more generally prospective structures, in orienting action is fully appreciated. This leads us to consider how agency and (emerging) structure influence each other, and build upon each other, so as to create a "spiral" of increasing articulation and stabilization.

Let us start with the voicing of promises: In and of themselves these will not be sufficiently powerful to establish requirements for technology development. This demands that the promise is taken up in an agenda (a local, a field, or a macro agenda). While an expectation statement is performative, i.e. it does something, it is an action — and because of its script, others may react — it only becomes an issue that demands consideration and action when actions and reactions converge sufficiently for it to become part of an agenda. While agenda building processes are shaped by many forces (and contingencies), as our discussion of strategic games around chips and HDTV made clear, they are the royal road for voiced expectations and are, at the same time, influenced by these expectations.

The notion of a sequence "promise — agenda — requirement" may suggest inevitability. This is clearly not the case: many promises remain just that, or remain unheard. The cases we know most about are the success stories. If one looks more closely into the conditions for success, or better, the steps that help to overcome contingency, the key factor is the presence (or creation, or emergence) of a protected space.[27] This point has been recognized in the literature on innovation (cf. promise champion plus protection from a higher level) and in technology studies (Law and Callon 1992; Rip 1992); what we add here is that the protected space is not a static phenomenon, but is linked to and adjusted to the agenda-building process. In all three cases discussed in this paper, one sees such links. Membrane technology presented as a promising research area gets support (and a protected space) because it fits the general promise of strategic science and technology. Technologies presented as the next generation almost create their own protected space; they are self-justifying because the notion of next generation is widely accepted.

Protected spaces, in their turn, may be created and maintained through expectations (Van Lente 1993). Thus, acceptance of promises (besides leading to activities to meet requirements) also helps to maintain a protected space. Within this protected space, subsequent promises are accepted more easily, provided they have the right shape. Thus, a promise-requirement cycle or spiral emerges.

We can now draw up a scheme for the dynamics of the conversion of promises to requirements: opportunities (generated within — or without — the protected space) presented as promises, get accepted and become part of an agenda; and are subsequently converted into requirements that guide the search processes. In the same movement a protected space is maintained (see Figure 1).[28] The outcome of a promise-requirement conversion (in terms of activities or a created space) is the background for a new one. In this way, promise-requirement conversions are superimposed upon each other. And, in fact, become a promise-requirement *spiral*.

BACKGROUND	# OF POSTERS	# OF PERSONS	RATIO
University	26	72	0.36
Research Institute	19	40	0.48
Firm	6	82	0.07

Figure 1: Conversion of Promises to Requirements

Figure 1 can be read as a technology-dynamics version of the co-evolution of structure and agency, and also the other way around: co-evolution of structure and agency can be seen as a spiral of promises and requirements (see Section 3). Instead of promises, we now have explicit and implicit story-telling (the latter through positioning), shaped by agendas and repertoires. Requirements are action-orientations taken up creatively by agents. A spiral of structure and agency is the result: social structure as the context of action, and (reinforced or transformed) social structure as the outcome of action. Figure 2 gives an impression of this dynamic.

Figure 2 has a lot in common with recent attempts (Giddens, Burt, Shibutani) to overcome the duality between structure and agency. So what is the added value of our analysis? What do our examples from technology studies have to offer, compared with the general attempts to create a coherent picture of structure and agency? The important contribution is that we can highlight coordination, as a process and as an outcome, and specify mechanisms which are more visible in technological developments because of the immanence of novelty.

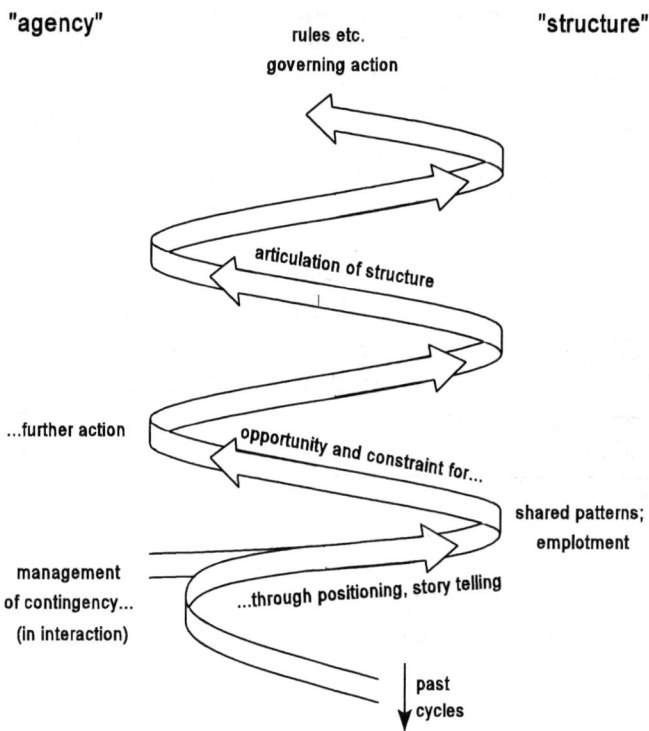

Figure 2: Co-evolution of Agency and Structure

Firstly, the promise-requirement conversion in its various forms indicates how structure emerges and how it shapes action in reference to specific *content* rather than to more diffuse traditional sociological categories like socialization, interests, or group. It is the *content* of orientations and scripts which pulls actors together, and when they act in terms of the script/scenario, a structure emerges which shapes further action. This is coordination through content (as in a story, as noted in Section 3).

7 Expectations in Technological Developments   225

The phenomena analyzed in this chapter relate to expectations that help to interlock activities and to build up agendas. In other words, expectations structure activities, but in a different way than structures normally do. The shared expectations are structures to be realized, actor-worlds as Callon (1986b) has called them. They do not *yet* exist, but nevertheless exert force. The mechanism by which they exert force is the script allocating roles; the overall movement is a promise — requirement conversion. So new social order is possible on the basis of the (heterogeneous) contents of collective-level projections of the future.

This leads us to the second contribution: how structure can be seen as a context of action. It is not because the structure offers models, roles, or other clues for behavior. (This would neglect the creativity of actors.) It is because the structure (i.e. the scenario, the blueprint) has to be specified in order to become forceful. Actors have to articulate specifications — which in itself is a creative act — and act accordingly.[29] Only through this detour can structure influence subsequent actions.

So one should recognize that structures can be prospective, and still be influential.[30] To capture this point we introduced the (paradoxical) term prospective structure. A prospective structure is made up of links which can appear in texts. In subsequent actions and reactions, the structure is filled in, modified, reshuffled — and becomes social structure, in its various forms, i.e. emphatically including new technological artifacts.

One can now understand why the division of labor in sociology as discussed by Giddens (see Section 1), is too static: retrospective structure is contrasted with voluntaristic actions. Prospective structure should be added to the ontology of the social world: i.e. a not-yet-existing structure that is to be filled in by agency, and which precisely in that movement also redefines and modifies that agency. Prospective structure hence has the same power as "forceful fiction," which is the power of opening up space for action.

Notes

* An earlier version of this paper was prepared for presentation to the XIIIth World Congress of Sociology, ISA, Bielefeld, Germany, 18-23 July, 1994.
1 In the opposition between structure and agency, "structure" refers mainly to the constraints which individual action faces. The idea is that these constraints stem from the forceful coherence of an aggregated level. In this section, we will use this sense of the term, which has, in other contexts, other connotations; Boudon and Bourricaud (1989) give a list of about twenty meanings.
2 One might expect questions of coordination processes and their outcomes to be a core sociological activity. Indeed, mutual accommodation of activities and the role of

institutions as governance structures are basic issues in sociology. However, the term "coordination" is not a very common one in the sociological literature. In the indexes of handbooks on sociology 'coordination' does not appear. We search in vain for a discussion on coordination in books like *The Encyclopedia of Social Sciences* (Shills 1968), *A New Dictionary of Sociology* (Mitchell 1979), *Handbook of Sociology* (Smelser 1988), *Sociology: Themes and Perspectives* (Haralambos and Holborn 1991), *A Critical Dictionary of Sociology* (Boudon and Bourricaud 1989). While even chemistry has an international journal entitled *Coordination Chemistry Review*, there is no sociological journal on coordination. This indicates that in sociology coordination — the mutual taking into account by actors and a key feature of social life — is not explicitly discussed. What sociology deals with instead is "social order," "structure," "institutions," "rules." Our point is that while some coordination is obviously implied and understood in these concepts and insights, it is not explicated as such.

3 The three examples are from earlier work. The first example is discussed in Rip (1992). The second and third case are discussed and analyzed in Van Lente (1993).

4 An example in the Netherlands and Germany is the Megachip project, in which Philips and Siemens received a few hundred million dollars each to develop the one and four Megabit memory chip. Although Philips abandoned the project in 1990, partner Siemens continued, as did the Japanese and American competitors The Japanese reached 16 megabits in 1992 and went for 64 in 1994.

5 This was the challenge taken up by the Siemens-Philips collaboration in the second half of the 1980s: to go for a second generation beyond the current state of the art. It turned out that they could not develop all the necessary production technologies, and had to conclude a strategic partnership with Toshiba. When they were finally able to make the leap to the 1 and 4 Mb memory chips — at great costs — they found they could not sell them profitably.

6 Wafilin is a subsidiary firm of Wavin, a producer of plastics; AKZO and Shell are multinational chemical and oil companies, respectively.

7 TNO, NIZO and KIWA are partly government-subsidized research institutes for applied research, dairy research and water purification, respectively.

8 In the preliminary study of 1982 (A) the umbrella term was not yet used. The author was content to list functions and processes: membrane filtration, dialysis, gas separation with membranes, pervaporation, electrodialysis, liquid membranes, membrane bioreactors.

9 We use "macro-actor" here in the sense that Callon et al. (1986) speak of "macro-terms," and refer to Callon and Latour (1981) for the original concept and argumentation.

10 Quotes are from participants at the IOP day on October 5, 1990, Ede, the Netherlands.

11 Remarks by G.W. Meindersma, the membrane (promise) champion at the Dutch multinational DSM, who now finds people coming to him and asking these same questions (C).

12 In membrane technology, one can see attempts to specify the characteristics of an "ideal membrane," and the mutual articulation of specifications for competing types of membranes (organic polymer membranes versus ceramic membranes).

13 The CCIR meeting can be seen as an instance of what in political analysis has been labeled the opening of a "window of opportunity," or "policy window." "The policy window is an opportunity for advocates of proposals to push their pet solutions, or to draw attention to their special problems" (Kingdon 1984: 173).

7 Expectations in Technological Developments 227

14 Participants in the HDTV business continue to urge its priority. The Project Manager of the Dutch Platform for High Definition Television: "[...] the future is already here because decisions have to be taken now if such a service is to be available in a few years time. Participation in Eutelsat [...] for instance, requires an investment decision of some 16 million Dutch guilders [about 9 million dollars] before the end of 1990. And that with the term HDTV as yet far from commonplace in Dutch politics!" (B: 10).

15 (J: Appendix V.) The report continues: "These shortcomings [of the present system] are due to the following choices:
   - the low screen frequencies;
   - the choice of interlining in scanning;
   - the limited number of lines for each image [...];
   - the screen ratio of 4:3;
   - the limited possibilities for sound, in quality and quantity."

16 Philosophers like Dumouchel (1979) and Achterhuis (1988) argue that a persistent feature of modernity is an allergy for limitations. A perceived limitation provokes action to transcend the limitation, instead of cultural arrangements to deal with the limitation, the pre-modern solution. Scarcity, they argue, does not refer to (the amount of) things, but to this cultural response to limitations. In the case of HDTV we see how scarcity of, say, the number of transmission channels, is constructed. Moreover, the case shows that the solution to scarcity may be prior to the scarcity itself.

17 The two-headed monster is comparable with the Janus-head of ready-made science and technology ("we are believed because it is true") versus science and technology in the making ("we must get our claims accepted as true"), as Latour (1987) has described it. The focus is different, though. Latour tends to focus on accounts that actors make afterwards, accounts that are different from what actually happened. So the ready-made part is later than the in-the-making part. In the case of our two-headed monster, the ready-made part is earlier: It is a projected future that, by implication, sets the boundaries of the in-the-making present. Correspondingly, Latour's emphasis is on epistemological outcomes: How the constructions of science and technology become "true" or "working," i.e. how they are given a life of their own due to the activities of scientists. Here the interest is in research agendas and interlocking activities as the outcome. Thus, the focus is on the reverse movement: How things that have a life of their own, like HDTV being on the macro agenda, affect the activities of technologists.

18 Bögels: "After all, through Eureka, we have been able to make up for our knowledge deficiencies. Therefore, we must now continue European cooperation in the area of HDTV" (F).

19 If Philips invests heavily in HDTV, and the system turns out to be a market failure, then Philips Company is endangered, as well as possibly the entire Dutch economy. Philips knows this, other European television manufacturers know this, and the Ministry of Economic Affairs knows this. Once again, the question on which points HDTV is actually better than present television has become irrelevant.

20 Compare the way the enunciator of a story can show elements of structure and how they pervade what happens, while the characters in the story can also refer to structure as they see it — and this is part of the story as enunciated. In the "story" that is told jointly in the expectations and actions, there is no simple asymmetry between enunciator and characters. They both show the (minimalistic) structure and refer to it.

21 The script of expectation statements is comparable with Callon's (1986b) "socio-technical scenarios." The notion of script is used by Akrich (1992) and Latour (1992) to capture the (implicit) messages and guidelines in artifacts.
22 See Callon's (1986b) discussion on the "socio-technical scenarios" put forward by Renault and Electricité de France.
23 For instance at the conference about "Synthetic Membranes in Science and Industry" (Tübingen, 4-8 September, 1989), organized by the European Society of Membrane Science and Technology, and the Eighth Annual Membrane and Technology Conference, Boston, MA, 15-17 October, 1990.
24 At the IOP day on October 5, 1989, both firms gave a lecture; X-Flow under the title "New hydrophillic microporous membranes prepared from polymer blends," and HIC on "Specific advantages of ceramics as a material for membranes" (Both printed in *Membraantechnologie*).
25 A simple example from Van Langenhove and Harré (1990: 7) will make the basic idea clear. In the statement: "Please, iron my shirts" that Smith utters in the presence of Jones, both Smith and Jones are positioned: Smith as someone who has the moral right to command Jones, and Jones as someone who can be commanded by Smith. The positioning can be accepted ("OK, I'll do it this afternoon") or not ("Sorry, I am not your servant"). The example shows that in conversations selves and others are located in some moral order. "Within a conversation, each of the participants always positions the other while simultaneously positioning him or herself" (Harré and Van Langenhove 1990: 11). Van Langenhove and Harré's main effort is directed towards a typology of positioning.
26 "[...] man is in his actions and practices, as well as in his fictions, essentially a story-telling animal [...] a teller of stories that aspire to truth [...]" (MacIntyre 1981: 216).
27 E.g., a laboratory, where attempts to try out expectations are protected, i.e. failures are accepted due to the belief that in the end important results will be achieved. These protected spaces also occur within firms ("niches"), where an innovation project creates a (temporary) space for trial and error. At the level of society, the protected space might be created by technology policy measures to stimulate strategic research areas: for the time being, disappointing results will not affect the possibilities for research. The concept of protected space is elaborated in Van Lente (1993).
28 An analogy of the promise-requirement conversion is the *credibility cycle* in science and technology (Latour and Woolgar 1979), which focuses on the ongoing conversion of *resources*. Financial resources are needed to pay the salary of researchers and to maintain a laboratory; researchers and experiments lead to publications; publications add to the reputation of the scientists and the laboratory; reputation is a resource to get further funding and to attract new researchers, which adds to the capacity to produce new claims. A difference is that the credibility cycle is essentially an *ongoing* conversion, whereas the promise-requirement conversion may appear just once, without necessarily compelling a next round: it may remain a one-time detour. What they have in common is the appearance of Janus-head type statements: in the former we find science-in-the-making versus ready made science; in the latter, promising technology versus technology as (invented) means to (given) ends.
29 This point relates to the well-known discussion of rule-following, which, in spite of the connotations of the term "following," is a creative process (Wittgenstein 1972). Compare also Barnes (1974) on the creativity involved in elaborating a scientific paradigm.

30 In this formulation, there are close resemblances to the analysis of *social movements*. Indeed, it is possible and fruitful to study some aspects of the dynamics of expectations in the same way social movements are studied. There are, for instance, parallels between Becker's (1963) moral entrepreneurs and our promise champions. And Ron Eyerman and Andrew Jamison (1991) analyze social movements in a way that has some intriguing parallels with our account of membrane technology. The histories of the environmental movements in Sweden, Denmark, and the Netherlands show how environmentalism developed from a heated debate to functional arrangements: "In the course of the 1980s [... environmentalists] formed companies and consulting firms to market their new environmental products. What had been an integrated process of social learning in the heat of anti-nuclear opposition in the late 1970s, when environmentalism had been viewed as one of the 'new social movements,' had by the late 1980s largely fragmented into more or less separate and quasi-professional activities." (Eyerman and Jamison 1991: 69). Look at the interesting prophecy we get when we fill in "membrane technology" as social movement in this quote: "In the course of the 1990s the membrane technologists formed companies and consulting firms to market their new membrane products. What had been an integrated process of social learning in the heat of strategic science and technology in the 1980s, when membrane technology had been viewed as one of the 'strategic research areas' had by the early 1990s largely fragmented into more or less separate and quasi-professional activities."

Sources

A. Bargeman, D. (1982): *Innovatiegerichte onderzoekprogramma's. Voorstudie membranen*. Enschede: TH Twente.
B. Eupen, T. van (1990): "Six months of Dutch platform HDTV," *HDTV report*, 3, September: 10.
C. Meindersma, G.W. (1990): Interviewed on March 21.
D. Ministry of Economic Affairs (1982): *Membraantechnologie in Japan, verslag van een rapporteursmissie*. The Hague, October.
E. *Nederlandse Staatscourant*, December 28, 1983: 11–14.
F. *NRC Handelsblad*, January 28, 1989: 23.
G. *NRC Handelsblad*, March 28, 1990: 12.
H. *NRC Handelsblad*, May 11, 1993: 19.
I. Program Committee of the Innovation-oriented research programs (1986): *Membraantechnologie in Nederland*. Part 1: 1986; Part 2: 1987.
J. Slaa, P. and H.-P. Siderius (1990): *Hoge definitie televisie. Een overzicht van de huidige stand van zaken met betrekking tot ontwikkeling en introductie*. Den Haag/Amsterdam: SWOKA/VU.
K. Terpelle, B.J. (1991): "The retirement of Prof. Kees Smolders," in: *UT mediair*, 3, June: 8.
L. Interview by Harro van Lente with Frans Muilwijk (Research Manager at HIC), February 13, 1990.
M. Interview by Harro van Lente with Eric Roesink (Research Manager at X-Flow), December 7, 1989.

Chapter 8
Boundary Maintenance and Radioactive Waste Disposal Technology in the U.S., 1945-1970*

Adri A. de la Brühèze

1 Introduction

After the first wave of excitement about atomic power in the U.S. in the late 1940s, the fact that hazardous radioactive wastes were also being produced gradually came to be recognized as a separate issue that would have to be dealt with. On scrutiny, it appeared that at the various, geographically dispersed, laboratories of the U.S. Atomic Energy Commission (AEC), radioactive wastes were being produced and handled according to locally developed insights. That this might be or might become a problem was put squarely on the agenda in 1948 by the AEC's Safety and Industrial Health Board (SIHB), a body charged with reviewing the safety of operational practices. While the SIHB report had little effect on the actual operational waste handling practices, it was the signal for increasing attention to radioactive waste management, in particular, a search for ways and methods to handle radioactive waste more carefully and safely.

Two interlinked processes governed the development of radioactive waste management technology in the U.S. On the one hand, the "bureaucratization" of radioactive waste and, on the other hand its "technologization." "Bureaucratization" occurs when the emphasis is on classification into different categories, for example high-level versus low-level radioactive waste.

Each category in a classification is subject to specific rules stipulating how pertinent wastes are to be handled. These rules may be refined or changed, and the categories may be specified in time, but the basic point is that the problem is managed by relying on classification and rules for categorization. Besides the categories of "high-level," "low-level," "solid," "gaseous" and "fluid," the most important classification was the distinction between "military" or production wastes on the one hand and "peaceful" or commercial wastes on the other. Military wastes were generated in the "production work" aimed at making nuclear weapons (De la Brühèze 1992a,

1992b). Commercial (or civilian) wastes were generated in the peaceful nuclear power program.

"Technologization" occurs when technological means are developed to manage a categorized problem. Often, specific technical means are developed for each of the categories. In the case of U.S. radioactive waste management, "military" (or production wastes) and "peaceful" (or commercial wastes) were seen as different categories of waste and therefore different methods of treatment could be developed. As a result, the management of these two types of waste was split along two separate bureaucratic and technological trajectories.

The dual processes of bureaucratization and technologization acquired impetus with the question what to do with the expected large amount of commercial radioactive wastes in a future nuclear power economy. This entailed a clear mandate for the AEC to develop adequate solutions. Subsequently, over a period of at least ten years this agency developed a specific and most promising technology for the management of future commercial wastes. In so doing, however, it also created environments of actors included and excluded from what can be called a "radwaste world." While such exclusion may not be too great a risk for an R&D program aimed at developing a prototype, the actual implementation of a technology is a different matter. This entails the involvement of many more (and more heterogeneous) actors and requires considerably more managerial force and agility. When the AEC continued to keep actors excluded during the implementation of its commercial radwaste management technology, it drove many of the "outsiders" — unable and unwilling to understand and accept AEC waste classifications and waste management plans — into one oppositional camp. In fact, processes of inclusion and exclusion were the covert drivers of radwaste technology development. Sometimes, the inclusion-exclusion process worked primarily through the bureaucratization of radwaste, in other instances there was a direct effect on the technologization.

This chapter will highlight some of these inclusion-exclusion processes and their outcomes, from the early phase when the AEC was still in search of its organizational identity and role, through the 1950s, when a stable radioactive waste "world" (Strauss 1978) was created, including, by 1959, a dominant problem definition. By the 1960s, the AEC could concentrate on patrolling the boundaries of this accomplished radwaste world, only to be faced with an irresolvable controversy around 1970, the effects of which are still being felt. These inclusion-exclusion processes and their outcomes can be conceived as "antagonistic coordination," that is, as a socio-technical process shaping the mutual actions of actual or potential opponents and hence also the kind of technology that was developed.

2 The Atomic Energy Commission and the Construction of a Radwaste World

Even after World War II, the development, promotion and regulation of military and peaceful nuclear technology in the United States remained a government undertaking delegated to the U.S. Atomic Energy Agency (AEC). Although the responsibility of a civilian organization, nuclear technology served primarily military purposes until the early 1960s. These included producing plutonium in production reactors for atomic fission bombs and developing atomic power reactors for the propulsion of military ships, submarines and aircraft (Dawson 1976; Hertsgaard 1983). As a result, military and civilian nuclear technology were indistinguishable (Hilgartner et al. 1982). In fact, most radioactive waste was, in the first instance, the responsibility of the military. As a result of the prevailing secrecy about all military nuclear activities during the cold war years of the 1940s and 1950s, the existence, quantity, and chemical composition of these military wastes were closely kept secrets because they could "be of substantial assistance to a competitor nation" (C).

Although at the time nuclear technology was primarily a military technology, its future civilian (or peaceful) aspects were widely touted during the cold war in such a way that it gradually became the icon of future technical progress (Gandara 1977; Van Lente 1993; Corn 1987; DelSesto 1987). The peaceful atom was expected to provide the U.S. with electricity "too cheap to meter," a statement usually attributed to AEC chairman Lewis Strauss (Pfau 1984). Disturbing information about lethal and toxic radioactive wastes discorded with this future technological promise and was therefore not mentioned at all (Boyer 1985). Hence, from the very inception of nuclear technology, radioactive waste played a special role: in the military program it was a secret topic and in the peaceful program it was seen as an unwelcome dissonant that had to be kept hidden from the outside world.

But at the same time, the mundane operations in the local AEC laboratories generated all kinds of radioactive "ashes, poisons, scrap or waste" (Compton 1956). In the early phase (but persisting long thereafter) there were no general rules or procedures for classifying and handling these wastes. The AEC as an organization was still exploring its charter (Truman 1956). Within this context the management of radioactive waste was not considered a priority.

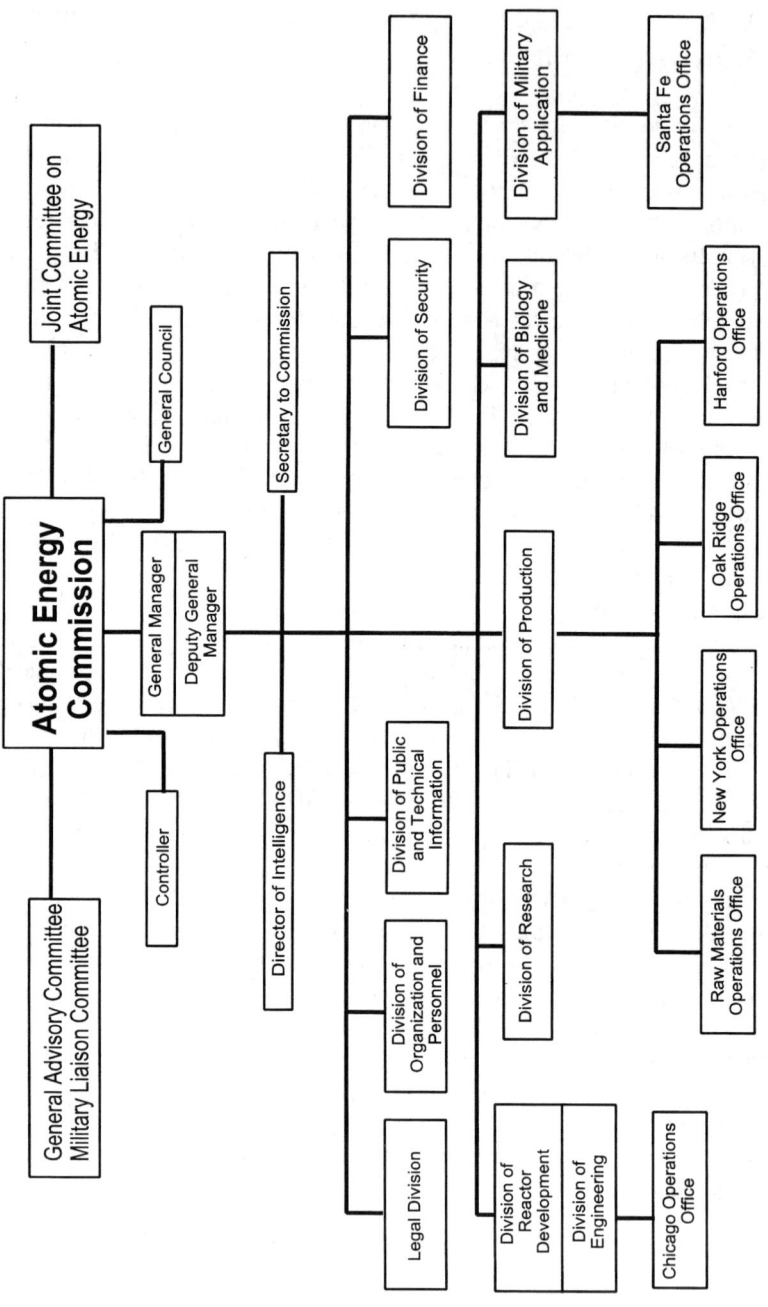

Figure 1: Organization Chart AEC, December 1948

The 1946 Atomic Energy Act statutorily divided the AEC headquarters organization into four Divisions: Research, Production, Engineering and Military Applications (Titus 1986). In 1949, a Division of Biology and Medicine and a Division of Reactor Development were added. The Division of Production was charged with the procurement, pre-treatment and production of weapon grade material, and supervised and controlled most of the local laboratories. The Division of Military Applications was charged with the engineering, testing and stockpiling of nuclear weapons. The Division of Engineering and the Division of Research provided supportive tasks on behalf of R&D programs carried out by the local laboratories.

Between 1945 and 1970, no headquarters division was exclusively responsible for radioactive waste management at the local laboratories. Rather, waste management was just one aspect of the overall operations of the laboratories. As a result, and in combination with the complex and often conflicting organizational responsibilities and the policy of "operational and managerial flexibility" (II; QQ) the local laboratories acted rather autonomously with respect to the definition, categorization and treatment of radioactive waste (Groves 1963). Hence, different kinds of radioactive wastes were distinguished in a range of different ways. Broadly speaking, however, two kinds of waste were generally distinguished: military wastes created in all kinds of operations at the military production sites and laboratory wastes containing all kinds of military and non-military wastes.

At the local military production sites and laboratories the military and laboratory wastes were broadly classified as "high-level," "low-level," "liquid," "solid" and "gaseous." But within this broad classification, local standards and rules for categorization differed. These local differences, and thus the existence of different wastes within one classification, were legitimated by claiming that the laboratories had different tasks, were involved in different operations and processes, used different materials for different purposes and (therefore) generated different kinds of waste with respect to their chemical composition and radioactivity.

Although the standards and rules for waste categorization differed at the local laboratories, the actual treatment of military and laboratory waste was similar. Radioactive wastes categorized as "low level solid" were buried on-site in the ground or off-site at sea, whereas liquid and gaseous "low level" radioactive wastes were diluted with water and air and dispersed into surface waters or the atmosphere. The existence of different local treatment practices (e.g., burial on site versus burial off-site) were legitimated with the argument that different kinds of waste and different environmental conditions permitted a variety of environmental dispersal and disposal practices.

Liquid residues which remained after the chemical reprocessing of irradiated uranium fuel elements from military or "production-purpose" reactors were commonly defined as (military) "liquid high-level wastes." The irradiated fuel elements were removed from the reactor, temporarily stored, and shredded and dissolved in nitric acid. Chemicals were added to extract plutonium and unburned uranium. The liquid residue that remained after extracting plutonium for nuclear weapons during reprocessing was called the "first cycle solvent extraction waste" and contained most of the radioactive fission products and toxic chemicals. These wastes were produced and stored in underground (steel and/or concrete) tanks at the Hanford (Washington) military production-site and, from 1954 onward, at the Savannah River (South Carolina) site as well. Smaller amounts of such military high-level wastes were produced and stored at the National Reactor Testing Station (Idaho), Oak Ridge National Laboratory (Tennessee), Argonne National Laboratory (Illinois) and Brookhaven National Laboratory (New York), which also produced non-military laboratory wastes. The military and laboratory wastes at the national laboratories were stored and treated as one.

Thus, while one can emphasize local variety in the definition and handling of military and laboratory wastes, one can also see the emergence of broad classifications ("high-level," "low-level," etc.) and similar waste handling practices at the local sites. This can be illustrated by comparing waste classifications and waste handling practices at the Hanford military production site and Oak Ridge National Laboratory in the period 1943-1949 (see Table 1).

The Hanford and Oak Ridge waste management practices show the existence of local technical regimes (Disco et al. 1992), i.e. similar waste classifications ("high-level," "low-level," etc.), views and management practices, but within these classifications locally varying rules for waste categorization, technical means and various technical means and "habits" to handle the wastes. These local technical regimes were to a large degree pragmatic and opportunistic because the reduction of waste management costs was the paramount guiding principle, and because the regimes were not seriously evaluated with respect to environmental safety and with respect to the amount and (chemical) composition of future wastes. Short and long term solutions for the growing amount of military and laboratory wastes were always sought in the construction of additional storage tanks and in developing means (e.g., temporary storage) to reduce the radioactivity of the wastes before they were environmentally disposed. Safety considerations played a role only in forms of operational safety measures for AEC personnel.

Organizationally, a key step in the management of radioactive waste — and especially of (future) civilian radioactive waste — was the establishment of the Sanitary Engineering Branch (SEB) in the AEC in 1948. With the activities of the SEB, new professional groups appeared in the world of radioactive waste management. These new groups helped shape the continuing bureaucratic and technological handling of civilian waste, rather than actually changing or standardizing the local military and laboratory waste management practices.

Table 1: Operational Waste-management Procedures at Hanford and at Oak Ridge National Laboratories, 1943-1949

Liquid high-level radioactive waste	Liquid low-level radioactive waste	Solid wastes (hl/ll)	Gaseous wastes (hl/ll)
Hanford 1943-1949:			
First solvent extraction waste alkaline stored in mild steel lined concrete tanks. Second solvent waste stored in mild steel lined concrete tanks. Tanks precipitation and storage of high-level boiling sludge in tanks.	Reactor cooling water discharged to Columbia River. Discharged to ground via "pits" and "cribs": all kinds of liquid wastes, including supernatants from precipitated wastes, and condensate from self-boiling precipitate stored in tanks.	Buried on-site	Filtered, scrubbed, and released through high stacks.
Oak Ridge National Laboratory, 1943-1949:			
Temporary tank storage of neutralized metal and radiochemical wastes for decay, followed by in-tank precipitation. Highly radioactive precipitate held in tanks. Supernatants diluted with large volume process waste (cooling water), and discharged to White Oak Creek and Clinch River	Discharged to White Oak Creek and Clinch River: "process wastes," including sanitary sewage, reactor cooling water, and slightly radioactive process waste from laboratory sinks. Also discharged to White Oak Creek: Supernatants from stored waste, diluted with reactive cooling water	Buried on-site	Filtered, scrubbed, and released through stacks.

Sources: Hanford: B; XX; III; LLL; Oak Ridge: A; B; X; NN; LLL; OOO.

A Reconsidering Existing and Future Waste Management: The Sanitary Engineering Branch

The Sanitary Engineering Branch was the brainchild of Professor Abel Wolman, a national authority on sanitary engineering and public health. As a result of his acquaintance with AEC Chairman David E. Lilienthal, Wolman became a member of the Safety and Industrial Health Board (SIHB) charged with reviewing local operations (Hollander 1981).

In April 1948 the SIHB published a rather provocative report (YY), in which it noted a variety of operational practices which had violated industrial hygiene and environmental sanitation rules and had paid insufficient attention to public health hazards. The SIHB report rejected the environmental disposal practices for "low level" military and laboratory radioactive wastes at the local laboratories because these practices lacked adequate supportive laboratory research, field tests and environmental data. As a first important step in the direction of an interdisciplinary problem-solving approach and the creation of a regulatory health protection unit in the AEC, the SIHB (and notably Wolman) recommended the assignment of a sanitary engineer to the Division of Engineering (YY: 75, 83).

In response to the SIHB recommendations, the AEC Commissioners appointed Arthur E. Gorman and Josef A. Lieberman as staff engineers in the Division of Engineering. Until the mid-1950s, these men constituted the AEC's Sanitary Engineering Branch (SEB), or "the Environmental Sanitation Branch." After 1949 the SEB was located in the AEC Division of Reactor Development.

The SEB considered the permanent storage of military liquid high-level waste in tanks as a public health hazard, and only acceptable if used as an interim solution prior to permanent and safe disposal. As possible permanent disposal solutions the SEB considered the concentration of the liquid high-level waste into a solid form, followed by geological disposal of the solids. The SEB was also concerned about the local management practices of the military and laboratory low-level waste, and investigated new (sanitary engineering) methods to confine radioactive and chemical toxicity to permissible levels prior to environmental disposal. Initially the SEB tried to get agreement on waste definition, classification and categorization, as well as on safe management of wastes. For this purpose, the SEB enlisted the cooperation of authoritative public agencies like the U.S. Geological Survey, the U.S. Weather Bureau, the U.S. Public Health Service, the Bureau of Mines and the U.S. National Academy of Sciences. Furthermore, the SEB set up an independent waste R&D program, funded local waste R&D work and developed "working relationships" in order to get access to the local laboratories and some measure of influence.

The SEB's influence increased with the passage of the second atomic energy act in 1954, aimed at the development of commercial nuclear power. AEC's Division of Reactor Development expected that within a future nuclear power economy large quantities of liquid high-level radioactive wastes would be generated by commercially built and operated power reactors and reprocessing plants. In this scenario, existing military wastes and their management served as an exemplar for the future management of commercial wastes: irradiated fuel elements from commercial power reactors would also be chemically reprocessed in order to recover unburned (and at the time scarce) uranium, and the resulting commercial liquid high-level wastes would have to be (temporarily) stored in tanks at the reprocessing sites. In order to avoid the safety and public health problems and costs of permanent tank storage which might hamper the growth of commercial nuclear power, the Division of Reactor Development considered the development of a safe and technically sound commercial waste disposal technology of crucial importance.

Therefore, the Division of Reactor Development charged its Sanitary Engineering Branch (SEB) with the evaluation and assessment of technologies for managing future commercial high-level waste. The 1954 atomic energy act thus offered the SEB an opportunity to play a major role in future commercial radioactive waste management. Via this role, the SEB hoped and aimed to change existing military and laboratory waste management practices at the local laboratories as well.

B   Geological Waste Disposal: The Hess Committee

As part of its strategy to assess future technologies for managing commercial radwaste, the SEB, in 1954, asked the Earth Science Division of the U.S. National Academy of Sciences (NAS) to study the feasibility of "land disposal" of future commercial high-level wastes. The NAS Division appointed a committee composed of non-AEC geologists chaired by Princeton University Professor Harry Hess. In September 1955, the "Hess committee" organized a conference at Princeton University to generate ideas on geological disposal of future commercial high-level radioactive wastes from power reactors. The sixty-five participants were for the most part geologists, but also included petroleum and mining engineers from universities, industry, and government agencies. The SEB and the Hess Committee were points of entry for new professional groups like geologists and petroleum and mining engineers into the field of radioactive waste management. As a result, new views, concepts and knowledge were articulated, encouraging new R&D in the process.

For example, the only way to assess the magnitude of the future commercial waste problem appeared to be an evaluation of the efficiency and technical suitability of present treatment and storage practices used for military waste. However, secrecy about existing military high-level radioactive waste and its management made such evaluation very difficult. As an alternative, the Hess committee proceeded to examine methods used by the oil industry to store liquified petroleum gas, petroleum, and oil in subsurface cavities and to dispose of oil and brine wastes in geological formations, hoping that these methods would offer solutions for the radioactive waste problem (UU).

Two subcommittees were set up during the Princeton conference to consider the different strategies of deep and near-surface disposal (6,000-7,000 feet). On the basis of the experience of petroleum engineers with geological brine waste disposal in their oil-mining operations, the deep disposal subcommittee concluded that future commercial liquid high-level radioactive waste might be pumped into the bottom of a geological structure from which oil had been or was being pumped out higher up. The near-surface committee recommended geological salt formations for the disposal of future commercial high-level radioactive waste. This recommendation derived in part from the awareness among petroleum geologists that there were many salt beds and salt domes in the U.S. that could be cheaply acquired. In addition, geologists knew that salt was an old, stable and dry geological formation that conducted heat well and had a high melting point, and that cracks in salt structures tended to seal themselves under pressure.[1] The near-surface group advocated the solidification of the future commercial liquid high-level waste prior to its disposal as a long-term goal.

These recommendations were presented in a Hess report in 1957 (LLL) as an "agenda," i.e. as a gradually realizable, practical approach to the disposal of existing military and future commercial high-level radioactive waste. Safe disposal was defined as complete isolation from the biosphere for at least 600 years. Direct disposal of commercial liquid high-level waste in geological salt formations was seen as the most promising method for the near future. In the long run, however, the emphasis should shift to the solidification of the liquid wastes followed by disposal of the solids in geological salt formations. Another important, but also controversial, recommendation was that the presence of suitable geological (salt) formations for waste disposal should determine the location of all future military and peaceful nuclear installations.

The Princeton conference and the 1957 NAS report strengthened the position of the SEB in the AEC, in part because the Hess Committee was considered independent and authoritative. Nevertheless, the Hess committee met increasing opposition from AEC's Division of Production (DoP), the

8 Boundary Maintenance and Radioactive Waste Disposal Technology 241

production sites and the local laboratories. These latter rejected the Hess committee recommendation that local burial of military and laboratory low-level waste and the local practice of storing liquid high-level military waste in tanks should be terminated and an alternative sought. Their opposition was understandable because the recommendation, if adopted, would have required a complete change of current management operations, installations and instrumentation and would have led to unacceptably higher costs (TT). The DoP and the laboratories were convinced that existing waste management practices could be improved and optimized so as to improve environmental safety and public health. In addition, the DoP and the production sites vehemently questioned the feasibility of solidifying liquid high-level military wastes into solids and the subsequent off-site transportation and geological disposal of the solids as recommended by the Hess committee. In the view of the DoP and the production sites tank storage constituted a safe and permanent management method for military liquid high-level radioactive wastes.

This emergent waste management controversy made it difficult for the SEB to impose its general waste management concept on the entire AEC and on all radwastes. This became clear as the Division of Reactor Development (DRD) decided that the new (general) concepts of solidification and geological disposal as recommended by the Hess committee and the SEB would be developed by the DRD for the management of future civilian/commercial high-level radioactive waste. This decision reinforced the formal bureaucratic distinction between (future) civilian/commercial waste and (existing) military wastes as the DoP continued the practices of storing liquid high level military wastes in tanks and of disposing low-level military waste in the environment. As a result, the (future) commercial and (existing) military wastes became embedded within their own separate bureaucratic and technical regimes. The Division of Production became responsible for the management of military wastes at the local production sites, while the Division of Reactor Development was charged with the R&D on a commercial radwaste management technology along the lines recommended by the Hess Committee and the SEB. This bureaucratic and technological distinction worried and increasingly frustrated the Hess Committee, because it gave the DoP and the local laboratories license to persevere in (military and laboratory) waste management and waste disposal practices which the Hess Committee considered hazardous.

C   The 1959 Hearings: Towards Stabilization

While the waste management dispute between the Hess Committee and the AEC was developing, an institutional basis for a shared commercial waste categorization and commercial waste management approach was being laid at the AEC. In 1958, the Congressional Joint Committee on Atomic Energy (JCAE) decided to hold hearings on radioactive waste management, chiefly because concerns about the disposal of low-level radioactive waste in coastal waters had placed waste disposal on the public agenda (Divine 1978; Mazuzan and Walker 1984). Despite some skepticism and reference to many difficulties and problems, the January/February 1959 hearings conveyed to the outside world that a safe and efficient technical solution for future civilian/commercial high-level radioactive waste disposal was possible and would be available in the near future. Many AEC witnesses stressed the progress and results of commercial waste management R&D at the national laboratories. Josef Lieberman of the SEB expressed the confidence of the SEB and the Division of Reactor Development that safe management solutions for commercial high-level wastes would be found before large-scale commercial nuclear power would have become a reality. As the most promising treatment and disposal solution for the liquid high-level commercial wastes, Lieberman mentioned interim tank storage, followed by the separation of Strontium 90 and Cesium 137 from the liquid waste, the subsequent solidification of the remaining liquids and the geological disposal of the solidified waste. The impression conveyed to the outside world was that there existed no disagreement within the AEC and other prestigious organizations on the classification, categorization and treatment of commercial high-level radioactive waste (DD). The existence of two kinds of high-level wastes, military and commercial, with their own characteristics and different bureaucratic and technical regimes, was hardly mentioned.

The consensus visible throughout the hearings reflected the fact that all divisions, production sites, laboratories and study groups in the AEC and (divisions) of other organizations had accepted the distinction between military and civilian wastes and that they shared, or at least did not reject, the view that the expected large quantities of future commercial radioactive wastes constituted a (public health, a financial and a public relations) hurdle for the development of nuclear power that needed to be overcome. Despite the fact that no long-term engineering solution for the ultimate disposal of commercial high-level radioactive waste was yet available, all radwaste actors subscribed to, or at least condoned, the new R&D program on the treatment of radioactive wastes and their geological disposal, methods that were considered technically feasible as well as desirable in terms of public and environmental safety and costs.

8 Boundary Maintenance and Radioactive Waste Disposal Technology

Now that a shared perspective on the problems of and solutions for future commercial radioactive wastes had been reached, both within and without the AEC, this perspective had to be protected and strengthened, so as not to impede the development of a suitable commercial radioactive waste management technology. A clear example of the outcomes of this protective strategy is what happened to the critical and dissatisfied Hess committee.

In 1959, and based upon the January/February hearings, the AEC and the JCAE proudly presented a reassuring view about the actual state and future of radioactive waste management. But in the course of the same year, under the influence of the raging debate about the public health hazards of radioactive fallout from atmospheric atomic bomb testing (Divine 1978; Kopp 1979), the dumping of low-level radioactive waste by private (licensed) companies in U.S. coastal waters again became highly controversial. The AEC was concerned that this debate might hamper the development of nuclear power (D; E; Mazuzan and Walker 1984). For much the same reason, the continuing — but only internally voiced — criticism by the Hess Committee of waste-disposal practices at AEC laboratories was not taken lightly by the AEC. In June 1960, Hess wrote a letter to AEC chairman McCone criticizing the ground-disposal practices at the AEC laboratories and recommending both that current local waste disposal practices be carried out at other more suitable geological sites, and that future commercial power-reactors and commercial reprocessing facilities be built at locations suitable for waste disposal. Hess' letter received much attention within the AEC; Hess was told that his committee had exceeded its charter by evaluating current practices (F; OO). To silence the Hess Committee, the AEC commissioners renamed it "the Committee on Geological Aspects of Radioactive Waste Disposal," implying that the committee was limited to advising on and evaluating engineering aspects of geological research and development work relevant to radioactive waste disposal.

So, at the threshold of the 1960s, the Hess Committee was silenced in order to preserve the general consensus on commercial radioactive waste and its future management in the AEC. The success of this effort would further stabilize the social network around the defined problems and their solutions as the radwaste world moved into the 1960s.

D Radioactive Waste R&D in the 1960s

During the 1960s, with the demarcation of military and commercial wastes and the emergent consensus on how to solve the future commercial radioactive waste problem so as not to impede the development of commercial nuclear power, radioactive waste management research was stimulated within the civilian nuclear program. Because commercial liquid high-level waste, i.e. reprocessed high-level radioactive waste from commercial nuclear power plants, did not yet exist, commercial waste management R&D had to be carried out with military wastes. This implied cooperation between the Divisions of Reactor Development and Production and even at times a certain R&D rapprochement.

In the wake of the Hess Committee report (LLL) commercial radwaste R&D was carried out by the Division of Reactor Development along three lines: first, direct discharge of liquid high-level radioactive waste into selected geological formations: second, the solidification of liquid high level radioactive waste and the subsequent long-term storage of these solids in selected geological formations: third, the separation and storage of highly radioactive fission products from the liquid high level radioactive wastes prior to the solidification of the remaining "supernatant" (FF).

Direct disposal of liquid high level waste into salt structures was emphasized in the first (direct disposal) approach. In 1959 ORNL contracted with the Carey Salt Company to conduct a series of field investigations in their abandoned Hutchinson, Kansas, salt mine. During the first experiment in 1960, commercial high-level radioactive waste was simulated by using acidic and neutralized (military) Purex wastes.

Gradually, the R&D emphasis shifted from direct liquid disposal to the second approach whereby solidified high-level radioactive wastes would be geologically disposed. As part of this shift in focus, ORNL carried out project "Salt Vault" between November 1965 and July 1967 in the abandoned Carey Salt Company mine near the city of Lyons, Kansas. In this field demonstration, canned, short-cooled irradiated fuel elements from military "production reactors" were used to simulate the thermal and radiation characteristics of reprocessed commercial power reactor wastes (N).

A third R&D approach involved the solidification of liquid wastes remaining after the extraction of highly radioactive fission products. The calcination of both liquid high-level commercial and high-level military wastes at the National Reactor Testing Station (NRTS) in Idaho was the main option being investigated.

Within the military program, R&D was carried out along the lines of solidification, separation of fission products and direct disposal. In an effort to replace the costly direct-ground disposal of its "intermediate-level wastes," Oak Ridge National Laboratory was investigating and testing Hydrofracturing, i.e. a technique of injecting a waste-cement-clay mixture under high pressure into artificially produced fractures in subsurface impermeable shale strata. From 1958 onwards, the Savannah River Plant in South Carolina investigated the feasibility of storing its liquid high level wastes in tun-

8 Boundary Maintenance and Radioactive Waste Disposal Technology

nels mined out of deep impermeable crystalline bedrock formations underlying the site at a depth of 2,000 feet.[2] In the 1960s the Hanford production site concentrated on the extraction of long-lived fission products (Strontium 90, Cesium 137 and 144, and Promethium 147) from the stored liquid high-level waste to reduce the radioactive-decay heat. The extracted fission products would be encapsulated in small containers as solids for long-term storage. The residual low-heating-rates wastes remaining after the fission product extraction would be concentrated into a "salt cake" by in-tank evaporation (BBB; H).

Table 2 shows the rapprochement of the commercial and the military waste R&D programs on finding engineering solutions for (actual and expected) radioactive waste management problems. Despite this rapprochement, it is still possible to distinguish separate global military and commercial technical regimes (Disco, et al. 1992). The research approaches are similar, but within these approaches there are clearly different purposes, aims and technological trajectories. In the commercial R&D program, emphasis was on temporary tank storage, waste solidification, off-site transportation and permanent off-site disposal of the solidified waste in selected geological (salt) formations. The military R&D program instead concentrated on improving permanent tank storage technology and on developing on-site disposal methods. Within the military regime, different local technical regimes can be distinguished, rooted in locally varying rules for waste categorization, and in various traditions, experiences and routines for handling the local wastes within prevailing environmental, organizational en technical conditions.

Table 2: Commercial and Military High-level Radwaste R&D in the 1960s. Funded by the DRD and DOP Respectively

		approaches		
		direct discharge to selected geological formations	solidification and storage of solids	separation of fission products and solidification of remaining liquids
emphasis in actual research	commercial H-L radwaste R&D funded by DRD in the 1960s	salt storage R&D carried out by ORNL in Kansas 1959-1965	salt mine R&D carried out by ORNL in Kansas 1965-1970	calcination notably at NRTS in Idaho
	military H-L radwaste R&D funded by DOP in the 1960s	- hydrofracking at ORNL - bedrock storage below SRP in South-California	in-tank evaporation and in-tank storage of residues at Hanford and SRP	- extraction and storage of isotopes at Hanford - calcination at NRTS in Idaho

3 Boundary Maintenance of a Radwaste World

In the early 1960s, a definition of the commercial radioactive waste problem and its solution had stabilized. This process of stabilization (Pinch and Bijker 1984) limited the number of participants involved and created what can be called an "inside" and an "outside" world. The "insiders" were chiefly located within the AEC, but also within the National Academy of Sciences (NAS), the U.S. Geological Survey, the Congressional Joint Committee on Atomic Energy (JCAE), private industry, universities, professional communities, the U.S. Weather Bureau, the U.S. Bureau of Mines, the American Association of Petroleum Geologists (AAPG), the Federal Radiation Council and the National Committee on Radiation Protection. Radioactive waste insiders were interconnected by institutional, programmatic, professional, scientific, technical, financial and ideological ties (Joerges 1988).

The stabilized problem definition of commercial high-level radioactive waste shared by the inside world, and the related R&D program, served as a frame of meaning and an orientation for action within this world. But the same problem definition, converted into a technological promise ("we will solve the radwaste problem") also served as a legitimization to the outside world, as a way to protect the AEC and the inside against meddlesome interference. This process of protection, or boundary maintenance, took the form of policing (Bijker and Law 1992), i.e. strategies and processes to keep actors, views and things in place. The social-cognitive process of boundary maintenance had an internal and an external impulse. Internally, it was directed at keeping the inside together in a social and cognitive sense, that is, in keeping actors, problem definitions, solutions and technical trajectories aligned (inside maintenance). Externally, it was directed at maintaining and augmenting the authority and credibility of the inside in the outside world (outside maintenance). The philosopher Derrida stresses the interconnectedness of the inside and the outside by noting that the outside plays a necessary constituting role in the formation of the inside, simply by virtue of the fact of being branded by insiders as supplementary, as inessential and even as parasitical. In Derrida's view, the outside, far from being a mere accessory, is thus a central feature of the inside. The outside can thus be seen as the intrinsic feature of a system, displacing the inside (Cooper 1989).

Because of its legal and political status, its authority and its resources, the AEC was the most visible actor in and representative of the radwaste world, also acting as its main "policer." For all atomic matters, the AEC was the "obligatory point of passage" (Law and Callon 1992). However, this by no means implied that other inside actors were just passive yes-men or dedicated followers. Each time the AEC had to take into account the positions, interests and views of the other inside actors, leaving traces, and sometimes

8 Boundary Maintenance and Radioactive Waste Disposal Technology   247

even scars, as existing bureaucratic arrangements and technical trajectories had to be adapted and changed. This, as well as the overall process of inside maintenance, can be illustrated by the continuing activities of the NAS Committee on Geological Aspects of Waste Disposal (the former Hess committee) and by the involvement of the U.S. General Accounting Office.

A Inside Maintenance: The U.S. National Academy of Sciences

In May, 1966, the NAS Committee on Geological Aspects of Radioactive Waste Disposal, or the "Galley Committee" after the chairman who had succeeded Harry Hess, submitted a report to the Division of Reactor Development (MMM), in which it evaluated the geological aspects of the commercial radwaste R&D program of the Division, as well as the geological waste disposal practices at the local laboratories and production sites.

The committee praised the salt and solidification R&D of the Division of Reactor Development but was negative about military high-level waste management R&D and the disposal plans being worked out at production sites (see Table 2). The bedrock disposal concept being investigated at the Savannah River Plant was evaluated as "in its essence dangerous" because it was doubted whether dry caverns in crystalline bedrock below the site could be found and whether the nearby and much used Tuscaloosa acquifers could be protected. The hydro-fracting disposal tests at Oak Ridge National Laboratory were evaluated as unsafe and not duplicable at other AEC sites. The Galley committee also opposed operational waste disposal practices of low and intermediate liquid and solid radioactive military and laboratory wastes at the local laboratories, because the laboratories were all located over some of the largest fresh-water acquifers in the U.S. (MMM). To change waste handling practices at local laboratories and "production" sites the Committee recommended the use of a new and uniform waste definition for commercial, military and laboratory wastes based on the longevity of radioactive nuclides and the concentration of radioactivity (MMM). This recommendation, if adopted, would require a complete change of waste classification and categorization, of actual local waste management operations ('practices'), of installations and of instrumentation, all of which would naturally lead to increased costs.

While the substance of the advice might well have been correct, the Galley committee had sought to interfere with the military waste management practices of the Division of Production. It had thus not only exceeded its charter, but had expressed views which, if made public, could easily have aroused public concern and suspicion. Therefore, the Division of Reactor Development decided not to publish the report and to terminate the advisory work of the Galley Committee. Because the National Academy of Sciences objected, and because the AEC Commissioners valued good relations with this prestigious body, the establishment of a new advisory waste committee was considered (RR; FFF; KKK). It was agreed that the AEC would have a crucial say in its membership, and that the new committee would concentrate on the management of future commercial high-level radioactive wastes generated by a commercial reprocessing industry (G; JJ; SS; GGG). In February, 1970, the new NAS Committee, the Committee on Radioactive Waste Management (CRWM) published its first report in which it wholeheartedly concurred in existing and planned R&D programs for the management of both commercial and military wastes (NNN; Boffey 1975). Unlike the Hess and the Galley committees, the CRWM did not express its opinion of military waste management practices at the local laboratories.

The Galley committee history shows that by excluding certain actors and views, and including others, social and cognitive stability was restored. Still, the movement had left technological traces. In the R&D program pertaining to future commercial wastes, the salt disposal approach had become stronger and more dominant. A 1969 evaluation by ORNL even indicated that it would be possible to establish a salt mine repository in the Eastern part of the U.S. within short order to serve the waste disposal needs of the commercial nuclear program over the next 20 to 30 years (N; FF). At the local production sites and laboratories, the ground disposal of low level military and laboratory wastes were continued, although now there was greater emphasis on methods for concentrating and containing radioactivity, as well as on reducing the activity levels of waste streams before discharge into the environment.

B   Inside Maintenance: The U.S. General Accounting Office

The issue of managing military high-level liquid wastes was unexpectedly raised by the U.S. General Accounting Office (GAO) in response to a request by the Congressional Joint Committee on Atomic Energy (JCAE) to evaluate the AEC's operational waste management practices as well as its waste R&D programs, chiefly because policy evaluation had become a common method for evaluating governmental performance in the late 1960s. Besides pointing at the local variety of waste categorizations and waste management practices, the May, 1968, GAO report also revealed that at least 10 of the 149 underground carbon steel waste storage tanks at Hanford had leaked an estimated 227,400 gallons of high-level waste into the ground (HH).

8 Boundary Maintenance and Radioactive Waste Disposal Technology 249

In response, the AEC classified the GAO report as "secret restricted data" because it contained information that might reveal production rates of plutonium and tritium (PP). At the same time, however, it established waste management consultant groups for the production sites, and an informal Task Force (I) to investigate the alleged leakages and to propose (future) solutions. On the basis of recommendations by the consultant groups and the task force, the AEC Commissioners informed the JCAE that there would be adequate time to develop permanent and safe disposal methods because the AEC had already adopted a policy of interim tank storage, solidification and long term storage that would be effective as of December 31, 1975 (J; S).

This bureaucratic and technical promise, accepted on credit by the JCAE, put pressure on military waste management at the production sites. Due to repeated tank failures, attention at Hanford was devoted to the construction of additional tanks and to the evaporation of wastes in the waste storage tanks. As a result of political pressure and of financial and technical problems, the proposed policy of interim tank storage, waste solidification and long term storage was transformed into a program of interim tank storage followed by in-tank evaporation ("solidification") and long term in-tank storage of the remaining "salt cake." This approach was presented as the end-goal and the solution for the management of liquid high-level radioactive military wastes. Solidification of military wastes and shipment of the solids to an off-site repository as recommended by the Sanitary Engineering Branch, the Hess Committee and the Galley Committee, was defined as unacceptable in terms of costs and safety (L; EE; EEE). As a result, the options of on-site disposal of liquid and solidified high-level military wastes in basalt below the Hanford production site, as well as disposal in crystalline bedrock below the Savannah River production site gained new impetus.

As in the case of the Galley Committee, the JCAE–GAO history shows that insiders can become troublesome because they work from their own specific perspectives and positions. By classifying the 1968 GAO report, and by defusing its criticism through bureaucratic and technical promises, the AEC was able to manage GAO criticism and to reassure the JCAE. Hence, social and cognitive stability was once again achieved in the radwaste world. But this outcome also left bureaucratic and technological traces. The AEC was forced to articulate its military waste management policy and to determine an administrative target for its implementation. This put pressure on military waste management and military waste management R&D, which culminated in the unintended articulation of a high-level military waste management solution.

The bureaucratic and technical traces left by inside maintenance became bureaucratic and technical legacies the moment new critical insiders and outsiders started questioning the rationale of past decisions. Besides defending and rationalizing past decisions, the AEC had to cope with new demands put forward by these critics. The effects of these exclusion legacies are visible during inside maintenance (i.e. in the Galley Committee and the GAO histories), but especially during outside maintenance, as classified reports, excluded critics and views were rediscovered, interconnected and put on the AEC-agenda by critical outside actors.

C Outside Maintenance: Plutonium Wastes in Idaho and Bedrock Disposal in South Carolina

Bureaucratic classifications and technical promises are not impervious to events in the real world. Confronted with the AEC's concrete waste management plans and activities, outsiders became restive and started asking questions. This alone would have gotten the AEC into hot water, but it also happened to coincide with the rising tide of "environmentalism" (DelSesto 1979; Hays 1987; Ford 1984). Due to the increased vigor and meddlesomeness of outside actors and views, and in interaction with "environmentalism," heterogeneous (and formerly excluded) actors, views, interests, local events and classified reports became linked up in unexpected ways. The demand that military plutonium-contaminated wastes be separated from other military wastes prior to burial was, for instance, more an outcome of events in the outside and inside world, than of any risk assessment within the AEC's Division of Production.

In 1969, plutonium-contaminated wastes from a fire at the Rocky Flats military plutonium plant in Colorado were being buried in the waste burial grounds at Idaho's National Reactor Testing Station (NRTS) (Metzger 1972). After coverage by local news media, Governor Samuelson mobilized State agencies to investigate the potential hazards, and requested the AEC to provide information about possible public health hazards. Idaho senator Frank Church did the same on the Federal level by requesting federal agencies, such as the Public Health Service, to evaluate the public health hazards of storing radioactive waste above the Snake River Plane Aquifer at the NRTS (Y; Z; MM; ZZ). Moreover, when Church became aware of the NAS-Galley committee's unpublished 1966 report he (and many with him) requested a copy of this report and an AEC explanation of its suppression (CCC; DDD; Cook 1972). With the limited distribution of the 1966 NAS report, concerns in Idaho were not laid to rest: the AEC had to find other ways to allay public health concerns. In an attempt to manage public concern, an AEC task force recommended that plutonium-contaminated wastes should be kept segregated from other wastes when buried, and that these wastes should be retrievable after burial (K; EE; LL; M).

A January, 1970, report by the Public Health Service (PHS), requested by Senator Church, made the same recommendations. The rapprochement between the AEC and PHS recommendations was, however, nullified by another report requested by Church in April, 1970, on waste disposal practices at NRTS. In this Federal Water Quality Administration (FWQA) report it was recommended that existing waste burial practices at NRTS be abandoned, and that the already buried wastes (including those from Rocky Flats) be removed to another more suitable location (BB; GG). Church urged the AEC to implement the FWQA recommendations (AA; BB). Although the AEC strongly disagreed (O), it realized that it had to take action. In the event, the AEC connected military and commercial wastes by proposing that the transuranium Rocky Flats military wastes buried at NRTS be excavated and transported to a salt mine repository which the AEC Division of Reactor Development was designing for commercial high-level wastes. Church gladly and publicly accepted this proposal, translating it as an AEC offer. AEC Chairman Seaborg, fearing a major controversy in Idaho, could only confirm Church's "translation" (P; DDD).

With this outcome the AEC Commissioners had violated the bureaucratic and technical boundaries between military and commercial waste and the segregated bureaucratic and technical trajectories carefully constructed and successfully maintained in the preceding years. This muddied the distinction, both in the inside and outside radwaste worlds, between the management of military and the management of commercial wastes, and, as an unintended outcome, helped to link heterogeneous actors, views, events and reports. For example, in the early 1970s, Southern States opposed AEC plans to store military high-level radioactive wastes in mined bedrock caverns below the Savannah River Plant (SRP) in South Carolina. Anticipating local South Carolina opposition after the widely publicized "Idaho affair," and also compelled by environmental policies, the AEC Commissioners charged the Division of Production with informing Georgia and South Carolina public officials about the waste storage plans at the Savannah River Plant. Despite this effort, the military high-level SRP wastes destined to be stored in bedrock became linked to commercial radioactive wastes to be stored in geological salt formations, as well as to classified reports, critical views and excluded actors.

Taking up the thread of the Idaho controversy and its outcome, the incipient Lyons controversy and the classified 1966 NAS-Galley report, the environmental Natural Resources Defense Council Inc. (NRDC), for example, wondered why the AEC was investigating bedrock storage "[...] when all the literature suggests that salt mine storage is the preferred method," and "how can bedrock storage at Savannah River and Hanford possibly be safe when the 1966 NAS study concluded that both these sites were unsatisfactory?" (HHH). Despite AEC reassurances, opposition grew as more actors and views got involved and became connected (Cook 1972). In March, 1972, Georgia Governor Jimmy Carter asserted that his state opposed the bedrock storage plans in a neighboring state, citing the 1966 NAS-Galley report's negative evaluation of bedrock disposal. Carter preferred surface storage of (military and commercial) high-level radioactive waste because this would allow for more effective surveillance and monitoring (JJJ). Carter's opposition was supported by South Carolina senator Ernest F. Hollings, who also emphasized that an earthquake might break open bedrock storage vaults (LL). In addition, suspicion grew in South Carolina that the Savannah River Plant would become a federal waste repository for military and commercial radioactive waste. This suspicion was confirmed as the activities of a special Nuclear Study Committee appointed by the State legislature to investigate the plans of Allied-Gulf Nuclear Service to build a commercial reprocessing plant at Barnwell (SC), became linked with the plans to store high-level military wastes in bedrock caverns below the SRP-site. This linkage between civilian and military wastes, discovered and articulated by local nuclear opponents, gradually increased anti-nuclear sentiments in South Carolina. Ultimately, in September, 1972, in the wake of the Lyons controversy, the AEC decided to postpone the Savannah River Bedrock disposal plans indefinitely.

The Idaho and South Carolina case histories show that outside maintenance left bureaucratic and technical scars which necessitated implementation of new bureaucratic arrangements and rationalizations as well as radical changes in existing and planned activities. The case histories also show that boundary maintenance processes, i.e. processes of inclusion and exclusion as part of inside and outside maintenance, can encourage alignment of heterogeneous and excluded outsiders. Unable and unwilling to accept AEC waste management plans and waste classifications, most notably the demarcation between commercial and military wastes, excluded outsiders tended to link radioactive waste issues, local events, (classified) reports and views which the AEC would have preferred to keep separate. This process of issue linkage in fact constituted an empowering and aligning factor for the outside world.

Boundary maintenance legacies thus not only had a direct effect on the bureaucratization and technologizaton of U.S. radwaste technology, but also on the kind and character of the interactions between inside and outside actors. In fact, the processes of boundary maintenance, i.e. inclusion and exclusion processes, and their sociotechnical outcomes shaped U.S. radwaste management technology. This process became strikingly evident the moment the AEC tried to implement its commercial radioactive waste management technology at Lyons, Kansas, in the early 1970s.

D Outside Maintenance: The Implementation of AEC's Commercial Radioactive Waste Disposal Technology at Lyons, Kansas

In May, 1970, the AEC promised Senator Church to remove the plutonium-contaminated Rocky Flats wastes buried at Idaho to a planned Federal salt mine waste repository. Subsequently, in June, 1970, the AEC publicly announced the "tentative selection" of the abandoned Carey salt mine near Lyons, Kansas, as the site where America's first commercial radioactive waste repository, pending "confirmatory geologic testing," would be built (Q; W). This announcement marked the start of AEC's effort to implement the commercial radioactive waste disposal technology it had worked on for nearly fifteen years within its charmed circle of insiders. On the basis of 10 years of research in the abandoned Lyons salt mine, AEC's Division of Reactor Development was convinced that this was the perfect site to demonstrate the feasibility of this technology and its underlying premises. The ensuing implementation entailed a further expansion of the already large number of outside actors the AEC had to deal with.

In November, 1970, the AEC published its commercial radioactive waste policy in the Federal Register (R). The policy implied that commercial liquid high-level radioactive wastes generated during the reprocessing of irradiated power reactor fuel, would be solidified at the reprocessing plants within five years after their generation and transported off-site to a Federally owned waste repository within ten years after their generation. Reprocessing of irradiated power reactor fuel was considered necessary because uranium supply was believed to be scarce, and because reprocessing was seen as a proven technology developed within the military program. Tank storage of commercial liquid high-level wastes was ruled out as a long-term solution, because this waste (due to the longer irradiation in power reactors) would be thermally hotter, more radioactive and more corrosive than high-level military wastes.

In Kansas, most actors were completely surprised by the AEC's public announcement in June, 1970. Confronted by what they saw as a *fait accompli*, they demanded explanations why Lyons had been chosen, and asked for more information and scientific data. Important actors were Kansas representative Joe Skubitz, the Kansas Geological Survey (KGS) under the direction of William Hambleton and Kansas Governor Docking. These actors urged that the safety of waste disposal in the Lyons salt mine be demonstrated before disposal operations could start. Although they, and especially the KGS, never explicitly opposed the salt disposal concept, the technical issues they raised (heat flow, heat dissipation, surface subsidence and radiation damage) in fact questioned the AEC's self-assurance about salt disposal, past radwaste R&D and the wisdom of the Lyons site selection.

Other events and legal requirements, such as *Environmental Impact Statements* (EIS) that had to be reviewed by relevant State and Federal Agencies, and congressional Authorizing Legislation hearings, introduced even more actors and positions. New actors were, for instance: the Southern Interstate Nuclear Board (SINB), the Louisiana Board of Nuclear Energy (LBNE), and the Southern Governor's Conference (SGC) which urged investigating the suitability of southern domed salt formations for holding radioactive waste before selecting bedded salt formations in Kansas as the preferred formation; private companies actually engaged in or planning to start commercial reprocessing, which urged the revision of the AEC radioactive waste policy requirements; the National Academy of Sciences Committee on Radioactive Waste Management (NAS-CRWM), with an ad-hoc salt-mine panel including Kansas Geological Survey director William Hambleton; the Department of the Interior (DOI); the Environmental Protection Agency (EPA) and Kansas State Agencies. The new issues these actors introduced were: the lifetime of waste containers embedded in salt, the impact of radioactivity and heat on salt, the leachability of the solidified wastes, earthquake hazards, transportation safety, nuclear theft, retrievability of the buried waste in case of emergency situations, post-operational and environmental monitoring, and the environmental effects of radiation. Despite the different issues raised, nearly all these actors shared the view that additional confirmatory research had to be performed before the planned demonstration waste repository could be built.

During and after the Environmental Impact procedures and the JCAE appropriation legislation hearings, the AEC faced an enormous proliferation of requests for the most diverse kinds of information. The Lyons EISs were the first ever written and every actor had additional questions to ask. Because of the heterogeneity of the actors and their interests, no single fact, problem, or solution could be defined so as to satisfy all the audiences even tolerably well.

The AEC, however, remained convinced that the Lyons site was suitable for the radioactive waste repository. In an attempt to manage opposition, the AEC therefore promised that if substantive problems arose during construction, work would be halted until they were solved and, if solutions could not be found, the project would be canceled. While promises and other such statements can of course be mere rhetorical and political flourishes, they also create a decision agenda: in the event problems and safety hazards were actually encountered, work would indeed have to be stopped.

This scenario was in fact mobilized when an ORNL consultant, and a recognized expert in this field, discovered a great number of unrecorded and unplugged former bore holes at the Lyons site. In addition, the American Salt Company (ASC), which had announced plans to expand its salt solution activities near the repository mine, disclosed earlier unexpected water flows and large losses of water (175,000 gallons) during solution mining activities without knowing either where the water had come from or where it had gone to. These findings and announcements created so much *prima facie* uncertainty about the geological integrity of the Lyons salt mine, that a discussion whether the reported problems were "substantial" or not did not even get off the ground.

About the same time, i.e. August, 1971, the Kansas Congressional Delegation had acquired broad support from a congressional majority unwilling to act against the unanimous opposition of the elected representatives of that state (Lear 1972). As a result, Kansas Senators Dole and Pearson were able to move amendments to the AEC authorization Act of 1972, stipulating that the AEC had to lease the Lyons site for three years before it was allowed to acquire the land, and that the burial of wastes would be postponed until complete safety could be affirmed by an independent advisory council appointed by the President (VV). In addition, the approval of the AEC plan was linked to the provision that the disposal technology should satisfy strict safety requirements.

The technical problems and the Congressional amendment strengthened the position of Kansas opponents. As a result of the technical problems, the AEC halted its Lyons activities, as it had promised to do, and asked the Kansas Geological Survey to undertake a paper survey of other Kansas areas with geological formations similar to those at Lyons (AAA). In its January, 1972, survey the Kansas Geological Survey concluded that the Lyons site constituted the poorest candidate for future study (Angino et al. 1972).

While the AEC may not have been convinced that the salt disposal concept had been refuted, it had come to recognize the potential impact of public concern and public misunderstanding. Convinced that its problem definitions and solutions were correct, the AEC redefined the radioactive waste disposal problem as a public relations and a public credibility problem (T; U). The problems encountered in Kansas were defined as "specific to the Lyons site and in no way negated the original generally accepted conclusion that burial in bedded salt offers the greatest currently known potential for ultimately safe disposal of radioactive waste" (V; WW).

Internally, however, the AEC shifted to another technical option. After evaluating the Lyons experience, the AEC's Division of Waste Management and Transport concluded in February, 1972, that geological disposal should be abandoned for the time being and that, as outsiders had been urging, retrievable and controllable surface storage should be explored (Downey 1985). Besides these "engineered surface facilities" with an expected life time of 100 years, the division proposed continuing evaluation of and R&D on bedded salt options.

It is important to recognize that the Idaho, South Carolina, and Lyons controversies did not emerge as a result of scientific uncertainties, as is often claimed (Rolph 1979; Nelkin 1992; Fallows 1979), but rather that controversiality lead to the articulation of scientific uncertainty. The controversies expanded further as issues, views, events and actors became linked. This process of *issue linkage* occurred as an overall socio-cognitive process, as well as an actor strategy. The AEC strategy, for instance, was defensive: it tried to uphold the social and cognitive boundaries between inside and outside worlds by claiming that technology and politics were (and should be) separated, and that political considerations were constraining the implementation of a sound and safe technology worked on for many years by many experts. This projected boundary can be seen as an attempt to maintain inside-outside boundaries: the AEC tried to mold contrasting views into a coherent perspective. The AEC acted accordingly, and so helped to align outside actors as well as helped to build a stabilized picture of a polarized radwaste controversy in the outside world. This is how the bipolarity to which we have become accustomed was socially constructed. At present this bipolarity is not only a socio-political conflict as many claim, but also a structural frame within which nuclear technological developments are shaped (Rip and Talma, this volume).

4 Discussion and Conclusion

The emergence, and to a large extent active building, of a radioactive waste world and the accompanying processes of bureaucratization and technologization of radwaste, is interesting in its own right (De la Bruhèze 1992c). With respect to the theme of this volume, this chapter shows that social-cognitive (institutional, professional, scientific and technical) links aligned actors, views and things, and thereby provided an orientation for action and interaction (laid down in programs, plans, goals and contracts.) In fact, these links soon gelled into a socio-technical structure providing opportunities and constraints for waste world actors (insiders), and which ultimately defined possibilities for change. The efforts and investments carried out within and on behalf of the salt-disposal research, for instance, made the salt trajectory for commercial high-level wastes become irreversible both in and outside the AEC.

It was also shown that this social-technical structure or order (the inside) required maintenance in order to develop a radioactive waste management technology along the lines set out. This maintenance took the form of "policing," i.e. processes and strategies to keep actors, views and things in place. Boundary maintenance was initially quite successful but began to take its toll as processes of inclusion and exclusion left bureaucratic and technical traces where new arrangements had to be implemented and planned work and research had to be altered. The bureaucratic and technical traces became burdensome bureaucratic and technical legacies the moment new critical insiders and outsiders started questioning the rationale of present and past decisions, and, in so doing, started rediscovering and interconnecting excluded actors, views and reports. As a result, inside and outside actors were increasingly able to influence the inside agenda.

A major point is that boundaries were never unambiguous, or at least never static. For insiders to pursue their central concerns, they had to accommodate to issues raised by others and to positions taken by important actors. Who was "in" or "out" depended on the content and outcomes of interactions. This chapter showed that U.S. radwaste technology was shaped by inclusion and exclusion processes. Besides the AEC, many other actors were linking up or parting ways (issue linkage) by the 1970s. These processes and their outcomes are summarized in Figure 2.

The socio-technical processes portrayed show that guiding waste management concepts like "retrievability" and "surface storage" had their own specific history. They cannot therefore simply be attributed to the influence of the environmental movement as is still often, but mistakenly, claimed. The patterns portrayed also show that changes and adaptations occurred in the domain of technologization, while the bureaucratic waste categorizations and classifications remained impervious to change. Thus, while radwaste plans, policies and practices were contested and adapted to changing circumstances, definitions and classifications of radwaste remained unchanged throughout the entire period under study. The preservation of waste definitions and classifications was the central element in boundary maintenance processes because changes in this domain would have implied complete changes in the management of actual waste-handling operations and would have led to unacceptably higher costs. This successful conservation of the original definitions, despite early efforts on the part of the sanitary engineers, Hess and Galley, to change them, was the root cause of the irreversibility of the U.S. radwaste technology trajectory, as well as of the radwaste contestation that lingers on to the present day.

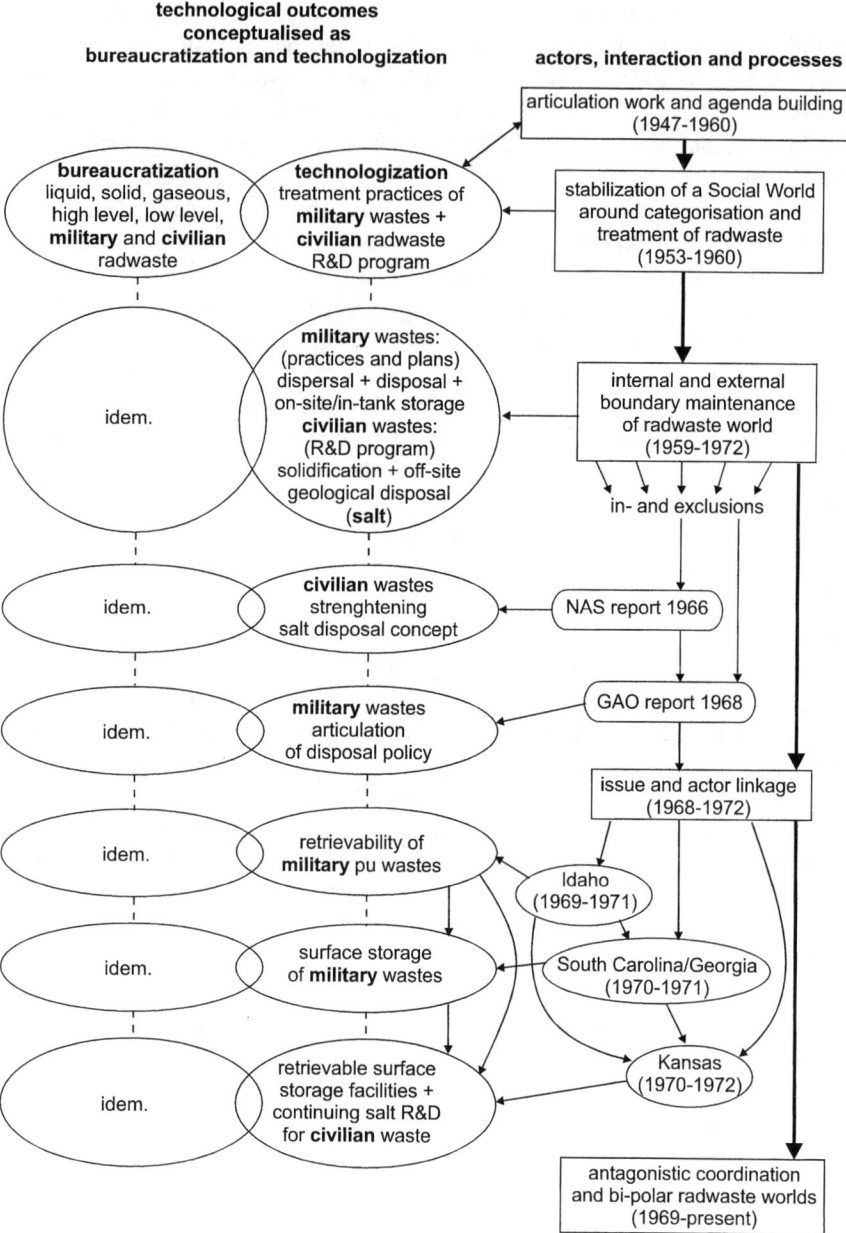

Figure 2: Sociotechnical Processes and Outcomes in the Development of Radwaste Technology in the U.S.

8 Boundary Maintenance and Radioactive Waste Disposal Technology 259

The conceptual (and ironic) lesson of this chapter is that boundary maintenance processes can in fact facilitate the alignment of heterogeneous and excluded outsiders into one oppositional camp. The coordination that then actually shapes technological developments is not coordination from one center, nor is it emergent coordination among more or less atomistic entities, but rather "antagonistic coordination" in which the interaction between actual or potential opponents determines what kind of technology will be developed (Rip and Talma, this volume). In this chapter antagonistic coordination has been described at the level of events, where interactions were usually very specific and focused on particular technical issues, for instance, the effects of temperature on the plastic movement of salt formations. While participants tend to think of such struggles as scientific or technical controversies (or at least maintain that this is what they should be about), the outcomes of the interaction processes — especially when there is no clear winner — can lead to the sedimentation of arguments and views at the collective level of institutions and organizations where they are available and operate as collectively shared perspectives or paradigms. Continuing interaction can then be characterized as "manoeuvering" because it is linked with potential inclusion or exclusion in the mainstream of the development. Scientific and technical points are then important because they enable or prevent access.

Antagonistic coordination leads to specific outcomes. The AEC had to improve (and sometimes modify) its arguments and its practices in order to successfully include or exclude particular actors and issues. Exclusion subsequently reinforced the logic of antagonism as beleaguered outsider groups began to link issues across sites and through time. Thus we see the AEC's exclusionary strategies producing and defining its own opposition as these groups linked issues like AEC plans to bury plutonium-contaminated wastes in Idaho, to bury military high-level waste in bedrock below the Savannah River Plant, and to use the bedded salt formations at Lyons for disposing solidified commercial high-level wastes. Such compilation of formerly isolated episodes into a (negative) image of what the AEC was about fed the flames of protest and produced strong polarizations. As a result, salt mine disposal was scrutinized very closely, and further work was expended to counter possible criticism. In fact, the technological option had to become perfect in order to pass muster.

Given the generic nature of such processes, one can argue that such outcomes can be generally expected. Antagonistic coordination increases demands on the technology being developed. It ups the ante, even though the actors do not start out with any such intention.

Notes

* Empirical sources are primarily archival materials of the US Atomic Energy Commission (USAEC), the Congressional Joint Committee on Atomic Energy (USJCAE), and the U.S. National Academy of Sciences (USNAS), which are located at the U.S. National Archives, the Archives and the History Division of the U.S. Department of Energy (USDOE), the Archives of the U.S. National Academy of Sciences, and at the U.S. Library of Congress.
  The abbreviation NA-326 AEC refers to records of the U.S. Atomic Energy Commission (AEC), record group 326 located in the U.S. National Archives. NA-128 JCAE refers to the records of the Congressional Joint Committee on Atomic Energy (JCAE), record group 128, located in the U.S. National Archives. DOE-AEC, SECY..-..,/CONTROL OF ATOMIC WASTES/ MATERIALS 12 / MATERIALS 20 refer to AEC radioactive waste documents located in the Archives and the History Division of the U.S. Department of Energy (DOE), categorized as "Secretary Files" (SECY) and subcategorized as "Control of Atomic Wastes," "Materials 12" and "Materials 20."
  I want to thank Arie Rip and Nil Disco for their many useful and stimulating comments on earlier drafts of this chapter.
1 In other words: at a depth of 3,000 to 4,000 feet the plastic flow in salt formations was thought to prevail over fracturing. For this reason the salt would effectively seal liquid or solid wastes and isolate them permanently from circulating water.
2 The bedrock disposal concept had originally been developed by the oil and chemical industry for storing hazardous waste materials in caverns excavated in impermeable rock. Dupont Corp., operator of the Savannah River Plant and well acquainted with this storage concept, quite naturally had an interest in testing its suitability for radioactive wastes.

Sources

A. Albee, H.H. (1995): "Problems in the burial of solid wastes at Oak Ridge National Laboratory," Proceedings of a Seminar Sponsored by AEC and the Public Health Service, 6-9 December, Cincinnati, Ohio.
B. Arnold, E.D. (1957): "Compilation and analysis of waste disposal information," Oak Ridge National Laboratory, Oak Ridge, Tennessee: Interdivisional Committee on Waste Disposal, Divisions of Chemical Technology and Health Physics, ORNL-57-2-20, February.
C. AEC 180-1 (1949): "Reporting of the handling of radioactive waste materials in the United States Atomic Energy program." October 17, Appendix C, DOE-AEC, SECY-47-51, Box 1232, Control of Atomic Wastes-I.
D. AEC (1960): Minutes Commission Meeting no. 1617. May 6, DOE-AEC, SECY 58-66, Materials 12.
E. AEC (1960): Minutes Commission Meeting no. 1630. June 20, DOE-AEC, SECY, Box 1359, "Waste Processing."
F. AEC 180-13 (1960): Appendix C. Minutes AEC Commission meeting 1675, 23 November.

8 Boundary Maintenance and Radioactive Waste Disposal Technology 261

G. AEC (1968): General Manager to John T. Conway, Executive Director JCAE, March.
H. AEC (1968): AEC 180/30. April 5, DOE Archives, SECY 66-68, Materials 20.
I. AEC 180-43 (1968): "Report of Task Force on AEC operational radioactive waste management." August 12, DOE-AEC, SECY 68-69, Materials 20-V1.
J. AEC 180-47 (1968): "Siting of commercial reprocessing plants and related waste management facilities." October 9, DOE-AEC, SECY 68-69, Materials 20-V1.
K. AEC 180-72 (1969): "Status of Task Force on AEC operational radioactive waste management," December 2.
L. AEC (1969): Meeting with AEC and GAO representatives concerning radioactive waste management, December 8.
M. AEC 180-75 (1970): "Task Force on radioactive waste management," January 5, DOE-AEC, SECY 69-70, Materials 20-V3.
N. AEC 180-81 (1970): "Solid radioactive wastes: Salt mine storage," April 23, Appendix A, DOE Archives, AEC-SECY 69-70, Materials 20-V3.
O. AEC 180-84 (1970): "Comments on FWQA report on waste management at NRTS," May 26: 2, 11-14, DOE-AEC, SECY 69-70, Materials 20-V4.
P. AEC 180-86 (1970): "Idaho press report on new waste disposal area,." June 3.
Q. AEC (1970): AEC press release N-102. June 17, DOE-AEC, "AEC-Press releases."
R. AEC (1970): Commercial radioactive waste policy. November, Appendix F to 10 CFR (Federal Register) 50.
S. AEC (1971): "Statement on 1968 GAO waste report," in: Ramey, AEC Commissioner, to Bauser, executive Director JCAE, January, DOE-AEC, SECY 71, Materials 20-V3.
T. AEC-SECY 2371 (1972): "Management of commercial high-level radioactive wastes," March 17, DOE-AEC, SECY 71-72, Box 7820, "Radioactive waste disposal."
U. AEC-SECY 2373 (1972): "Proposed press release on expansion of waste management program," March 20, DOE-AEC, SECY 71-72, Materials 20-V6.
V. AEC Fact Sheet. (1973): "Commercial high-level radioactive waste," June 21, NA-128 JCAE, Box 706, Vol.5.
W. AGMO (1970): "Fact sheet -- salt mine repository," in: Erlewine, AGMO, to Bauser, Executive Director JCAE, June 16, NA-128 JCAE, Box 707, Vol. Kansas Site.
X. Browder, F.N. (1951): "Liquid waste disposal at Oak Ridge National Laboratory," Industrial and Engineering Chemistry 43, 7, July: 1502-1505.
Y. Church (1969): Church to Gottschalk, Director Bureau of Sport Fisheries & Wildlife, September 12.
Z. Church (1969): Church to Seaborg, September 13.
AA. Church (1970): Statement of Church in: AEC 180-83, "Waste management at NRTS: Guidance and disposal operations at the NRTS, Idaho Falls, Idaho," April
BB. Church to Seaborg, May 1, DOE-AEC, SECY 69-70, Materials 20-V4.
CC. Church (1969): Church to Seaborg, October 7.
DD. Congress, the 86th (1963): "Hearings on industrial radioactive waste disposal," First session, January 28-30 and February 2-3. Washington D.C.: Government Printing Office.
EE. Costagliola (1969): Memo from Costagliola, JCAE staff, to (JCAE) record. December 8, NA-128 JCAE, Box 706, Vol.5.

FF. Division of Reactor Development and Technology (1967): "Long term high-activity waste management," Statement of the Division of RD&T, September 8, DOE Archives, SECY 66–68, Materials 20.
GG. Federal Water Quality Administration (FWQA) (1970): "A report on the examination and disposal operations at the NRTS," April, Idaho Falls, Idaho.
HH. GAO (1968): "Observations concerning the management of high level radioactive waste material. Report to the Joint Committee on Atomic Energy by the Comptroller General of the United States," Washington, D.C.: USGAO, May, B-164052.
II. Hewlett, R.G. (1978): *Federal policy for the disposal of highly radioactive Wastes from commercial nuclear power plants — an historical analysis.* March 9, Washington D.C.: Department of Energy, DOE/MA-0153 Draft.
JJ. Hobbs, F.T. (1968): Acting Commission Secretary, to file, March 13.
KK. Hollings (1972): Hollings to Schlesinger, Chairman AEC, March 3, DOE-AEC, SECY 71–72, Materials 20-V6.
LL. Hollingsworth (1969): Hollingsworth, General Manager AEC, to Commissioners and Field Office Managers, October 10, DOE-AEC, SECY 69, Materials 20-V2.
MM. Jordan, L.B. (1960): Senator Len B. Jordan (Idaho) to Seaborg, September 15, DOE-AEC, SECY 69, Materials 20-V2.
NN. Joseph, A.B. (1955): *Radioactive waste disposal practices in the atomic industry: A survey of costs.* Johns Hopkins University, written under AEC contract.
OO. Luedecke (1961): Letter from Luedecke to Hess, January 4, DOE-AEC, SECY 58–66, Materials 12.
PP. Marshall, C.L. (1971): Director Division of Classification, to Bauser, Executive Director JCAE, January 6, DOE-AEC, Materials 20-V3.
QQ. Metlay, D. (1985): "Radioactive waste management policy making," *Managing the nation's commercial high-level radioactive waste.* Washington, D.C.: Office of Technology Assessment (OTA): 199–244.
RR. McCool (1967): Memo from W.B. McCool to File, June 12.
SS. McCool (1968) Memo from W.B. McCool, Commission Secretary, to File, January 22.
TT. Parker, Morgan, and Western (1956): Report of study group g, "Proposals of panel on genetics on limitations of radiation exposure for general population and for occupational personnel," April 7, Records of the Committee on Disposal and Dispersal of Radioactive Waste (part of BEAR Committee), USNAS Archives, Washington, D.C.
UU. Parker, F.L., L. Hemphill and J. Crowell (1958): *Status report on waste disposal in natural salt formations.* September, Oak Ridge National Laboratory, Oak Ridge, Tennessee: ORNL-2560, Health Physics Division.
VV. Pearson, J.B. (1972): "Radioactive waste management: An international environmental challenge," Statement of Senator James B. Pearson (R-Kansas) in the Senate of the United States, February 22, NA-128 JCAE, Box 706, Vol.5.
WW. Pittman (1972): Speech at the American Nuclear Society Meeting, Washington, D.C., November 16, NA-128 JCAE, Box 16, "Speeches."
XX. Rhodes, D.W., J.R. Raymond, and H.V. Clukey (1955): "Operational experience in Hanford liquid waste disposal," *Sanitary Engineering Aspects of the Atomic Energy Industry*, Proceedings of a Seminar Sponsored by the AEC and the Public Health Service, December 6–9, Cincinnati, Ohio.

8 Boundary Maintenance and Radioactive Waste Disposal Technology 263

YY. Safety and Industrial Health Advisory Board (1948): "Report of the Safety and Industrial Health Advisory Board." April, NA, 326–AEC, Lilienthal Office Files, 1946–1950, Box 17, Folder: "Safety and Industrial Health Advisory Board."

ZZ. Samuelson (1969): Samuelson, Governor of Idaho, to Seaborg, September 11.

AAA. Schlesinger (1971): Schlesinger to Docking, November 19, DOE–AEC, SECY 71, Box 7820, "Radioactive waste disposal."

BBB. Seaborg (1965): Seaborg to Seitz, including: "Management of radioactive wastes from the nuclear power industry," February, (1965), November 1, DOE Archives, SECY 58–66, Materials 12.

CCC. Seaborg (1969): Seaborg to Church. October 17, DOE–AEC, SECY 69, Materials 20–V2.

DDD. Seaborg (1970): Seaborg to Church. June 9, DOE–AEC, SECY 69–70, Materials 20–V4.

EEE. Seaborg (1971): Seaborg to Senator Hatfield. May 28, DOE–AEC.

FFF. Seitz (1967): Seitz to Commissioner Tape, August 30.

GGG. Seitz (1968): Seitz to Seaborg, including: "Proposal for Committee on Radioactive Waste Disposal Advisory to the Atomic Energy Commission," February 29, DOE–AEC, SECY 66–68, Materials 20.

HHH. Speth, J.G. (1971): Speth, Chairman of the NRDC, to Schlesinger, Chairman AEC, November 4, DOE–AEC, SECY 71, Box 7820, "Radioactive Waste Disposal."

III. Steele, K.D. (1988): "Hanford's bitter legacy," *Bulletin of Atomic Scientists*. January/February: 17–24.

JJJ. Stetson and Carter (1972): Stetson, Manager SRP, to Jimmy Carter, February 1. Carter to Stetson, March 14, DOE–AEC, SECY 71–72, Materials 20–V6.

KKK. Tape (1967): Tape to Seitz, October 17, DOE–AEC, SECY 66–68, Materials 20.

LLL. USNAS (1957): "Proceedings of the Princeton Conference on Disposal of Radioactive Waste Products," September 10–12, 1955, Princeton University, New Jersey, in: *The Disposal of Radioactive Waste on Land*. Washington, D.C.: U.S. National Academy of Sciences.

MMM. USNAS (1966): *Report to the division of reactor development and technology, United States Atomic Energy Commission*. May, Washington, D.C.: U.S. National Academy of Sciences, Division of Earth Sciences, Committee on Geological Aspects of (Radioactive) Waste Disposal.

NNN. USNAS (1970): *Radioactive waste management: An interim report of the Comittee on radioactive waste management*. February 17, Washington, D.C.: U.S. National Academy of Sciences, Committee on Radioactive Waste Management (CRWM).

OOO. Witkowski, E.J. (1955): "Operational experience in the disposal of radioactive wastes in open pits," *Sanitary Engineering Aspects of the Atomic Energy Industry*, Proceedings of a Seminar Sponsored by the AEC and the Public Health Service, December 6–9, Cincinnati, Ohio.

Chapter 9
Meaningful Boundaries: Symbolic Representations in Heterogeneous Research and Development Projects

Elke Duncker and *Cornelis Disco*

1   Introduction

Research and development, whether in the context of large manufacturing companies, in the engineering faculties of universities, or in scientific research laboratories, is increasingly becoming a collaborative enterprise. This can be put down to ever increasing specialization as well as to the increasingly stringent and diverse requirements which products, prototypes and experimental apparatus must satisfy. So, as more — and more diverse — expertise is required, that expertise itself is chopped up into increasingly smaller disciplinary and sub-disciplinary bits. The upshot is that R&D now tends to be organized in projects which are elaborate concatenations of different scientific disciplines and professionalized practices like engineering, fabrication, marketing, management, jurisprudence, etc.

While such heterogeneous projects may have a central management structure, R&D work on the shop floor evolves chiefly as a conversation among the groups and practices involved. This is because R&D (and *a fortiori* multi-disciplinary types of R&D) is a highly indeterminate search process (Bodewitz et al. 1988; Van den Belt and Rip 1987). Participants must constantly attune their partial search processes to those of others by mutually articulating findings, needs for information, constraints, reports, plans etc. as the project proceeds. Such communication not only has a particular content, it also has a specific form like texts, equations, drawings, and other types of "symbolic representations."

Given the large and obvious differences among disciplines and practices in the way they order their phenomenal worlds and in the "symbolic repertoires" they use to describe and manipulate them, it might seem that such complex communication would be difficult, if not entirely impossible. In fact, as Star and Griesemer (1989) Star (1995a) and Fujimura (1995) have shown and our present research confirms, communication among disciplines and practices is rife with misunderstandings and other pitfalls. This suggests that in order to collaborate effectively the different disciplinary and professional groups must make extra efforts to understand one another. We are not asserting that collaborators have to achieve anything like deep mutual understanding, or that they in fact seek any such thing, but only that they have to be able to understand enough of each others' work to contextualize their own products and to steer the work and resource flows in the collaboration as a whole. As we shall see, superficial "black-boxed" comprehension will often be enough. Nonetheless, even this is not a matter of course in communications among heterogeneous practices and in fact it requires special effort.

These efforts are shaped by the basic properties of symbolic repertoires. On the one hand, symbolic repertoires structure unique universes of meaning and practice and so maintain the ontological and methodological coherence of scientific and technological practices (and the boundaries among them). Thus they help produce and maintain more or less self-referential worlds of meaning and practice. On the other hand, the system of symbolic repertoires taken as a whole exhibits a hierarchical and nested structure based on the differentiation of current specialisms from more generic ancestors and the widespread use of certain basic symbolic infrastructures like Applied Mathematics and graphical conventions. This provides substantial links and overlaps among symbolic repertoires and makes boundaries among practices more permeable than at first sight might seem the case. While ongoing conversations among groups in heterogeneous R&D collaborations therefore have to overcome the apparent incommensurability of different self-referential symbolic universes, they can do this by exploiting various ancestral relationships among contemporary symbolic repertoires — manifest, for example, in the widespread use of Applied Mathematics and Applied Mechanics.

Nonetheless, despite the shared resources of generic and semi-specific repertoires, communication among different technological and scientific practices can be a serious problem. All participants in one of the collaborative projects we have been studying, the Twente micro-optics project, pointed repeatedly to the different "languages" they spoke and to the time it costs to discover the differences and the misunderstandings. The following remark is typical of how participants in this collaboration described this problem:

Respondent: "In this sense you can be easily at cross-purposes. They (another participating group, ED&CD), and the same goes for us too, use concepts which are completely obvious for them, very well defined, and they know exactly what they mean, but the other group doesn't know that at all."
Interviewer: "Couldn't you ask these things?"

Respondent: "Yes, but it must first occur to you, that they mean something different. And that's the problem. It takes time to discover that."

It is quite evident that the depicted problem is due to esoteric parts of two disciplinary repertoires which cannot be satisfactorily linked up, at least not on an ad hoc basis. It costs extra time and effort to realize that this rupture exists and to assemble materials with which to bridge the symbolic divide. A salient point is that insofar as actors can only comprehend each other in terms of their own symbolic repertoires, they will tend to normalize strange expressions so as to fit them into their own repertoires. Native symbolic repertoires form, as it were, the semantic frame through which the other gets understood. Only extraordinary effort — perhaps following on extraordinary failure — can uncover such normalizations and the misunderstandings they hide. In the end, however, the success of R&D collaboration would seem to depend not only on uncovering embedded misunderstandings as in the above example but, as noted above, on instituting an ongoing conversation. For this, routine attitudes and procedures must be developed so that groups can assess the implications of one anothers' activities. This means that R&D success can only be guaranteed if: a) groups can succeed in transforming their own esoteric work into terms comprehensible to the others or, b) groups can learn to transform the esoteric symbolizations of other groups into terms they can understand, that is, transform them into the terms of their own symbolic repertoires.

The upshot is that collaborative R&D projects are not only interactive settings in which actors non-reflexively normalize one anothers' strange expressions into familiar — but often erroneous — terms, but also settings in which reflexive learning takes place and in which rules for transformations and possibly new hybrid repertoires can emerge. To put it generally, collaborative R&D projects are not only arenas in which existing symbolic structures are played out, but also arenas in which new symbolic structure is produced.[1] In terms of the general agency-structure dynamic (Van Lente and Rip, this volume) we can see collaborative projects as interactive settings which can improve the fidelity and quality of their own communications (and those of future projects) by dint of learning processes coupled to achieving practical goals.

Our empirical findings, based on a study of two collaborative research and development projects, suggest that distinct symbolic repertoires and their associated practices can become pragmatically coupled, and hence coordinated, via a number of rather distinct mechanisms. We now simply enumerate them and will describe them in more detail below:

0) Use of very generic (everyday) language, hardly any esoteric symbolic representations. (representational null-option);

1) Coupling by using generic and semi-specific repertoires as a lingua franca;
2) Coupling by passive transformation of esoteric representations. (Listeners' Dictionaries);
3) Coupling by active transformation of esoteric representations. (Speakers' Dictionaries);
4) The development of hybrid repertoires associated with a specific heterogeneous practice, or type of practice.

Two remarks are in order here. The first concerns the conditions under which coupling strategies may be initiated. Briefly, actors can initiate such efforts either as repair or as anticipation. Efforts at repair can only follow on the perception that communication has in fact broken down in whole or in part. This is itself a difficult accomplishment (as the above citation from the Twente project suggests). However, when actors realize that something is symbolically awry, they may attempt to repeat the unsuccessful communication by using one of the modalities from the list above. In this case a specific misunderstanding becomes perceived as an obstacle to continuing coordination, followed by negotiation about the meaning of the symbolic representation. Hence the two repertoires get linked up at a specific symbolic junction. On the other hand, actors can also anticipate symbolic difficulties — on the basis of past experience, for example — and can take what are essentially prophylactic measures against mis-communications, again by utilizing one of the coupling modalities from this list. In this case, potential mis- or non-understandings are anticipated by both speakers and listeners evaluating each others' expertise and their possible familiarity with different symbolic repertoires. Linkages between different repertoires are laid in advance and irrespective of an actual misunderstanding. In practice we will see a mixture of repair and anticipatory measures, including failed attempts at both.

The second remark is that the five coupling and transformation mechanisms may represent a hierarchy of effort — in the order in which we have enumerated them. Hence, as we go down the list, the modalities require an increasing amount of prior effort before they can be applied. Whereas falling back on very generic shared repertoires (which are already in position) doesn't require any extra effort besides its actual performance, the use of dictionaries presupposes prior experience or some education. Mechanisms (0 and 1) which do not require prior efforts are applicable ad hoc; others (2-4) need (planned) efforts, which means that the collaboration may have to span a longer period in order to make this worth the participants' while.

The two empirical cases we have studied differ on at least two main counts: the time perspective (their expected duration) and the disciplinary heterogeneity. Hence, an overall difference can be observed in the types of coupling and transformation mechanisms employed in the two projects. While the Muon g-2 project is formally monodisciplinary and is, moreover, embedded in a context of repeated and continuous collaborations in High Energy Physics over a span of 50 years or more, the Twente micro-optics is nothing if not disciplinarily discontinuous and is only the fifth annual collaborative project supported by a special university research stimulation fund. Examples of ad hoc repair and anticipatory modalities were chiefly found in the Twente micro-optics project, while a reasonably stable hybrid repertoire was visible only in the g-2 project. In general, the rather different contexts and aims of the projects can serve as a preliminary indication of the generalizability of our conclusions, though we obviously cannot make methodologically compelling claims.

The two projects on which the research for this chapter are based can be described as follows: The Twente Micro-optics project is a collaborative R&D project based at the University of Twente in the Netherlands. It is the fourth project in a series of multidisciplinary collaborations specially supported by the university council to encourage interdisciplinary technological and applied science R&D at the university. The explicit aim of the project is to design two prototype devices which exploit the optical non-linearity of certain polymers. The devices are a "blue laser" (based on the non-linear optical (NLO) effect of frequency conversion from red laser light into blue laser light) and a "soliton switch," which is intended as a light-based analogon of the switching function of electronic transistors, using solitons (ultra short light pulses) instead of electrons. Aside from these devices, the program aims to improve the quality of available NLO-active polymers as well as of procedures for thin film deposition of these materials and their use in optical structures. This means that aside from the applied physicists and electrical engineers working specifically on the blue laser and the soliton switch, the project also includes physical and organic chemists involved in calculating, designing, and synthesizing new NLO-active molecules and monolayer configurations.

The Muon g-2 project (see also Disco, this volume) entails the design and construction of a superconducting magnetic storage ring and detector system for use in a high-energy physics experiment aiming to measure a theoretically relevant anomaly in a fundamental parameter of the Muon (a very heavy and unstable electron-like particle). This project is being carried out by a collaboration of physicists and engineers based at universities and research sites around the globe. Although small by current physics standards, the technological challenges of the project are formidable. This is reflected in the duration of the R&D effort. The experiment was approved in 1986 and will hence be almost a decade old by the time this text is published. It is expected to take another two years to complete the instrumentation, with another two years to run the actual experiment. Tasks have been assumed by the various collaborating groups and are coordinated at Brookhaven National Laboratories on Long Island (U.S.) where the experiment will also be run. As the various components become more tightly specified, problems of interference and compatibility have been increasing. Various modes of transformation and coupling of symbolic repertoires play a role, both in the process of specification and in solving interferences and ensuring compatibility of functions and dimensions within the overall goals of the experiment. Heterogeneity of symbolic repertoires here is rooted mostly in the specialized physics and technology which have to be mobilized by different working groups as well as the gap between physics repertoires on the one hand and engineering repertoires on the other.

In Sections 2 through 5 we will use data from these two projects to examine and to illustrate the four coupling and transforming mechanisms (items 1-4 in the list above). In Section 6 we discuss the coordinating role of symbolic representations in hybrid R&D settings as mutable "immutable mobiles" (Latour 1987; 1990) and relate this to the concept of "boundary objects" (Star and Griesemer 1989). We also draw conclusions about the nature of the coupling and transforming mechanisms and we consider whether they are a simple taxonomy or evidence for a secular process of innovation in symbolic repertoires taking place over decades.

2 The Use of Generic Semi-specific Repertoires

One way the different groups involved in collaborative R&D projects can carry on their coordinative conversation — in spite of differences in their specific symbolic repertoires — is to revert to a commonly shared more generic repertoire. This is a fairly universal strategy, one for example that travelers employ when they revert to gestures to make themselves understood by natives in spite of an inability to speak the local language. In R&D contexts, useful generic repertoires are natural languages and, particularly, forms of applied mathematics. The charm of such linguae franca is that they are available to a wide variety of scientists as a pre-existing common symbolic "infrastructure." No extra effort has to be expended to produce them — they are already part of the symbolic *habitus* of every scientist — and they can readily be mobilized to mediate, more or less transparently, among a variety of specific repertoires. In this section we shall show how such semi-specific repertoires were used in the early stages of the Twente micro-optics project to establish a common definition of the R&D effort.

9 Meaningful Boundaries

The basic strategy in using semi-specific repertoires to close communicative gaps can be described as "symmetrical transformation." For example, in a conversation, a chemist utters X (using chemical symbols) and a physicist utters Y (using physics symbols). Neither understands what the other means. Then, the chemist says that (for specified conditions) $X=p$, where p is some generic mathematical expression, while the physicist says that (for specified conditions) $Y=q$, where q is some other generic mathematical expression. Subsequently a conversation can ensue about the differences and similarities between p and q (because both the chemist and the physicist are at home in applied mathematics) from which symmetrical conclusions can be drawn about the meaning of X and Y.

One striking form of the use of semi-specific repertoires to maintain conversation was repeatedly observed at the Twente project's twice-yearly general meetings. During the breaks, small groups of participants would typically stand about discussing some particular point of a given presentation. In the course of these discussions, questions would arise about particular physics concepts. As it was apparently very confusing to talk about complex concepts without seeing them represented in some way or other, the conversation sometimes stagnated. Discomfort was expressed by non-verbal signs like knitting the brows or by asking the same question repeatedly, both clear indications that something was going wrong. In such situations somebody invariably went to the blackboard (and the others followed her) to write down some pertinent mathematical formula. Subsequently, everybody participating in this discussion referred to the formula on the blackboard, sometimes by pointing at relevant terms, sometimes by writing down modifications or transformations or by providing a similar formula from their own field.

In general, lapses in interdisciplinary conversations were repaired not only by means of mathematical representations, but also by the use of other symbolic representations like graphs or technical drawings and even sketches. However, certainly in the beginning stages of the Twente collaboration, the repertoire of applied mathematics was used by far the most often, not only in ad hoc repair work but also as an anticipatory strategy for preventing potential misunderstandings. The standardized symbolic representation proper to applied mathematical formulae was familiar to all scientists involved in the Twente project, because Applied Mathematics plays a major role in three of the four participating disciplines (Electrical Engineering, Physics, and Chemical Physics) and is embedded in the academic curricula of all four disciplines (thus also in Chemistry). At the very outset of the project, therefore, the following formula detailing non-linearity for optical phenomena was accessible to all participants and became a way for them to express both their common goal and, to some extent, a division of labor:

$$\vec{\mu} = \vec{\mu}_o + \alpha \cdot F + \beta : FF + \gamma :: FFF$$

In the context of the Twente micro-optics project it describes the several components of optical properties a substance can have. α is the first order (or linear) optical coefficient, β the second order (non-linear) optical coefficient and γ the third order (non-linear) optical coefficient. Substances can have optical properties of all orders, but the non-linear part is normally extremely small. Normal glass, for instance, has a high linear coefficient, but there is special glass which becomes opaque in the sun and transparent in the shade, which is a non-linear optical effect.

Figure 1: Equation for Non-linearity

The formula was transferable from one practice to another, because it was highly standardized. However, in spite of the standardization, it became obvious that mathematical expressions could also take on specific meanings in different contexts. For instance the coefficient β in Figure 1 was named differently in different contexts. The formula as it is written down above was the formula used by the chemical physicists. Their expression for the second order non-linear optical coefficient was β, which was related to one single molecule. As soon as the emphasis was on a group of molecules, the second order non-linear coefficient was called $\chi^{(2)}$. In order to synthesize good materials the organic chemists had to look for both the β of a single molecule and the $\chi^{(2)}$ of the substance.[2] The applied physicists made thin layers of this material and measured its non-linear optical properties. Under certain conditions the general second order coefficient $\chi^{(2)}$ was replaced by a more specific term d.[3] In the context of the specific application pursued by the applied physicists, the term d was used as a synonym of the second order coefficient. The component d_{33}, in particular — sometimes accompanied by d_{31} — was used as a relevant representative of d. While the overall structure of the formula remained the same, different actors imputed different interpretations to the same (syntactic) position in that structure. This was possible because of the intimate juxtaposition of the interpretative flexibility of the mathematical formulae with their standardized and robust meaning across boundaries. This produced a mutually accessible mode of representation and thus facilitated communication and coordination among the heterogeneous practices of the collaboration.

Using mathematical formulae as a shared system of representation not only allowed the work of the moment to proceed, it also stimulated mutual learning among the groups. The shared symbolic infrastructure of mathematical formulae provided a medium for interpreting the exotic symbolic representations of other groups in terms of native symbolizations or even for speaking to other groups in terms of their exotic symbolic repertoires. In this very process of reverting to semi-specific repertoires, actors learn to couple formerly distinctive symbolic repertoires into something of a shared cognitive structure. In other words, the use of semi-specific repertoires is also the basis for long term evolving common culture, and not merely a resource for ad hoc repair work. This entails the construction of what we call "dictionaries," that is, lists of more or less explicit instructions for transforming strange expressions cast in the framework of exotic symbolic repertoires into familiar expressions in the native symbolic repertoire — and vice-versa.[4]

These dictionaries are ongoingly constructed as a by-product of successful attempts at ad hoc couplings between specific repertoires, for example, via semi-specific repertoires. Whereas using semi-specific infrastructural repertoires as a lingua franca does not require prior investments by collaborators and can be applied ad hoc (the shared infrastructural repertoire is already programmed into the actors and is on-line as it were) this is not so with the use of dictionaries. Their use at any point in time is predicated on their having been built up in the prior course of the collaboration itself. Hence the use of dictionaries cannot be ad hoc in the sense that mobilizing semi-specific or generic repertoires generally is. To have dictionaries available, mutual learning must first take place in the context of anticipating and repairing potential and actual misunderstandings. Only then can dictionaries actually be used.

3 Coupling and Transformation via Passive Dictionary, Agency of the Listener

Reverting to semi-specific repertoires is a symmetrical way of dealing with mutual incomprehension, whether as ad hoc repair work or as long term anticipation. Parties communicate by transforming statements in their own "native" symbolic repertoires into the generic lingua franca and back again. This entails both passive and active transformation in a concurrent mode. Two kinds of assymetric strategies are common. Here the emphasis is either on passive transformation (I turn the statements made by you, which are strange to me, into ones that are familiar to me, but again strange to you) or on active transformation (I turn my familiar statements into what are for me strange statements but for you familiar ones). These strategies both rely on and further augment the dictionaries which actors have acquired as a result of special study or, more usually, in the course of maintaining the interdisciplinary conversations essential to heterogeneous R&D projects. In other words, because they do not revert to shared semi-specific repertoires, neither the ad hoc tactics of repair and anticipation nor the making of dictionary entries can rely on the prior competencies of the actors (like, for example, facility in applied mathematics); instead, transformations and dictionaries emerge in attempts at communication themselves.

It should be emphasized that the transformations are not, generally speaking, deep ones, i.e. the actors do not actually adopt or even comprehend one anothers' repertoires in any profound sense. What we are pointing to here are pragmatic strategies that actors use to get on with the work — in spite of very diverse cognitive frameworks and symbolic repertoires. Getting on with the work does not often depend on profound mutual comprehension; more usually it depends only on being able to assess what consequences the demands and actions of others will have for your own specific task. For this, relatively superficial routinized exchanges of symbolic black boxes is often sufficient. We speak of "black boxes" here because the listener or the speaker does not have to comprehend the full complexity (i.e. the complexity available to native speakers) of the representations he or she is hearing or speaking. It suffices to be able to understand or express those features which are relevant for the work at hand — the rest may conveniently be left in obscurity. The motto here is: "Don't bother me with all your details and I won't bother you with all of mine." Hence, there is communication of salient features rather than anything like full mutual comprehension.

These sorts of communications strategies can be illustrated by exchanges between engineers and physicists in the context of the muon g–2 experiment. Experimental HEP is a form of science which depends heavily on the in-house design of experimental apparatus (Disco, this volume). In order to accomplish the necessary design processes physicists must rely on engineers, not only because, as the "resident spokesman" of the g–2 collaboration put it, "there simply aren't enough of us," but also because (as he didn't say) physicists are rarely well enough versed in engineering to produce trustworthy designs unaided. On the other hand, neither are engineers, at least from a physics perspective. The machines in question have to produce physics data and getting them to do so requires constant attention to design details by physicists themselves. Although engineers can suggest alternative designs, only physicists are in a position to assess their physics effects. The upshot is an ongoing conversation between physicists and engineers about ways to achieve pragmatic designs for what is ultimately to be a machine for the production of physics data. This conversation often requires more than ordinary effort inasmuch as the symbolic repertoires of engineers and physicists, at least as disciplinary repertoires, are largely disjunct.

Of course, the practice of R&D in the context of HEP is itself a lifelong education for the engineers and physicists involved; in particular, engineers and physicists educate one another — at least to some extent — in the use of their respective symbolic repertoires. While engineers may not (and need not) develop facility in reading Feynman diagrams and while physicists may not (and need not) develop facility in controlling and compensating for welding stresses and deformation; engineers do learn to interpret representations of particles, of fields, of beamlines and physicists do learn to read mechanical drawings and think about stresses and dimensions. So there has emerged a shared pragmatic symbolic repertoire which provides resources for the ad hoc repair and anticipation which always remain necessary. In other words, HEP already has rich dictionaries mediating between physics and engineering — at least on a pragmatic "black-box" level. Any observation of the ongoing conversation reveals the existence of this dictionary and the fluency with which well-socialized participants can use it. We return to this point in our conclusions.

Our example of "passive dictionaries" is taken from observations of a "Workshop on Vacuum Chamber and Subsystems" held at Brookhaven on January 14, 1993. A major portion of the workshop was devoted to resolving a spatial conflict in the storage ring vacuum chamber, a slender toroid with a rectangular cross-section of 15x20cm and a diameter of 14 meters. The two subsystems were a magnetic-field measurement trolley and a system of "pick-up electrodes." The trolley had to be able to circumnavigate the storage ring while providing data on the magnetic field and hence required an "arm" for purposes of traction and communication. The "pick-up" electrodes were to provide data on the location and composition of the bunches of pions/muons orbiting the storage ring. The conflict was about the dimensions and strength of the trolley arm vs. the surface area and sensitivity of the pickup electrodes. It appeared that the stronger and thus bigger the arm, the smaller and less sensitive the pickup electrodes could be. In order to bargain in this zero-sum conflict, physicists and engineers had to be able to deconstruct one another's claims. This tactical interest, as we shall see, demanded transformations of symbolic representations by the listeners; i.e. the mobilization of "passive dictionaries."

The workshop proceedings were dominated by the presentation of mechanical drawings in the form of overhead sheets and by verbal exchanges about them. Figure 2 shows a drawing made by the engineer of the group responsible for the trolley (Yale). The circle in the center represents the cross-section of the trolley. Attached at the upper left corner is the arm leading to a rack and pinion drive and slip rings. This design for the arm is constrained at this point in time only by the (hatched) rail/electrode, which has resulted in a rather thin cross-section at the extremity of the arm. Toward the trolley, however, the arm becomes quite thick and mechanically robust — at least in the Yale version. Figure 3 is another drawing shown later on in the workshop by the Brookhaven project coordinator. It is similar but now includes the pickup-electrodes (the eight elements shown as heavy

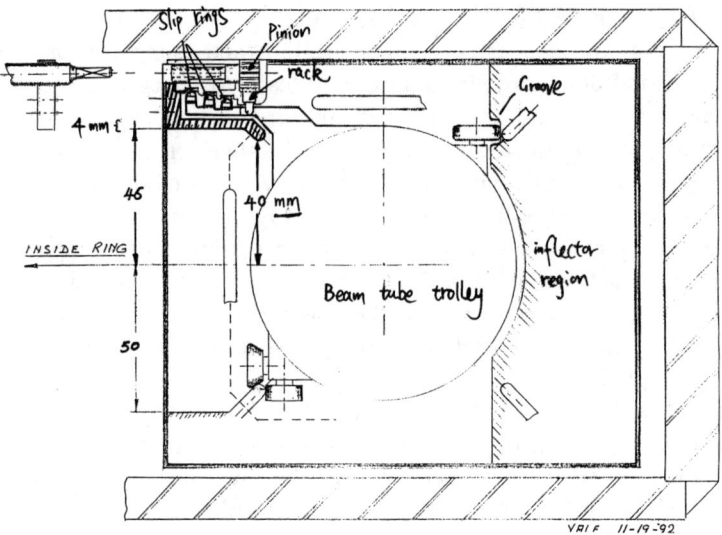

Figure 2: Cross-section of Beam Tube Trolley Showing Rack and Pinion Drive and Slip-rings (Yale Design)

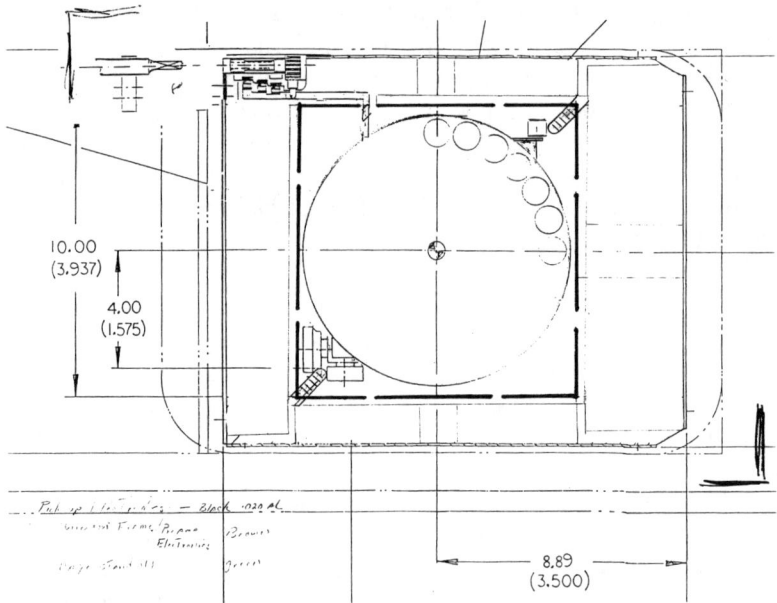

Figure 3: Cross-section of Beam chamber Showing Pick-up Electrodes and Trolley Drive Arm and Slip-ring Assembly (Brookhaven Design)

9 Meaningful Boundaries 277

lines and configured in a square) and shows the consequences of their geometry for the trolley arm. Not only is the arm now much thinner (and thus weaker) but also the dimensions of the rack and pinion and slip rings have been reduced so as to clear the pick–up electrodes.[5] Clearly, the meaning of Figure 3 from the Yale group's standpoint was to decrease their design margins considerably and in fact to imperil the mechanical integrity of the trolley arm. Their only defense lay in a deconstruction of the lines laid down by the Brookhaven group. But how much physics did they have to know in order to do this? As it turns out, not very much, but they did need an adequate dictionary.

In the protocol of the ensuing conversation, physicists are indicated in boldface, engineers in normal type, and a Brookhaven project coordinator in italics. YS is the Brookhaven physicist responsible for the "pickup electrodes." He wants to maximize their "azimuthal length" in order to have a maximum signal. This conflicts with the trolley arm that has to pass through the space potentially available to the pickup electrodes. XF is the Yale physicist formally responsible for the trolley and he is trying to nudge YS off his absolutist standpoint in order to gain more room for the arm. He is supported by one of the project mechanical engineers, JP, who, almost as a professional reflex, wants to know what the percentages really are, i.e. what are the allowable deviations from the ideal design.

JP: What test has been done on the pickup electrodes, or has the...
RL: Well, we're in the process of making the first...
JP: Yeah, but what I want to know is what's defined the shape so far and what's limiting us that we can't do what Andy is saying, open the gap or whatever we want to do.
YS: OK, what we are going to try to do is to set up, ah, a square of pickup electrodes, but, ah, you have to keep in mind this circle...
Voice: That's the parameter.
YS: That's the parameter, you are right. You have to, ah, every plate has to have the same solid angle... to some degree. Now..
GB: Is that really a solid angle? Technically...
YS: It doesn't ah, you can call it, ah, azimuthal length, right.
XF: So then, say, right now the gap is 6 mm. If you expand that then that means some part will be shortened. Then, say, what kind of error, you, uh, would introduce... If tolerable then you can expand that to 1 centimeter...
YS: Right, if you go to ah 4 centimeters it starts becoming a little, ah.... {silence}
JP: This is getting back to the point, ah, you've done some type of study or something you've done; you've run a model or... to do those percentages? "Cause I'm always looking for the percentages to how much I can deviate from the ideal design.
YS: If I make it half, I have half the signal. That's all I can say.
JP: OK, so we'll live with 90%, what do you think?
YS: OK, I mean... but you know how it is. It's a give and take game. So we won't give unless...
Voice: They give, you'll take, right?

YS is claiming a cognitively privileged position. He is using his authority as a physicist to lay claim to scarce space in the beam chamber for the subsystem which is his special responsibility. It is not clear on what his claim is actually based. The physicist XF can and does challenge his claim by suggesting he can in fact calculate the measurement error corresponding to specific dimensional compromises, implying that then other physicists can also participate in the evaluation and assess YS's real needs.

The engineer JP, on the other hand, will accept a black-boxed version of the outcome — after all only the physicists can decide what level of signal reduction is acceptable given the goals of the experiment, so that for JP a specific attack on YS' conclusions is not a real option. But he must design a beam chamber and so he needs a workable result. As a matter of course (engineers know that criteria are *always* negotiable) he does not believe YS has no room to manoeuver and tries for a feasibly small though utterly arbitrary reduction in length, to 90%. JP will be satisfied and can carry on if YS can grant that reduction — whatever the possibly arcane physics justifications might be.

The discussion refers to signal-noise ratios whose complexities are hidden behind YS's strategic positioning and which are symbolically represented only as lines on the drawing which indicate that surface area is being maximized. (and in the verbal statement that signal level is directly proportional to surface area). So what's happening? There is a symbolic representation of the beam chamber on the overhead. Both parties are perceiving this as a two-dimensional cross-sectional representation of a three-dimensional configuration. This is shared symbolic repertoire which provides the conceptual turf on which the conflict is fought out.

YS presents the pickup electrodes as a symmetric constellation of surfaces designed to measure particle impacts and thus to generate knowledge about the beam during experimental runs. He provides some "parameters" for the pickup electrodes (equi-azimuthal; signal proportional to surface area, hence to the sum of the length of lines on the sheet) and engineers are being encouraged to perceive these lines on the overhead as representations of specific spatial constraints — which as it turns out limits their freedom to optimize the trolley arm from a mechanical engineering perspective. Behind these lines YS has mustered all kinds of esoteric calculations culminating in judgements about surface area and signal levels. He presents this whole package as the legitimate symbolic deconstruction of these lines and in particular as constraints on proposed deviations. However, engineers' symbolic repertoire does not enable them to conceptualize (or challenge) the specific physics constraints imposed by these pickup electrodes. They are at the mercy of YS unless they can transform their perception of the lines into something that will save the trolley arm.

The engineers see the lines on the drawing as mechanical obstacles and in order to eliminate them as such have no other recourse than to transform YS's symbolic construction of these lines on the overhead into their own generic claims about electrodes and signal-to noise ratios in general. The overall movement is that what looks like exactly the same symbolic representation, i.e. the cross-section drawing, in fact represents two different objects (this is because the calculations by YS are hidden and are available only as constraining outcomes). So YS posits one representation and the engineers — via a dictionary rule that says all signals can always be weaker, hence surface areas can always be smaller (hence cross sections also) — transform the determinate lines into variable ones. So engineers come to see what elements YS presupposes in the lines, but actively and strategically try to transform the symbolic representation into a representation of something else — something they can at least work with by "playing the percentages." So this is not an example of understanding strange repertoires as a perfect mirroring, i.e. with the aim of "taking the role of the other," but in the process of transforming the strange into the familiar also reinterpreting it so that the other does not block your own progress.

What kind of dictionary might we say that JP is employing here to transform YS's physics requirements into constraints for his mechanical engineering specifications? It can be assumed that JP has only a shadowy knowledge of the role of these particular pickup electrodes in the physics performance of the overall detector. Nonetheless, he can transform these particular pickup electrodes into the more generic category of particle detection devices about which he knows enough to be able to bargain with YS over surface area on these particular ones. Hence, on analogy with many types of particle and radiation detection devices he can easily imagine a linear relationship between signal strength and surface area. He also knows that reduced signal strength always means that some information gets lost in background noise. Such is the way of all detection devices, and engineers, particularly those who have been working in HEP for a few years, know this as part of their own professional lore. What he also knows is that what is enough signal and what is too much noise is really a matter of convenience; those who have to work with the output from detectors always want the maximum (so they can be lazier on other counts), but very often they can make do with less. JP's transformations of these particular pickup electrodes (which YS wants to make as big as possible and about which JP as such knows next to nothing) into generic radiation detection devices (about which JP knows a lot) enables JP to bargain with YS *as if he knew about these particular pickup electrodes* and even to suggest an arbitrary reduction in size. The gambit works, among other reasons because the physicist *XF* backs him up in arguing that signal strength is a negotiable, not an absolute, quantity; the effects of deviations from the ideal can be "calculated" (and presumbably compensated).

The dictionary entry being used and reinforced in this case has a different structure than those based on recourse to a shared semi-specific repertoire. Instead of the transitive equivalences typical for the latter ($X=p$, $Y=p$, hence $X=Y$, where p is, for example, some generic mathematical expression) actors in this case exploit the relationship of member to a class: W is a member of class Z; what holds for Z thus holds for W. So in order to shorten the lines on the drawing which were presented as necessary outcomes of esoteric calculations, the Yale group and a Brookhaven engineer argued that whatever the specifics of the matter, electrodes generically could always make do with less gain than that desired by their users. In effect, therefore, the dictionary entry could be read as saying: "Detectors and electrodes can always be made less sensitive (and their properties modified accordingly)." Any specific detector can thus be made less sensitive, (and its properties modified accordingly). The strategic implication of the dictionary entry was therefore: "Always challenge those who claim that their detectors cannot be made less sensitive. It's a bluff."

4 Coupling and Transformation via Active Dictionary, Agency of the Speaker

Dictionaries are not only a basis for understanding strange repertoires, they are also used to transform "native" expressions into the repertoire of "exotic" audiences. This requires the construction of an identity for the audience and a partial modification of the speaker's own identity. In the Twente micro-optics project the groups judged their audiences in terms of a set of "threshold competencies" and accordingly re-structured their own presentations. Knowledge of "high school mathematics," "high school chemistry," or university level "Applied Mathematics" provided not only the basis for a lingua franca, it also provided criteria for constructing audiences in terms of their symbolic competencies as well as for modifying native symbolic identities.

The Twente physicists, for instance, normally constructed their symbolic identity in terms of the repertoire of applied physics. This meant that among themselves they communicated primarily in terms of mathematical formulae, supplemented by graphs, diagrams, and drawings. Doing physics, and non-linear optics in particular, was literally unthinkable for them without using mathematical expressions. Their meetings were saturated with mathematical expressions (even when they tried their best to avoid them). Graphs and dia-

9 Meaningful Boundaries

grams were used specifically to exhibit results of measurements and computer simulations. Drawings were used for communication with the machine shop and for documentary purposes.

However, the symbolic repertoire used by the physicists when representing their work to the other partners in the project was distinctly different. Only one complete mathematical formula was used and this appeared on overheads at all the meetings, supplemented by some names of coefficients and other fragments pertaining to mathematical formulae. In addition, they used a variety of symbolic representations, beginning with pictures, two-dimensional drawings, graphs, diagrams and tables, all of which were descriptive rather than analytical. Most strikingly, though, from the very beginning they used "exotic" representations — especially several kinds of molecular images and steric formulae — which are symbolic representations native to Chemistry and quite alien to normal physics.

The physicists had acquired their facility with chemical representations in a series of seminars with the physical chemists prior to the start of the micro-optics project. In these seminars they learned to use systems of representation for the spatial configuration of molecules: both two-dimensional images and three-dimensional models of molecules. In the absence of a concrete model or representation, they would resort to fingers and hands to portray the spatial structure of a given molecule. As a consequence they were able to express the import of some of their own work in terms of chemical symbolic representations as early as the first general meeting of the collaboration.

The following figure is (a part of) an overhead sheet accompanying the presentation of a senior member of the physics department during the third meeting of the collaboration. It shows the typical mixture of chemical and physical symbolic representations being used by the physicists at that time.

The handwritten title "4-nitro-calix[4]arene" and the sub-title in parentheses "(cone conformation)" are representations native to chemistry. The same goes for the handwritten remark above the graph "calix/chloroform". These terms are absent in normal physical repertoires. The remark "Spontaneous Raman" refers to a measurement method used in Twente by both the physicists and the chemists. The plot in fact shows the results of this measurement. The handwritten arrows point at the interesting peaks of the graph in support of the explanation given during the presentation.

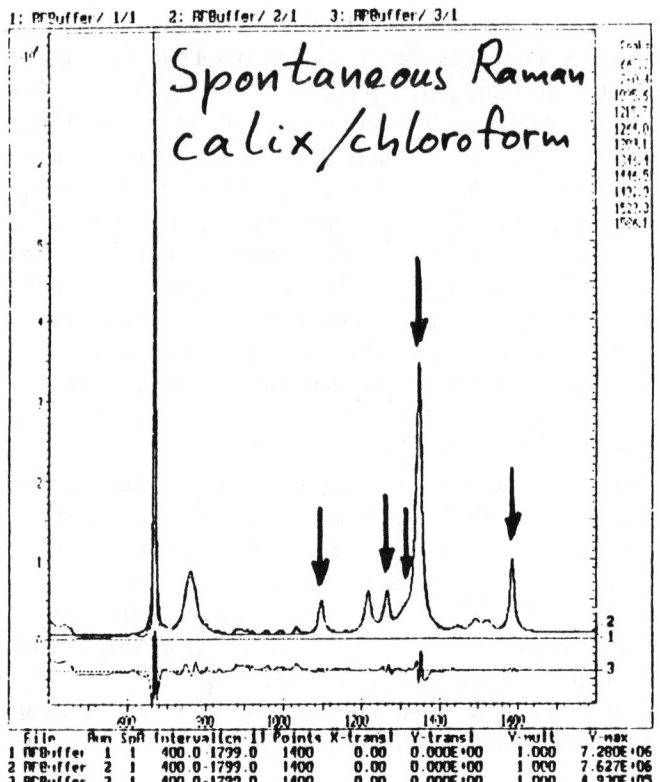

Figure 4: Mixed Symbolic Representations, Physicist, Project Meeting 3

While the physicists were clearly constructing their collaboration audience as interested organic chemists and hence dabbling in "exotic" symbolic representations native to chemistry, the same cannot be said of the organic chemists themselves — at least not at the outset of the collaboration. The latter persisted in using exactly what they used in their own labs: a large variety of native chemical representations of molecules, such as two-dimensional and three-dimensional images and various kinds of steric formulae. Their laboratory exhaust hoods, which functioned as a kind of impromptu blackboards, were covered with just such steric formulae.

By the third general meeting, however, the senior member of the organic chemistry department had begun to use symbolic representations native to physics. His presentation was formulated entirely in terms of mathematical formulae describing the physical concepts of non-linear properties as they were used by the Applied Physicists. Figure 5 is the first overhead sheet of this presentation:

The first and the second formulae show the two forms of the differential equations already discussed in Section 2 of this chapter. The coefficients are called α, β, and γ on the microscopic level and $\chi^{(1)}$, $\chi^{(2)}$, and $\chi^{(3)}$ on the macroscopic level. After some calculations and the application of further conditions, he arrives at the two lines of mathematical formula. The rest of the presentation is in the same style, one quite unusual among organic chemists at Twente prior to this meeting and indicative of an emergent ability to anticipate misunderstandings by actively transforming native findings into exotic representations aimed at a particular audience.

OSF Monday April, 11th 1994

A view from an organic chemist

microscopic:

pind = alpha·E + beta·E2 + gamma·E3 + ···

macrocopic:

Pol = chi1·E + chi2·E2 + chi3·E3 + ···

A simple model gives:

CHI2 = N · f · Beta · <cos3 theta>

$$<\cos^3 \text{theta}> = \frac{mu \cdot E}{m \cdot k \cdot T}$$

No dipole-dipole interactions in these models

Figure 5: Mixed Symbolic Representations, Chemist, Project Meeting 3

Another indication for the mutual appropriation of strange symbolic representations and the use of dictionaries can be found in the shared publications of the two departments. Several co-authored articles were published by both departments. In these articles the physicists describe substances with non-linear optical properties by means of symbolic representations proper to chemistry, whereas the chemists elaborate on the physical concept of non-linear optical properties by means of mathematical formulae typical of physics.

Two main strategies are visible in these examples:
1) The avoidance of esoteric symbolizations which will not be understood by the audience;
2) the partial appropriation of the "exotic" repertoires of other groups so as to speak to them in their own terminology.

In the first case the groups constructed their audience by anticipating which symbolic representations were likely to be misunderstood. Subsequently they modified their identities in order to avoid these particular symbolic representations. In the second case they also domesticated or appropriated "exotic" repertoires by taking over representations, which were typical for the other group. Such facility in the definition of an audience and in the symbolic adaptation of the self depended on such mutual learning as occurred in the process of cooperation itself, or at least in directly related activities like seminars etc.

The dictionaries being used and constructed here are more complex and specific than in the case of mutual reversion to semi-specific repertoires or in the case of passive transformations. This is because making meaningful statements in a strange repertoire generally requires more detailed mastery of its syntax and semantics than simply transforming strange utterances into meaningful ones in one's native repertoire. The situation is analogous to the difference between active and passive mastery of a foreign language. One would expect this increased complexity and difficulty to limit the scope of "active" transformations to collaborative settings of long duration or in which extra efforts have been made to expedite mutual learning — as was indeed the case in the Twente collaboration.

5 Hybrid Repertoires

Here we return to a point raised in Section 4 about emergent shared repertoires. Given enough time, recourse to the various transformation strategies outlined above (which inevitably involves mutual learning) may result in new and more encompassing, "hybrid repertoires." These can provide a solid basis for the coordination of longer-term hybrid practices.

The following example shows that the coherent practice of the design and construction of large detectors for HEP experiments in fact depends on the fluency of all participants in a hybrid symbolic repertoire which obviously antedates the g-2 collaboration itself (although the mutual learning taking place within g-2 also augments this fluency). This hybrid repertoire supersedes neither the repertoire of high-energy physics nor that of the various engineering disciplines involved. The hybrid repertoire is a means of getting on with the practice of defining physics requirements in terms of engineering specifications, but it cannot fully define the meaning of the experiment, either as a piece of physics or as a piece of engineering. The hybrid repertoire — as a standardized set of routine and symmetric transformations — has developed organically over time as a by-product of the more or less ad hoc solution of representational hassles within the particular activity of engineering design in a HEP context; but it itself is neither physics nor engineering. If anything, the hybrid practice (and the associated symbolic repertoire) could be described as the intersection of physics and engineering for the specific purpose of designing large detectors.[6]

The example is taken from the same meeting as was discussed in Section 3. Again, the physicist YS is the major actor, though now not in the role of direct contender for space in the beam chamber; here he is the responsible person for another subsystem, the so-called "electrostatic quadrupoles." His job now is to report on simulations he has done which show the electrostatic consequences of changes in the physical configuration of certain elements in the beam chamber. The elements in question are rails along which the NMR beam tube trolley is to travel. Because they intrude into the beam chamber they also act as elements in the electrostatic field and have in fact been incorporated into the current design for the electrostatic quadrupoles. As a result of Yale's dimensional requirements which were also at issue in the earlier exchanges, it has become desirable to move the support for one of these rails down by 4mm (in order to get more room for the trolley drive arm and slip-ring contacts). The question being put to YS is whether this will compromise the quality of the electrostatic field (whose uniformity is crucial for other physics reasons).

The issue can be settled to the satisfaction of both engineers and physicists thanks to the existence of a hybrid symbolic repertoire which allows them just sufficient insight into one anothers' particular concerns to get on with the business of designing the detector. Aside from a large collection of shared working concepts like "field," "particle," "6061 Aluminum," "shimming," "TIG welding," "vacuum feed-throughs," etc. the hybrid repertoire which has emerged out of the practice of experimental HEP over the past few decades depends heavily on mutually comprehensible diagrams and drawings. Henderson (1991) and Ferguson (1992) have shown how important visualizations are in engineering generally.

In this case, two classes of visualizations are used to resolve the problem. The first are traditional mechanical drawings made at the drafting-table. The second are computer-generated plots showing lines of potential in the electrostatic field for different configurations of rails. All the visualizations represent cross-sections of the beam chamber. The first, Figure 2, is the

Yale drawing featuring the trolley arm and the rack and pinion/slip ring assembly that also played a role in the struggle about the pickup electrodes. Note that the 4mm downward shift of the support for the upper left rail is emphasized. This is the drawing that sets the agenda for the configuration debate because it proposes changes in the existing quadrupole configuration. YS subsequently presents a second mechanical drawing, Figure 6, exhibiting that configuration in unmodified form.

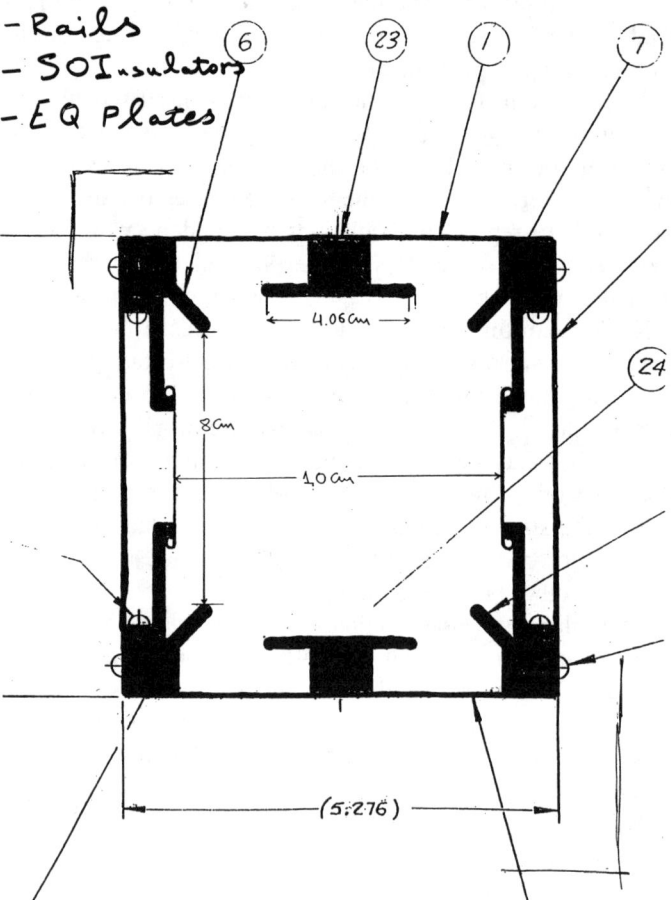

Figure 6: Ground Poles/Trolley Rails, YS's Configuration

Figure 6 could be seen by all participants as a basic statement of the ideal spatial configuration from the viewpoint of the electrostatic quadrupoles, that is, a design taking into account only the requirements of the quadrupole field itself and blissfully ignorant of impinging subsystems. It is a statement within a hybrid repertoire: it expresses physics parameters in the form of a mechanical drawing — a symbolic medium belonging to mechanical engineering.

While high energy physicists, through long experience, have learned to express physics requirements in terms of mechanical drawings, their engineers, through equally long experience, have learned to understand at least some typical physics representations in terms of their mechanical and spatial consequences. In some cases, both acquired aptitudes are exercised in the same visualization, i.e. both physics and engineering representations overlap in the same diagram. During the beam chamber meeting, subsequent to the projection of Figure 6, YS proceeds to project a series of figures which superimpose a physicists' representation of the electrostatic field on an engineers' representation of the cross-section of the beam chamber. The physicists' representation is the result of a so-called "Poisson Simulation" which is able to simulate the configuration of a field on the basis of pole shapes and the configuration of other susceptible matter in the space of the field. There are dedicated computer programs for running such simulations and producing plots which show the (in this case, electrostatic) "lines of force" in the field. This allows physicists to vary the composition of elements at will and to assess the results for field shape and uniformity. Figure 7 shows such a plot for the "Original, Symmetric Design." It will be seen that all the features of the beam chamber cross-section have been eliminated except for the electrostatic quadrupoles and the corner rails. In the Poisson Plot these features are basically points in the Cartesian space of the field, although the fact that they are, as it were, "aluminum points" makes them salient for the field's shape. In the original mechanical drawing (e.g., Figure 2) the same features are constructions which have to be machined, supported, and which stand in mechanical and spatial relations to many other objects in the beam chamber. Nonetheless, the two types of drawings connect unambiguously at the electrodes and the rails. Configurations that have to be changed because of mechanical considerations can be represented in the first type of drawing while the physics effects on the field can be assessed in a computer plot which overlaps in Cartesian space in the second type of drawing.

Physicists are quite at home with Poisson simulations and find computer plots of various kinds of fields utterly transparent. Engineers with experience in HEP have grown used to looking at them and reading them after a fashion, even though such plots were never part of their professional training. The drawings as such are analogous to many more generic and familiar kinds of plots like like contour lines on geophysical maps or isobars and isotherms on weather maps. Various kinds of analysis in engineering, for example stress-analysis, also use similar types of visualizations. So there is

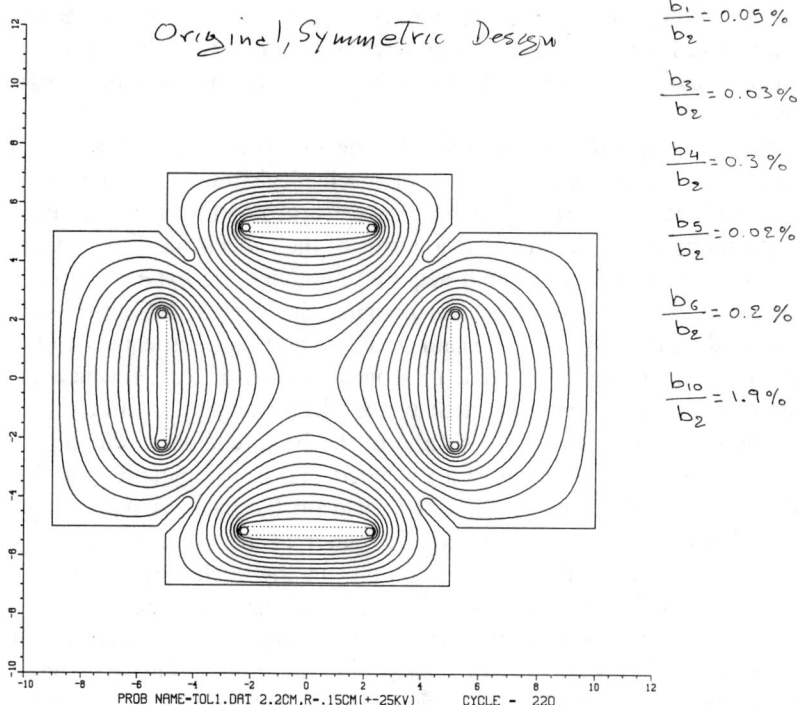

Figure 7: Poisson Plot Electrostatic Field, "Original, Symmetric Design"

background experience in reading such plots and in assessing such matters as symmetry, uniformity of gradients, and the overall shape of the patterns evoked. On this basis, it is possible for such plots to work as part of the hybrid symbolic repertoire native to the design of HEP apparatus.

Engineers need not formally, of course, read such plots with an eye to evaluating whether the physics effects of specific configurations are acceptable. That is, they do not have to understand the details or the critical parameters involved in such simulations. That is clearly the province of physicists. But in the culture of HEP design, openness and discursive inclusion are functionally important and highly valued, even if only at the superficial and pragmatic level of the hybrid repertoire. So YS presents these plots to his mixed audience of physicists and engineers as in part intuitive visualizations which everyone can "read." This is the style in which he presents Figure 8, showing the field "with corner rail base dropped 4mm" and everyone can see that in the region of the beam (the space between the quadrupoles) there is next to no visible effect. YS's confirmation that in his judgment as an electrostatic field expert this is indeed the case also confirms what everyone can (intuitively or otherwise) see for themselves.

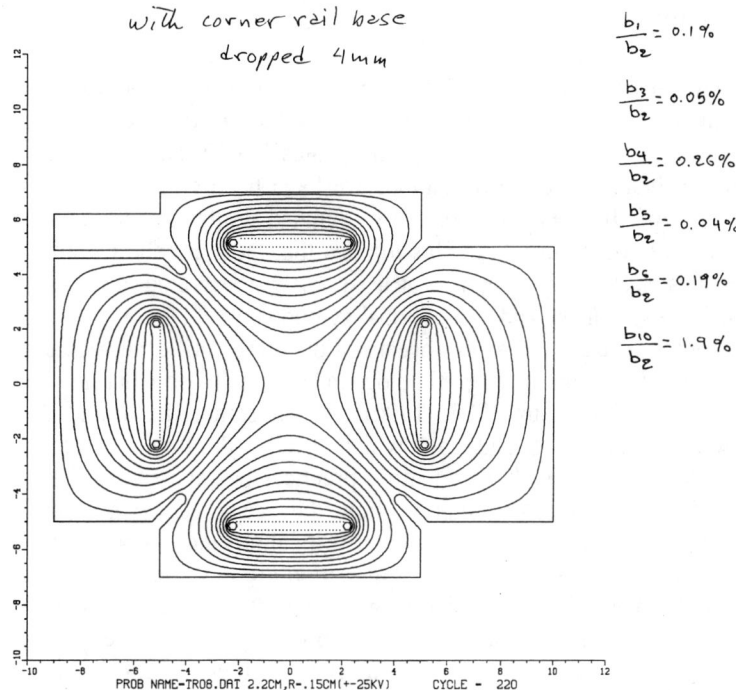

Figure 8: Poisson Plot Electrostatic Field, "With Corner Rail Base Dropped 4 mm"

On the other hand, there are physicists in the audience who have to be enrolled into YS's interpretation that the field is insensitive to the new mechanical requirements. Visual images, though suggestive, are apparently not enough here. For this reason Figures 7 and 8 also present a number of quantitative results of the Poisson simulations (the various ratios to the right of the plot). Because none of the engineers present, and probably not even all the physicists, are aware of the allowable tolerances (or perhaps what symbolic conventions are being followed) YS feels obliged to explain the meaning of the ratios. We may see this as constructing a temporary entry in the hybrid representational dictionary by means of which other physicists (and even engineers) can participate in the judgment that everything will be OK. In this particular case the judgment is sanctioned by the "resident spokesman" who intervenes in explaining the meaning of the numbers for the benefit of those present.

6 Conclusions

While we have described communicative practice at heterogeneous local R&D sites, the participant groups are always embedded in more cosmopolitan "fields," i.e. in various disciplinary and professional practices. Hence, the symbolic exchanges and confrontations we have been describing at the local level can also be seen as specifically situated confrontations among cosmopolitan symbolic repertoires. This implies that the symbolic dynamics of heterogeneous R&D projects can be at least partly interpreted in terms of the dynamics of disciplinary and professional symbolism.

Bruno Latour has argued that the organization of scientific and technological labor over space and through time depends crucially on what he calls "immutable mobiles" (Latour 1987; 1990) i.e. on standardized and decontextualized measurements and data which can be durably transported from site to site. Latour asks how scientific practices become stronger than competing orders of knowledge, that is, how they can become a macro-actor challenging and dominating other actors (Callon and Latour 1981). Part of the answer is that systems of representation are developed which allow local and idiosyncratic information to be represented in ways which are simultaneously immutable and mobile or, more precisely, immutable in spite of being highly mobile and transportable intact to "centers of calculation." Esoteric symbolic representations are clearly the most important of these immutable mobiles, precisely because they are the most abstract — and therefore the most immutable and the most mobile.

However in multi-disciplinary settings like those we have been investigating, disciplines can no longer complacently frame phenomena in terms of their own repertoire of "immutable mobiles," but must manage to relate their symbolic expressions to a variety of alien symbolic expressions. In short, disciplinary labor in such settings involves managing boundaries among heterogeneous practices and disciplines, it means exploring the *mutability* rather than exploiting the *immutability* of disciplinary "mobiles."

This involves the labor of producing what Star and Griesemer (1989) have called "boundary objects." Boundary objects organize shared but simultaneously distributed cognition among various "social worlds," the latter being defined by Strauss (1978) as constellations of groups and organizations committed to a particular activity and thereby sharing resources and building up shared ideologies. Academic disciplines and professions are examples of such social worlds. Boundary objects are used by different social worlds without presupposing a fully shared definition of an object. They are flexible enough, so that each social world can read a specific meaning from a boundary object sufficient to its needs. Simultaneously they are "robust enough to maintain a common identity across sites" (Star and Griesemer

1989: 393). As such they enable collaboration and communication among heterogeneous practices on equal terms, e.g., without recourse to a single dominant mode of symbolization.

Despite some differences in their sociological contexts, it can be argued that the concepts are two sides of the same coin. As Latour himself concedes, mobiles are never completely immutable (Latour 1987: 241). In order to be usable in a specific context, abstract and standardized immutable mobiles must again be adapted to the particular working situation. This of course is the inverse of the standardization and abstraction of representations which gives rise to immutable mobiles; it is in fact the "re-representation" (Star 1995b: 92) of immutable mobiles in a local context. According to Star (1995b: 91-95), a central tension emerges between static and abstract representations on the one hand and concrete, real-time work on the other:

> Taking this tension and its attendant tradeoffs into account, we can think of immutable mobiles as traveling along a path of work, where the tensions between mutability and immutablity are managed in each situation. This path is the re-representation path (Star 1995b: 92).

This (local) management of mutability and immutability assumes form as some specific way of using representations, including what we are calling transformations. In this chapter we describe the local management of mutability and immutability as specific strategies of transformation (that is, our various transformation mechanisms) used by actors to get heterogeneous collaborative settings working and to keep them working. In the process, novel dictionaries and repertoires are created which again reassert a certain immutability — at least for the duration.

When the collaboration emerges out of a context of repeated and durable associations among the same disciplines and practices (as in high energy physics) disciplinary symbolic representations can be an important resource for coordinating activities; but only insofar as they are routinely translatable on the basis of need by other parties. As the g–2 project shows, such routine translatability is already embedded in a hybrid repertoire, based on years of repeated co-operation in the building of detectors among the disciplines of high energy physics, mechanical engineering and electronic engineering. This enabled physicists and engineers to coordinate the actual R&D process and to ensure the physics performance and engineering integrity of the final detector.

This repertoire, as the account shows, is based on a complex set of well articulated boundary objects straddling the divide between physics and several engineering disciplines. These boundary objects are typically based on immutable mobiles specific to either physics or engineering (for example, Muons or 6061-T6 Aluminum) but which by dint of black-boxing — or selective perception of salient characteristics, which is the same thing — have become sufficiently mutable to become mobile even across the relevant boundaries. This means that while physicists and engineers may attach very different meanings to the concept "Muon," they can both work with the concept in a sense sufficient to their own practices — and to fulfilling their role in the collaboration.

At the same time, however, a new round of standardization sets in which tends to make the boundary objects which comprise the hybrid repertoire less mutable across the space and time of the high-energy physics world — without affecting their mobility. We can of course see only a little bit of this process at the level of a single experiment like the Muon g-2 measurement. At the cosmopolitan level of the design and construction of detectors, we could say that the erstwhile boundary objects are becoming a new set of immutable mobiles which may soon comprise an esoteric symbolic repertoire performable by designers and builders of detectors in HEP. This proto-discipline now already aggregates experiences worldwide, has its own journals and accumulates its own first principles and heuristics. Its adepts can function effectively — and without many contextualizing preliminaries — in any experimental HEP setting throughout the world and they have become as personally mobile as the modes of representation they embody. We may in fact be seeing the emergence of a new practice which, however tied to the apron-strings of mainstream HEP, increasingly has its own practical and symbolic fish to fry.

When the collaboration is multi-disciplinary, as in the Twente micro-optics project, there are also (disciplinary) symbolic immutables, but they are not mobile throughout the collaboration simply because disciplinary symbolic expressions cannot be (fully) retrieved in the symbolic repertoires of other disciplines. In order to create the necessary mobility throughout the collaboration, symbolic immutable mobiles had to be made mutable, that is, they had to become boundary objects coupling heterogeneous disciplinary practices. Mathematical formulae, for example, were able to become mutable and function as boundary objects thanks to the compilation of transforming dictionaries which allowed at least some of their meaning to be carried across disciplinary and practical boundaries. The possibilities for transformation depended on the inherent layeredness and flexibility of the "immutable" expressions: for example, the symbol of the coefficient $(b/c^{(2)}/d_{xy})$ in the differential equation of non-linearity and the concrete local meaning connected to it disappeared when the formula moved from one local practice to the other, whereas the structure and the general meaning of the mathematical formula remained the same. Each group interpreted the formula depending on their particular needs.

The dominant role of the symbolic repertoire of applied mathematics in the Twente collaboration, especially at the outset, when only a few shared cognitive structures were present, is also due to this layeredness and flexibility. Normally, the dominant role of mathematical equations in techno-scientific contexts is explained by extolling their virtues as immutable mobiles. Showing how they can also work as boundary objects reveals the local flexibility behind the standardized meaning and thus "ecologizes" the apparently natural hegemony of mathematical symbolic repertoires in these sciences. As such they exhibit some of the boundary dynamics prevailing between different scientific R&D practices and provide insight into how heterogeneous settings can remain coordinated in spite of the simultaneous use of divergent symbolic repertoires.

As basic modalities for the work that manages the tension between mutability and immutability we found five coupling and transformation mechanisms, four of which (1–5) we described on the basis of empirical examples:

0) Ad hoc use of very generic repertoires;
1) Use (ad hoc) of semi-specific repertoires as lingua franc.;
2) Understanding strange repertoires, dictionary, agency of the listener;
3) Talking in terms of strange repertoires, dictionary, agency of the speaker;
4) Emergent hybrid symbolic repertoires.

Which mechanisms are available at a given point in time depends on how much mutual education has already taken place. In other words, the repertoire of available coupling and transformation mechanisms in a given collaborative setting changes over time. What we are suggesting here is that emergent practices themselves produce these changes — which seem to be irreversible. Our data themselves suggest that the taxonomy of five coupling and transformation mechanisms can be conceived as a hierarchy in which the routinized accomplishment of each stage produces the conditions for the emergence of the next. This process can only occur when the collaborative activity persists over a long period of time, or when several collaborations emerge repeatedly within the same disciplinary context. Boundary objects emerge as the re-representation of immutable mobiles and — in the course of long-term and repeating projects within the same realm — become increasingly standardized across the participating practices, i.e. become increasingly immutable again. This dynamic is visualized in Figure 9.

Ad hoc collaborative settings like the Twente micro-optics project start out with very generic, semi-specific, specific and distinctive repertoires. Depending on the time perspective of the collaborative project, they may only apply ad hoc coupling mechanisms like the use of generic and semi-specific repertoires or they may start to compile dictionaries for listeners and speakers, which might eventually produce a new stabilized hybrid repertoire. In fields like high energy physics, where collaborations emerge repeatedly from the same disciplinary matrix, experience in accomplishing symbolic transformations is cumulated from collaboration to collaboration and ultimately a distinct culture of transformation repertoires emerges. Successive collaborations can routinely build on such interdisciplinary repertoires.

Hence, we are suggesting that in durable collaborations the different transformation mechanisms are not simply juxtaposed. Rather, we would argue that new structures emerge on the basis of irreversible mutual learning processes. These emergent shared socio-cognitive structures in turn allow for transformations of increased complexity, fidelity, and symmetry. This bootstrap operation seems to have two steps:

1) The first step includes the emergence of dictionaries as structures of semiotic coordination. Our first two mechanisms (that is, the use of generic and semi-specific repertoires) are types of ad hoc efforts to couple different esoteric repertoires at specific, pragmatically relevant, points. As an unintended consequence of these efforts, entries accumulate in a dictionary which, over the longer term, begins to couple the repertoires at more and more points. Hence meanings can be shared, even if specific cognitions still remain distributed. That is, the dictionaries remain asymmetric in the sense that expressions in strange repertoires remain largely "black-boxed" and can hardly be opened by the one who transforms them with the aid of dictionary entries. "Opening the black box" can only be performed by the natives, and they do so only when actual or expected misunderstandings are perceived as potential obstacles to the ongoing coordination of the project.

2) The second step in the bootstrap operation is the conversion from dictionaries into hybrid symbolic repertoires. This amounts to an emergence of cognitive coordinating structures on the basis of the semiotic dictionaries. Our next two mechanisms (the agency of the listener and the agency of the speaker) presuppose the prior existence — and hence prior fabrication — of semiotic dictionaries. As such they cannot be ad hoc strategies but must build on prior interactions and learning. In an advanced stage when the heterogeneous repertoires get coupled at more and more points, the use of these dictionaries encourages the opening of black boxes. Thereby, cognition becomes shared and the dictionaries become more and more symmetric, until, in the very long term, a new hybrid repertoire may emerge. The related standardization process turns former boundary objects increasingly into immutable mobiles specific to the type of collaboration.

9 Meaningful Boundaries

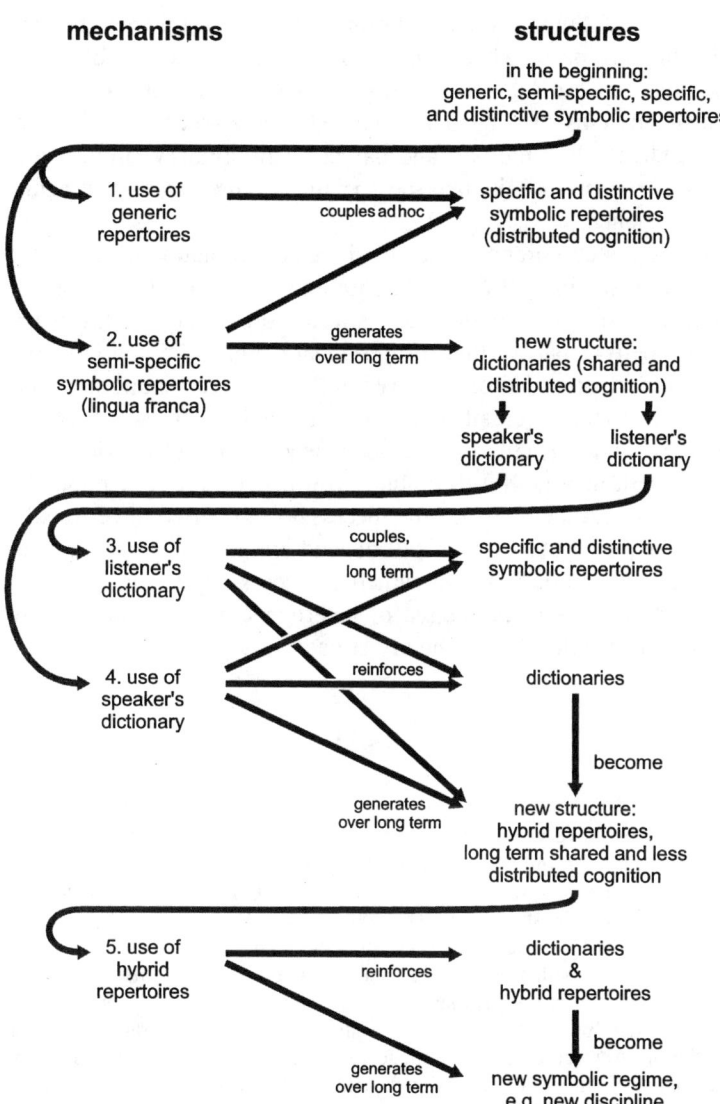

Figure 9: Dynamics of Emergent Dictionaries and Hybrid Repertoires

Substantiating these preliminary findings will require additional research. However, if the taxonomy of coupling and transformation mechanisms can indeed be shown to be steps in a hierarchy of symbolic structuration, this provides a scheme for symbolic innovation based on the active bridging of symbolic difference under conditions of limited mutual comprehension. Insofar as the ad hoc solutions of the collaborators become stabilized and routinized by dint of long or repeated association, conditions would be established for the emergence of shared repertoires of symbolic transformation. Stabilized transformations then become a shared second-order communicative repertoire and may become part and parcel of disciplinary cultures into which, for example, new recruits may be socialized quite apart from actual collaborative experience.

This means that at a certain point the cognitive co-ordination necessary to proceed with interdisciplinary R&D collaborations no longer has to be actively constructed by the collaborators but is now, to a greater or lesser extent, available to them as part of their professional culture. Semiotic dictionaries can at a certain point become shared means of coordination, superseding the ad hoc strategies entailed in stages 0) and 1) of our typology. How far this goes, and particularly whether it leads to novel hybrid repertoires associated with new (sub-) disciplines will depend on perceptions by actors of the costs and benefits of the necessary cognitive investments. Among other things, such perceptions will depend on the time perspective of the collaboration (i.e. the period over which the necessary semiotic investments can be amortized) and the degree of heterogeneity of the constituent repertoires — which is indicative of the effort necessary.

Notes

[1] Clearly our argument is very congenial to Peter Galison's concept of "trading zone" (Galison 1997) with its proliferation of "interlanguages" in the form of "pidgins" and "creoles." We share Galison's affinity with Star and Griesemer's (1989) idea of "boundary object" when he states that "The notion of cooperation through heterogeneity is key for their project and mine" (Galison 1997: 48n).

We disagree, however, with Galison's insistence that boundaries *always* have size and extent, i.e. that they are always sites of local interlanguages and hence of situated and coordinated "pidginized" practices. Galison seems to reject the notion that simple translation *ever* serves to cooordinate heterogeneous practices. In our view the relative content-richness of boundaries depends empirically on the effort that actors are willing (or forced) to put into getting their acts together. Our scheme is in fact a hierarchy of increasing linguistic coordinative effort (and hence costs) which tends to assume a particular form at the point when the perceived costs of coordination threaten to outweigh its expected advantages. This may never involve "pidginization" (let alone "creolization") but may consist only of recourse to a lingua franca or to versions of active or passive transformations and the creation of tacit "dictionaries".

Galison is too categorical for two reasons. First, physics collaborations, however ad hoc, are always embedded in longer term practices in which it is economical to develop collaboration-transcending "pidgins" and "creoles", what we call "hybrid languages." Other fields frequently exhibit more ephemeral collaborations in which this "economy of scale" is absent. Second, he is too literal in his reliance on linguistic anthropology. We think the notion of physics as a "culture" and of its components as "subcultures" and hence the direct equation of "trading zones" in physics with those found between "natural" cultures with their pidgins and creoles ought to be taken with a metaphorical grain of sand. Sometimes collaborative efforts are mere arrangements of convenience among heterogeneous practices which barely merit the appellation of "cultures" themselves — let alone the "trading zone" of their pragmatic and ephemeral intersection.

2 In order to achieve a high $\chi^{(2)}$ in a substance, a high beta in the single molecule is necessary, but this alone does not guarantee a high $\chi^{(2)}$.

3 Under the conditions of the second harmonic generation (a light beam with a certain frequency goes into the material and two light beams come out, one with the same frequency and another one with double the frequency) d is defined as half of $\chi^{(2)}$ and determines the efficiency of the frequency conversion. Since the applied physicists are pursuing a frequency converter application, d is very important for them. $\chi^{(2)}$ and d are mathematically speaking (third rank) tensors, which means that they are not one value, but consist of a number of components. $\chi^{(2)}$ has 27 components and d has only 18, which can be further reduced under certain conditions. Depending on the orientation of the molecules, certain components of the tensor d are expected to be the largest. In the case of corona-poled thin films, which were used in this project, the element d_{33} was the largest and d_{31} at most a third of this value. All others were either about the same as d_{31}, negligible, or zero.

4 In the Twente collaboration we found that such "dictionary"-based transformation was actively pursued by the participating groups. Interviews in which the question of mutual (mis-)understanding came up betray a keen awareness of divides between specific repertoires as well as numerous examples of the application of transformation dictionaries based on the use of semi-specific repertoires as semantic common ground. The following account by a senior member of the department of Organic Chemistry is a typical example:

"Scientific discussion with other groups is not a simple question. Everybody has his own jargon. I can give an example: Chemists talk about the extinction coefficient. Physicists talk about the absorption cross section. I never grasped the full meaning of it, but it was obvious that the concepts were associated in some way. What has recently become clear is that the absorption cross section is simply the area under an absorption peak, whereas we give the extinction coefficient for a certain wave length. Thus the physicists just integrate over the whole peak and that's all. Sometimes it takes years before you discover such relations. In this sense you can easily be at cross-purposes."

5 The project manager (who generated the drawings) stated at the workshop that the reduction was no problem. He had simply photocopied the slip ring region from Figure 2 at 70% reduction and pasted it into the drawing of Figure 3.

6 We speak of a practice here rather than a discipline because what we see lacks both the organizational inbedding and scientific standing of a discipline.

Chapter 10
Antagonistic Patterns and New Technologies

Arie Rip and *Siebe Talma*

1 Introduction

New and promising technologies are not born in a social vacuum. When entire domains of technological options are involved, as with nuclear fission and fusion, and now with genetic engineering and other forms of modern biotechnology, society responds to the promise, and to the uncertainties, of the unknown. In doing so it reflects patterns and attitudes available in its cultural repertoire, while also augmenting and changing these patterns.

The evolving socio-technical order has strong historical roots. The patterns in our present-day culture are themselves outcomes of earlier interactions, including experiences with new technologies. Societies, and groups and actors in societies, have learned how to deal with technology, but often unreflectively, and without any assurance that achieved arrangements are indeed good, or productive, or societally desirable. In recent years, the responses to new technology have been more explicit, and also more critical, and they have been voiced more broadly. The question whether the domestication of technology in society works well can now be discussed as such; the emergence of a technology assessment discourse is one indicator (Rip et al. 1995). Technological development and its embedding in society have become reflexive, and the new socio-technical order is subject to debate, and to some extent, is being consciously constructed.

In these multi-actor and non-coordinated processes the typical management approach of intentional action to optimize goal-achievement cannot be followed. For one thing, the discussions and responses are shaped by earlier patterns, and, while learning occurs, there is no central instance to absorb the lessons and apply them. It is distributed learning, and in the case of new technology, often cast in an antagonistic mold. It is particularly these antagonistic patterns, and the opportunities they provide for learning and for productive management of technology in society, that interest us here. These are complex questions, and it is wise to approach them historically, by asking how patterns relevant to new technologies arise and evolve. That there are controversies about new technologies is now an accepted fact of life, but we contend that what is played out in these controversies are patterns at a deeper (and less visible) level.

Take the Brent Spar controversy, a big issue in June, 1995. The eventual decision not to sink (or "dump") the oil platform in the Atlantic Ocean reverberated in the board rooms of industry and on the opinion pages of newspapers. The storyline was one of "David" Greenpeace against "Goliath" Shell. But it was not just Greenpeace that vanquished Shell, it was another Goliath, constructed ad hoc but no less powerful for that. The German motorists were part of the Goliath, and Chancelor Kohl, and wavering governments, and internal difference of opinion in Shell. And arguments about risk and costs. And professional activism, having a chance to score and reinforce the support of its constituencies. And culture. People inculcated with habits like "cleaning up after you" and recently also "separating different kinds of waste," saw that a big oil platform was to be dumped. So it became a symbol of what you should not do.

While the controversy was about a specific issue, its dynamics drew on available cultural and structural patterns. Its conclusion reinforced some strands in the patterns, and weakened others. And learning occurred: the various actors, in board rooms and on opinion pages of the newspapers, tried to draw out the relevant lessons.

How do antagonistic patterns around new technologies emerge? At first, there is recognition of novelty, and attempts to name it. When radioactive waste was suggested as a possible weapon, it was called "deathly sand" (De la Bruhèze 1992a). The "mushroom cloud" became the sign of the atom bomb. Such metaphors are one way of naming. Labels, like "the atom," around which actors can assemble, and with which they can link up, are the main route through which new technology acquires a social and cultural presence. Such labeling occurs at an early stage, before there is much experience with the new technology. But the label need not be very precise, nor need its contents be known and accepted, in order to connect different actors. Again, "the atom" shows this well. The labeling may be contested, and shift after a time (as when "atomic" became "nuclear," or when "genetic engineering" became one of the defining characteristics of [modern] biotechnology). The process of naming sets the scene, creates associations, and shapes learning about the new technology.

Within the set of possible and actual linkages, there is a difference between those connected with the development and introduction of a new technology, and those which are not. Obviously, different interests are involved (even within the group of introductors there will be such differences). But one also sees a particular pattern emerge: a difference between insiders and outsiders. The introductors see themselves as insiders who know much more about the technology and therefore position themselves as also more knowledgeable about its potential embedding in society. At first, actors not involved in the new technology need not consider themselves excluded, or as being outsiders — but insiders will nevertheless define them as such. Such labeling by insiders, and associated behaviors, can in fact create coherent groups of commentators/critics which were not there before, as De la Bruhèze has shown for the U.S. Atomic Energy Commission and radioactive waste disposal technology (De la Bruhèze, this volume).

The combination of labeling and diffuse group formation leads to situations where stereotyping and inclusion/exclusion behavior becomes self-reinforcing. Such situations are increasingly common, and this has given rise to a further, and reflexive, type of labeling, that of "proponents" and "opponents" of new technologies. Neither "proponents" nor "opponents" are simple categories, but to think in those terms seems natural. And the labels are used prospectively: introductors of new technology now expect that there may be contestation, and watch out for "opponents." The opponent/proponent dichotomy has become a pattern in our culture, and it serves actors in their attempts to order a complex environment. In that sense, the dichotomy is now a fact of life. (This goes so far that a whole secondary business has emerged around mediation and conflict-resolution, of attempts to increase acceptance of new technology, of public-opinion surveys and other monitoring exercises, and of training, on both sides, in strategies for effective contestation.)

The dichotomy of proponents and opponents is the strongest antagonistic pattern around new technologies, and its importance is related to longer-term features of our societies, and is reinforced by recent developments. One such development is the increasing role of mass-media, which claim to need dramatic story lines to draw attention — clearly the storyline of proponents and opponents is one of the basic available narrative forms.[1] Another development is the professionalization of the environmental movement and other critical movements, so that there is personal and professional continuity of opponents across different technologies.

Other patterns occur, because concerns about one technology, and promises about the same or another technology, are not separate issues anymore. They are connected, for example through shared labels of "risk" and "promise" of technology. Or through referencing back: "Make sure that recombinant DNA technology will not suffer the same fate as nuclear technology."

The central topic of our chapter is exactly that: Existing and evolving patterns define the situation for new technologies, partly independently of the specifics of the technology. Among these patterns, the metaphor of "contest," and the storyline of proponents and opponents have been particularly salient since at least the 1970s. After drawing out the longer-term and structural aspects of this salient pattern, we look more closely at the evolution of the socio-technical orders of nuclear technology and modern biotechnology, using our studies of the situation in the Netherlands to identify interesting further patterns. This allows us to introduce the necessary complexity into an otherwise overly simple juxta-position of proponents and opponents. Antagonistic interaction then becomes one variant of a wider range of what one can call "agonistic" interaction: Struggles, contrasts, tensions, and difficult assessments, but also complementarities and recognition of the roles of the different parties (and aspects of new technology).

The next question then is what sort of learning — that is, learning how to handle technology in society — can occur? For learning to occur, it is not necessary that shared views and common interests emerge. In contests and in battles, there is an element of agonistic coordination. The simple fact of knowing that it is a battle (and not a military exercise or a fox hunt) implies specific coordination: there must be some rules of the battleground, there is recognition of who are the opponents. Even in the antagonistic variant, the labeling of "proponents" and "opponents" thus serves a purpose: it reduces social uncertainty about who is who in the (presumed) battle.

Antagonisms provide coordination, but can lead to an impasse when, instead of learning, only mutual labeling occurs. Often, however, antagonistic struggles force actors to articulate the merits of their position, to search for arguments and counter-arguments, and to commission special research. So after some time, there is better understanding of the issues, and potentially, resources for more adequate action.

Our discussion of patterns around new technology, in the Netherlands and more generally, allows us to evaluate the possibilities of learning how to handle new technologies in society through agonistic, and also antagonistic interaction. We shall turn to this question briefly in the concluding section.

2 Agonistic and Antagonistic Patterns Around Novel Technologies: Their Shape and their Long-term Evolution

Conflicts, including conflicts over technologies, can be analyzed as such, as is done in the literature on controversies about technology (e.g., Nelkin 1992) and in the various versions of conflict sociology. Our approach is to see conflicts as part of evolving socio-technical orders. The emphasis is then on novelty, change and continuity, rather than on oppositions and conflicts per se.

It is not that novel technologies and the dichotomies that emerge cannot be dealt with in general sociological terms, as conflicts, and analyzed in terms of opposing groupings and the processes that maintain such distinctions. One could apply the general sociological argument about the "objective complicity" involved in antagonistic struggles, where both parties depend on their continued re-enacting, and thus re-inforcement, of a joint game. Bloomfield and Vurdubakis (1995: 535) speak of "constitutive complicity" here. One can also look at strategies, as in Bourdieu's general field theory, with the key distinction between the strategy of conservation (by the possessors of cultural capital), and the strategy of undermining (by those who do not (yet) possess such capital) (Pels 1985).

But there is an additional element. What is specific (and what thus makes our analysis of general interest for sociology as well) is that it is *new* technology, with its promises and uncertainties, that is at stake. There is not just an establishment fighting newcomers. Rather, the possessors of earlier technological capital are themselves challenged by the possibility and actuality of new technological developments. Thus, the ambiguous linkages between technology, modernity and progress will create a complex pattern. Technological innovation, especially when seen as an explicit goal — as was the case after the second world war — is a destabilizing factor in relations between technology actors (firms, government agencies, research groups/ institutions). Schumpeter spoke of "creative destruction," and it is now generally recognized that market equilibria are the exception rather than the rule.[2] New products are opportunities for growth, but they also generate uncertainties, for the innovators no less than for other actors. One way to reduce uncertainty is to set up strategic alliances. Besides the direct economic and strategic benefits, an important effect is that an in-group is created, of those in the know, distinct from an out-group that is hopefully less important, and need not be taken into account.

The arms race is another example of the dynamic of new technology and hoped-for reduction of uncertainty, but now at the level of states and military-industrial complexes rather than firms. The Manhattan Project to develop an atomic bomb, often considered to be a watershed in global history, is also interesting as an example of connections being made without actual sharing of purposes. Participants in the project did not have a common definition of the project. They linked up because the possibility of an atomic bomb — an open promise, and in that sense like an unarticulated label — was a way of pursuing their own separate goals: Doing "sweet" physics, meeting a technological challenge, showing the competence and value of the Army as against the Navy or the Air Force, beating the Germans in the race to an atomic bomb, showing the Russians that Americans can do better. When linked up, participants with different backgrounds gain common experiences, and shared legitimizations emerge which leads to reification of the value of the project, both at the time and afterwards. Thus, direct and indirect network formation will occur, the latter because of the shared reference to "the" project, here the Manhattan project.[3]

Along the trajectory of development, introduction, adoption and diffusion of new technologies, mutual dependencies among actors increase, and we see strong formations emerge around very practical concerns: say, the linkages between nuclear technology, electric power production and its distribution, and a number of industrial suppliers. Or in the case of the motor car, the whole assemblage of delivery, maintenance, and supply to users. The actors involved in such technologies see themselves, in a typically modernist stance, as the carriers of progress. This overall cultural legitimization creates a link across various technologies, in addition to the specific cross-linkages that occur; technological, financial (banking and venture capital, insurance) managerial, e.g., through mobility of managers, and through double functions of board members.

In the final analysis, a social world will emerge in which promotion of technology is a practice as well as an ideology. From within this world, what is outside appears only as an opportunity or barrier, and if actively resistant, as an opponent.[4] This is not a watertight boundary, but it is continually reinforced by a division of labor with this world's alter ego, a bureaucratic-professional world of technology regulation, as well as by the cultural dynamics of the introduction of new technology in modern society. We shall briefly discuss what overall patterns and formations can be distinguished. And we introduce another point: there is the backdrop of a technological culture native to industrial society: domestication of new technology and adaptation of practices and culture, i.e. the ways in which novel technologies become embedded in society, are shaped by, and reinforce, a cultural pattern with antagonistic flavors: "You had better watch out for chemicals, or nuclear plants, or genetically modified organisms; or for new technology generally."

The bureaucratic-professional world of regulation is the most easily recognizable. At the level of government agencies, various forms of regulation and control of technology have emerged: Regulation for mines, steam engines, and chemicals during the 19th century, for gas and electricity at the turn of the present century. By now, regulation occurs, and is expected to occur, across the board. At the same time, government agencies have been organized to promote technological development; their bureaucratic separation from the regulatory agencies is related to, and reinforces, the dichotomy of proponents and opponents. While the distinction between promotion and control is not absolute — think of the importance of standards for developers as well as regulators, and of the role of Food and Drug Acts since the late 19th century, enforcing quality and honest trade at one and the same time — their separation has become the obvious way to handle technology. And it has been underpinned by a general political argument about the need to separate conflicting interests; see for example how the U.S. Atomic Energy Commission was split up so as to create a separate Nuclear Regulatory Commission.

The way technology in society is managed at the level of governments is related to the broader socio-cultural development of modernity, with its interest in novelty and progress, and with technology (with a capital T) as one of its icons, and with its concerns about the smooth and optimal functioning of societal arrangements. Each particular case of promotion or regulation and control has to be understood against this background.

Similarly, each particular case of introduction of new technology and of response, especially critical response, is embedded in broader patterns. Proponents of a particular technology ally themselves with the labels of "progress" and "modernity," and opponents are forced to declare themselves against the presumption of modernity, or develop another view on progress. However, we are running ahead of ourselves here and presuming the existence of proponents and opponents as such, which is exactly what is at issue. The dichotomy, and hence the actual existence of proponents and opponents, is to some extent an outcome of the dynamic, rather than its origin. So we should rewrite the point, and also take the opportunity of locating criticism of technology more satisfactorily.

Introductors will foreground the promise of a new technology, and act to specify and realize the promise. They seek to overcome barriers against their project, and to label the hesitant and the doubtful as blind or irrational, and thus not deserving of serious consideration. Technological modernism gets articulated in this way.[5] Others can speak for constituencies outside the charmed circle of technological modernism (including future generations and the environment) and in doing so will link up with the discourse of danger and risk. A recurring point is that the unknown may harbor danger (unknown but undesirable effects of novel technology), so that precautions are in order. Almost unavoidably, the value of existing, embedded, socio-technical order is emphasized and sometimes alternative non-technological options are outlined. Alliances are wrought in those terms, and the strong populist overtones create an uneasy mix of conservative-romantic reaction and alternative modernism.

Looking back, one can see that a culture of concern and criticism about technology is visible from at least the late 19th century onwards, and probably earlier, if one takes the Luddites seriously (Schot 1991; Staudenmaier 1989). Such concerns are often related to the disturbing effects of new technology on existing social relations and societal patterns, and therefore labeled "conservative" or "romantic." The 1960s and 1970s saw an important change, in the sense that technology criticism was adopted in the politically liberal/progressive repertoire. One effect was the articulation, in public debate, of technology (in general) as intrinsically linked to domination and control (of individuals, of groups, of countries). Another effect was that by the 1980s, substantial arguments by technology critics, no less than their simple political presence, had become regular inputs into government regulation. Introductors of technology expected concerns, at least concerns about risks, and began to take these into account in shaping the technology.

While the distinction between promotion and control, between proponents and opponents, is deeply embedded in our culture, and one can encounter typical proponents and typical opponents of technology as (often self-styled) spokespersons, the actual patterns and practices are more complex. We have used the idea of intentional or emergent socio-cultural reduction of uncertainty about new technologies, as the key to understanding these patterns. In doing so, we have focused on the introductors, sometimes accepting the storyline about the heroic fights of technology's champions. But it is as important to understand the actual embedding of technologies in society, and how these too are shaped by socio-cultural patterns. Again, naming and other ways of routinizing situations occur.[6] This has led to a socio-cultural pattern which one of us (A.R.) has called the "emerging danger culture of industrial society," which shapes daily life as well as influences regulation and control of technology.

Think of the rules for our association with chemicals as an addition to the arsenal of cultural means of survival in industrial society. Rules like washing fruit, but also standards for maximum allowable concentrations of chemicals in the workplace and for acceptable daily intakes of food additives together form a net to catch and contain the dangers of chemicals. The rules may be justified in terms of toxicological data, but their effectivity is assured through the relevant cultural transformation.

Cultural transformations, as we use the term, take place around specific cognitive and technological novelties emerging in industrial society, for example, microbes or new chemicals (Rip 1991). In the 19th century, microbes were introduced as a medical invention. Around 1900, the idea that diseases could be caused by "germs," tiny organisms invisible to the naked eye but dangerous intruders all the same, took hold in our culture. The idea of health changed in the same movement (this was preceded by comparable changes within the medical community). Good health was no longer a matter of internal and external balance. Instead the body came to be seen as a fortress that had to be defended against intruders, specifically against germs, but generally against contaminants, toxins. Thus, new behavioral rules emerged: from mundane prescriptions like wash your hands, don't sneeze near others; to neurotic cleanliness "in the name of health." The symbolic element, in the sense of the importance of the rules for the way we order our world, is apparent not only in the ritual way in which the rules are followed, but also in the *anomie* resulting from their contradiction. People are devastated (or just disbelieving) when they are told that germs are everywhere, even in the air we breathe. "Can't we be safe anywhere, then?" is the reaction.[7]

The way we handle chemicals is also embedded in a cultural transformation, covering a period of a century or more. The associated practices indicate something of how our kind of society handles potentially dangerous technologies in general. For example, they suggest that there is a concern about chemicals even when there is no actual sign of danger. This implies that there is diffuse political support for stringent measures: we spend resources and political energy to check for risks of chemicals, (including the risk of carcinogeneity, which is a social and political mobilizer) and devise systems to set adequate standards. Here, the cultural backdrop links up with, and reinforces, the structural distinction between promotion and regulation.

The danger culture of industrial society, together with the vicissitudes of modernization and the contrast (or division of labor) between promotion and control, set the stage on which the dramas of new technologies are played out. Each unfolding story has its own specific plot, shaped by the particular technology, the actors, and the circumstances involved. But each story is also shaped by specific patterns that have emerged and are already in place. For novel technologies, these patterns may be strongly antagonistic. In this section we have shown, at least globally, how this has come about. The next step is to look more closely and on the basis of experience with two novel technologies, nuclear technology and modern biotechnology, to detail three specific patterns. We shall focus on data and experiences from the Netherlands, but will emphasize the generic features of the patterns rather than the specifics of the Dutch situation.

3 Promises and Risks in the Cases of Nuclear Technology and Biotechnology

Promises and risks of a new technology can be contrasted, and are, in fact, pitched against one another. In some cases, this gets framed by the proponent–opponent dichotomy to such an extent that it produces only grandiose declarations (as in the early days of the recombinant DNA controversy) and little learning. In other cases, weighing promises against risks leads to mutual articulation and a better understanding of the value of a new technology. What we want to show is how, over time, repertoires of promise and risk have emerged which allow such articulation of the value of a technology — without necessarily producing a consensus. In particular, a socio-cultural pattern for addressing novel technologies has become established since the 1970s.

Historically, promises came first. Electricity and synthetic chemicals, when they appeared at the end of the 19th century and came into their own at the beginning of the 20th century, were surrounded by a halo of promises about their contributions to modern life. This was not only salesmen's hype and social legitimization. Promises lead to articulation of further specifications and actions to realize them, and thus create certain paths of technological development and social embedding. A later example is the promise of nuclear technology: if electricity is to become "too cheap to meter," nuclear power plants must be designed to deliver. This promise-requirement cycle (Van Lente 1993; Van Lente and Rip, this volume) orients the actions of technologists, and legitimates them at the same time (Jelsma 1991).

Classically, the darker side of technological progress has been conceived of as only specific side-effects, to be handled by inspection and regulation. By the 1930s, in debates about mechanization and employment, and especially in the 1940s, with the early debates on nuclear arms, the idea emerged that novel technologies might be assessed at an early stage. The whole notion of control of technology became a recognized issue because of the example of nuclear technology. In spite of concerns about its use for bombs, and then about proliferation, it was pushed by short-term geopolitical considerations, and protected from criticism by exhortations to "develop your bomb first, then think about consequences." By the late 1950s, the consequences (actual and potential) had become visible, but by then the development appeared to be irreversible. In response, concerned scientists, intellectuals and politicians began to raise the issue of control of technology in general, and to identify the problem of (irreversible) momentum acquired by promising technologies.[8]

It is the issue of control, and partly the specific responses of the technology promoters to the possibility of control, that has shaped the risk repertoire characteristic of discussions, decision-making and practices since the early 1970s.[9] The new risk repertoire was rooted in the novelty of the technology (and thus in the unpredictability of the risks), and in the irreversibility and macro-character of the effects once they occurred (and thus in their essential "unmanageability"). In the Netherlands, a new term, macro-risk, was coined to denote such risks, the prime examples being the possibility of nuclear pollution as a genetic threat to a substantial part of the world population, or even the survival of the human race. Such risks were widely recognized in North-West Europe and the U.S., and the argument of concerned groups was that taking such a macro-risk could not be justified by whatever benefits might be ascribed to the technology. The promise-repertoire of promoters was powerless against such an argument. The socio-cultural effect was a gap between spokespersons for the promise of technology and spokespersons for concerns about technology, in particular nuclear technology, which made mutual labeling in terms of proponents and opponents an easy and obvious strategy. In order to persuade the other side (and as yet uncommitted publics) arguments would be inflated, with the result that rhetorical battles between big promises and doomsday scenarios were the order of the day.

In the Netherlands, as elsewhere, it became almost impossible to maintain an intermediate position about nuclear energy. As one of the proponents phrased it, nuclear technology was portrayed by the critics as the "new original sin," and thus in absolute contrast to the shining example of technology for the good of mankind that it really was. The recognition of the polarized situation led to attempts at partial resolution. In 1974, almost sixty scientists, intellectuals, industrialists and politicians were prepared to sign a "proposal to reflect on nuclear power," which urged a five-year moratorium on the building of new nuclear reactors in the Netherlands. This fueled public discussion, so that the labor unions were also moved to endorse the moratorium. Other groups aligned themselves in the emerging discussion as well. The Fall of 1975 saw further action (inspired by an initiative of the U.S. Union of Concerned Scientists) in the form of newspaper advertisement saying "Give us the benefit of the doubt," signed by over a thousand prominent citizens (with academic degrees) and asking Parliament to vote against further expansion of nuclear power in the Netherlands. Not only positions for or against nuclear power, but also roles and responsibilities, were being articulated. For example, proponents of nuclear power attacked the expertise of some of the signatories, arguing that theologians had no right to pronounce on nuclear technology, or any kind of technology for that matter.

Thus, in the first half of the seventies, nuclear energy was the prime *topos* for playing out, and thus reinforcing, antagonistic interaction with respect to new technology. A repertoire became available, including ways of making risk arguments; roles were articulated and legitimated, with some professionalization of opponents *vis-à-vis* proponents, and with continuity of individual opponents across issues as one indicator (Molenaar 1994). This culture and structure subsequently became the mold within which debate and action around the issue of recombinant DNA experiments were carried on. This research, a stepping stone to genetic engineering, emerged in the second half of the same decade, and has since become a second *topos*. To a certain extent, events were predictable. Promises would be voiced and would be countered by articulation of risks. The associations of concerned and critical scientists (VWO and BWA) would send the same spokespersons into the arena that had struggled against nuclear energy a little earlier. The cultural and structural alignments were already in place; the plot would unfold of itself. But there were also differences.

For one thing, the recombinant-DNA debate had different origins. The molecular and other biologists involved, at first sight unencumbered proponents of a brave new world, themselves drew attention to (certain classes of) risks cleaving to the new technology, and proposed a temporary moratorium in 1974 (A). Shortly thereafter, however, the existing patterns reasserted themselves. Concerns were voiced about unknown effects, and about macro-risks like runaway organisms and the effects of tampering with evolution. In response, molecular biologists augmented their promises, and sometimes depicted themselves as a Gideon's band oppressed by opponents ignorant of science and technology, even going so far as to compare their situation with Galileo's fight against the forces of darkness.

In the case of recombinant DNA, risk repertoires had structured the debate from the beginning (Krimsky 1982; Jelsma and Smit 1986). This took place in ways strongly reminiscent of the risk debate between proponents and opponents of nuclear technology. In the U.S. nuclear debate, the establishment approach to the risks of nuclear accidents had been set out in the so-called Rasmussen study. This report was released in 1975, and was hailed as a landmark achievement in risk analysis, more specifically, in probabilistic risk analysis (Rip 1986). The Dutch *Commissie Reactorveiligheid* (Committee for Reactor Safety) subsequently produced a study along similar lines. As in the U.S., the approach was heavily criticized. In the first place, the risk of nuclear accidents was perceived as being confined to that of the "runaway reactor." Secondly, the magnitude of this risk was established by using probability calculus. Risk was defined as the product of the chances — calculated on the basis of fault trees — and the effects of (elements of) nuclear accidents. Hence, in the case of very small chances, the product (the level of risk) would always be very small, even for very large effects. On this basis, the Rasmussen study concluded that the risk of nuclear power installations was acceptably small. The VWO experts used the alternative risk repertoire, and argued that in the light of meso- and especially macro-effects, even a very small chance is too large to be socially acceptable. It was bad (and deceitful) policy to multiply away the possibility of irreparable damage.

The same risk repertoires came to blows in the recombinant DNA debate in the late 1970s. The dominant conceptualization was that of the escape of a dangerous recombinant organism. This was the point of departure for the risk assessment produced by the U.S. National Institutes of Health (in 1976), which also proceeded probabilistically. The calculations again demonstrated that the risk was acceptably low. In the Netherlands, the recombinant DNA advisory committee pursued a similar line of reasoning, and this triggered the following critical response from VWO: "[Here we see] this defective reasoning being employed all over again in a new domain of technology."

Across the domains of nuclear and recombinant DNA technology, there was continuity in the type of risk conceived (runaway reactor/runaway organism, both with a potential for large-scale effects), in the type of arguments by which these risks were depicted as acceptable, and in the subsequent counter-arguments. Add to this the promise-requirement dynamics, where the claims about applications in medicine and agriculture shaped research agendas and venture-capital interest in new R&D firms, and it is clear that a socio-cultural mold with a definite antagonistic component had become available. This pattern was visible in other domains as well. Siting of chemical plants became increasingly controversial, and probabilistic risk assessment criteria were applied to resolve the conflicts.

The continuities also worked in the other direction. Experience with chemicals and chemical technology was incorporated in debates about recombinant DNA (as well as other technologies and products): One should be concerned about long-term adverse effects on the human and natural environment (with DDT, vinylchloride and chlorofluorocarbons as exemplars), about possible large-scale accidents (the chemical plants at Seveso and Bhopal), and about military applications (the development and use of chemical weapons). For chemicals, strict regulation was in place by the late 1970s and early 1980s. The regulatory culture for chemicals had become more rigorous, and this had implications for regulatory culture generally. It was easier now to propose strict regulation of recombinant-DNA experiments. Many governments employed a combination of restricted licensing, with liberalization after sufficient experience had accumulated.

4 Networks of Actors Linking up with Biotechnology (and Other Labels)

The history of recombinant DNA technology and modern biotechnology allows us to identify another type of coordination in evolving socio-technical orders: labels attached to a new technology are also occasions for actors to position themselves as connected, or separate. As far as labeling practices of this kind have accompanied the emergence of "genetic engineering" we can see a distinct transition. Genetic engineering and (modern) biotechnology are now often assumed to overlap, or even to be synonymous. Biotechnology "often refers vaguely to technologies associated with genetic engineering" (Bud 1991: 416). In the 1970s, however, genetic engineering and biotechnology were two separate domains, differing in respect to the actors involved, the topics of discussion, and the promises being made. Alignments formed at the time flagged either one label or the other, but not both. Mutual reference to a label allowed actors to interact without becoming specific about the content of such interaction; in that sense, the labels worked as "boundary objects."[10]

The content of the label gradually evolves. In the late 1970s, biotechnology was defined as the integrated application of biochemistry, microbiology and process technology.[11] Genetic engineering was considered an external (but important, sometimes even essential) input into biotechnology. Since the risk discussion was associated with the label "genetic engineering," risk did not at this stage become linked to biotechnology, while genetic engineering was already surrounded by (proposed) regulation and advisory committees, biotechnology was still regarded as an ordinary branch of industrial enterprise. By the early 1980s, however, molecular biology and molecular genetics gradually came to be seen as a fourth basic discipline in biotechnology.[12] And now, in the 1990s, biotechnology and the use of genetically modified organisms are often equated, not just as labels but also in the practice of firms and research organizations and in technology policy.

The promise of recombinant DNA technology as developed in the 1970s, and of genetic engineering or genetic modification techniques more broadly, was taken up by technology promoters, and by policy makers eager to fill their portfolio of strategic technology stimulation programs, and subsumed under the label "biotechnology." It was clearly important for these promoters to position themselves as advantageously as possible, and thus to align themselves with (modern) biotechnology, and not with genetic engineering (and related labels). Conversely, critics would emphasize the key role of genetic engineering in order to avail themselves of the risk repertoire that had resurfaced around it.

10 Antagonistic Patterns and New Technologies 313

We are suggesting that networks of labels emerge as a corollary of the positioning and alignment of actors, and that such networks, and especially their being partitioned (in this case, into promise + biotechnology, and risk + genetic engineering), enable and constrain the further strategies and alliances of actors. Hence there emerges an antagonistic socio-cultural pattern, in which border-crossings are guarded against. In this case, for example, biotechnology proponents do not want to be associated with labels like "genetic engineering" or "genetic manipulation," and insist, when cornered, on the less threatening term, "genetic modification."

We can test this suggestion with data from an analysis by one of us (S.T.) of associations of actors (institutions, agencies, groupings) with each other and with the several labels, using data for the Netherlands in the late 1970s and through the 1980s. Figure 1 shows a multi-dimensional scaling map of network linkages of actors, with differently shaded areas according to the labels with which these actors associate themselves. There is a clear division into three areas: to the left actors dealing exclusively with genetic engineering (or recombinant-DNA) activities; the area to the right has actors associated with biotechnology only; while in the middle actors are linked to genetic engineering as well as to biotechnology.[13] Classifying actors globally, one can say that technology promoters concentrate on biotechnology, but also relate to genetic engineering, and some of the regulatory actors connect to both labels — but social interest groups, the typical location of concerns about technology, link up exclusively with genetic engineering. Clearly, the term "biotechnology" figured in the repertoire of groups of scientists, industrialists and government officials, but was absent from the repertoire of social interest groups — otherwise they would have been located differently on the map in Figure 1.

The network in which social interest groups figure is separated from the network of technology promoters because of their distancing themselves (at the time) from biotechnology. Were one to go into historical and sociological detail, one would see that the boundary object maintaining their network is "risk" rather than "genetic engineering." This relates to the fact that a variety of terms is used in genetic engineering networks, such as recombinant DNA (technology), genetic manipulation, and genetic modification. Using the specific term "genetic engineering" (or any other specific term) is not necessary, because that is not the label that creates coherence. On the other hand, there are no alternatives to the term "biotechnology," but definitions of biotechnology abound. So it is a boundary object, which is indeed unique, but also "weakly structured in common use," so that the different user groups can have their own definitions.

The networks of labels, interlinked with alliances and strategies of actors, are recognized as important by actors themselves. Government agencies, both the regulators connecting to biotechnology and genetic engineering, as well as technology policy agencies interested in bridging the gap between proponents and opponents, actively try to influence labeling and alignments. By now, "biotechnology" and "genetic modification" have merged, and promises and risks are weighed against each other, rather than contrasted — although the contrasts and struggles can easily surface again.

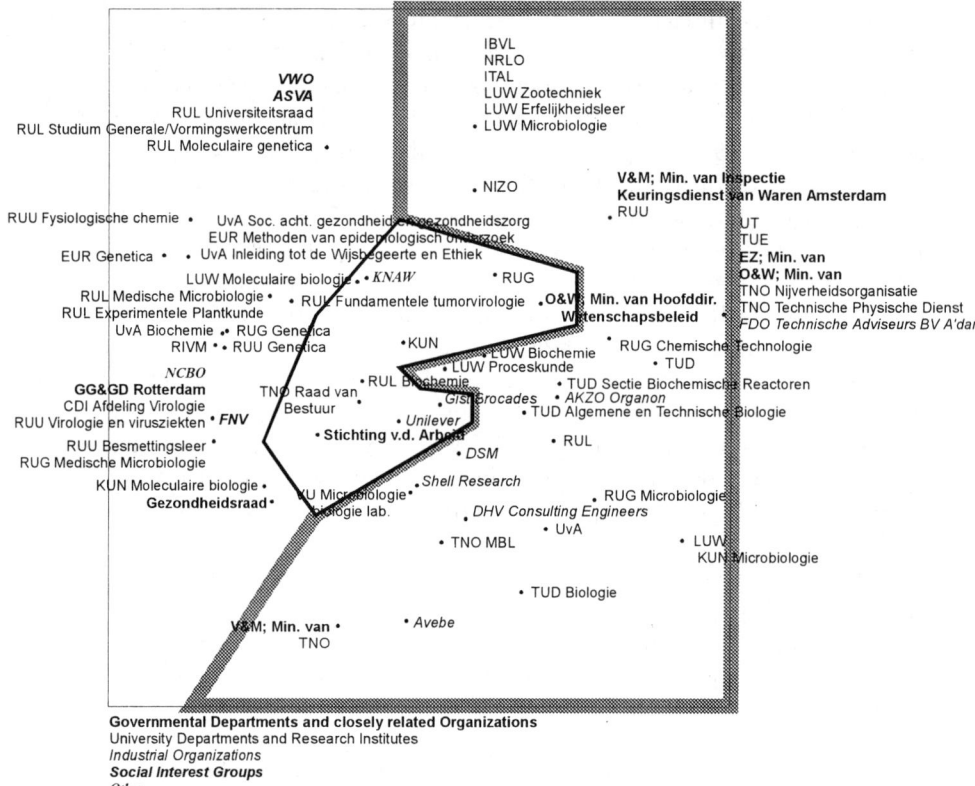

Figure 1: Network of Institutes Connected with Persons on the Basis of Shared Projects and Committees in the Period 01-01-78 to 05-30-80, Stress is 0.08491

An interesting example of the merger is provided by the *Consumentenbond* (the main Dutch consumers' organization) which linked up with the label "biotechnology" in 1990, when it wrote in its magazine: "Like every new technology, biotechnology offers opportunities as well as threats. Therefore, it is well that not only experts and interested firms think and decide about its desirability, but that the voice of consumers is heard as well. To achieve this a "Biotechnology Platform" will be established, partly on our insistence" (B). The entry of the *Consumentenbond* into the "biotechnology" network is through the label of "consumers," and this is related to the marketing of products in which genetically modified organisms, i.e. biotechnology, are involved.

While biotechnology is now the overarching label, concerns and opposition are still focused on genetic engineering — or, as is now common parlance among proponents, "genetic modification." Environmental release of genetically modified organisms, and the ethics of modifying animals have become the issues around which the basic struggle is being played out. While border crossings have become common, there still is something of a no man's land between the worlds of the modernist promoters of biotechnology and the concerned or merely reluctant receivers of biotechnology.

The patterns created by networks of labels and networks of actors aligned under such labels can be observed for other technologies. "Nuclear" became a contaminated label in the 1980s; hence the acronym NMR, for Nuclear Magnetic Resonance, has been replaced by MRI, Magnetic Resonance Imaging — even if this usage of "nuclear" had little to do with "nuclear" power. A formula with chemical symbols is suspect, in spite of campaigns by the chemical industry. "Natural" is a potent label, and actors will struggle to appropriate it for themselves. The antagonistic element in these sociocultural patterns derives from the association with "good" or "bad" that certain labels carry, and that actors will want to capture, or avoid.

Positioning, and the battle of labels, has implications for evolving sociotechnical orders. The impasse around nuclear power, and the attempts to create an inherently safe nuclear reactor, are one example. Labeling has become a down-to-earth issue in modern biotechnology, as it did in the case of the "scratching" egg (see Van de Poel, this volume). It is now being debated whether products like cheese or tomato sauce that have been produced with the help of genetically modified organisms, have to be physically labeled as such for retail sale. Whatever the outcome of this debate, it will certainly have consequences for the way such products are made. Our analysis indicates that the outcomes do not derive in a linear way from decisions (with ancillary argumentations and predictions) but also depend on networks and alignments, and on facile dichotomies between what is "good" and what is "bad."

5 Compartmentalized Cultures

The third socio-cultural pattern that we discuss has less to do with spokespersons, and their key role in the early stages of technological developments, than with ongoing practices. Biotechnology has evolved just far enough for us to explore how it might be embedded in practices, how it might, so to speak, be "domesticated." Synthetic chemicals have already become part of our daily life. This has forced people to evolve routines and other cultural ploys for handling chemicals. An overall danger culture has evolved, which is productive, even if full of tensions and concerns about the hidden dangers of chemical compounds (Rip 1991). Similarly, the appreciation of computer technology has been transformed now that personal computers are part of daily life. In the 1950s and 1960s, the computer was often depicted as the new Moloch which had come to replace/destroy us. But at the same time, there was the counter-myth of the computer as the willing slave or *Golem*. Our present personal computers are such slaves (even if the software tends to discipline the user, rather than the other way around). In cartoons, nuclear plants are always depicted as monstrous. This betrays exactly how nuclear technology has (failed to) become embedded in our society.

Some of the dangers of radiation have been known since the time X-rays were discovered, and the specific dangers of radio-active materials since the 1920s, with the use of radium to illuminate the hands of watches and alarms. When it turned out that nuclear technology could serve important social and political goals, like demonstrating American military superiority, and opening up the possibility of "electricity too cheap to meter," a specific occupational culture was called into being. In the 1940s and 1950s, nuclear technologist and operator were feared and respected occupations, as the blacksmith with his power over fire used to be feared and respected in earlier times. But by now, the occupation is clearly a beleaguered group. Why has this come about? We would suggest that this shift, and the dynamics of the preceding examples as well, all result from the *compartmentalization* of the danger culture.

The principle of handling the danger of nuclear technology was containment: physical containment, but also social and bureaucratic containment, with separate organizations, with special access regulations, with checks both on radiation exposure and on political reliability. Given such containment, with nuclear energy enabling the production of electricity, but at a separate location, the actual practices of handling danger, with the attendant daily routines, were limited to an inner circle. The outer circles knew about the danger, could trust safety measures and/or be afraid, but could not integrate dangers and routines in their daily practices, except in the form of images and myths; the image of the deadly nuclear plant is one example and the idea (myth) of the runaway nuclear reactor which features in risk discussions, another. Such practices would be expected to lead, and have indeed led, to a structural gap between the inner and outer circles.

In other words, the development of nuclear technology created its own separation of proponents and opponents. The coupling of risk and responsibility in a coherent danger culture, which takes a productive form in the occupational culture of miners and mountaineers, cannot, therefore, be expected in the case of nuclear technology. The danger culture of nuclear technology is compartmentalized. A semblance of a working relationship between the compartments can be created by public participation, and by inner circle receptivity to outside comments. But this is a precarious relationship. Nuclear technology would have bogged down completely in many countries had not the risk of global warming through the greenhouse effect been discovered in time.

There are similarities with the development of modern biotechnology. The genetic engineering part of biotechnology is novel and strange, and must be domesticated somehow, that is, become part of our culture. The main route that has been followed is, again, containment: physical and biological containment and bureaucratic containment through guidelines. And again, an "inner circle" has emerged, and it alternates between seeing itself as a Gideon's band, working on a bright future for mankind ("what's good for biotechnology is good for society," to paraphrase the saying about General Motors) and as a beleaguered group that was almost ousted out of Denmark and Germany, and fears similar ostracism everywhere.

So there is some compartmentalization of the danger culture of biotechnology. (It might have been less if key groups had been more reflexive, and less concerned about political legitimization. But it is humanly impossible not to react that way.) Ironically, though, what is happening now is that the actual implementation of biotechnology is breaking through the containment principle. Environmental release is a striking example: physical containment is impossible, full biological containment is inconsistent with the need to create robust rather than crippled micro-organisms: the new organisms must be able to survive in a hostile environment. The only containment left is bureaucratic containment.

Risk and promise become entwined when biotechnology becomes part of daily practices in agriculture, in animal husbandry, in the health sector. As we have chemicals, we will now also have biologicals. Routines will emerge: for instance, after the first experience with immunological side-effects of biotechnologically prepared insulin, people may well go on using such preparations, but be on the look-out for side-effects.

The working out of the danger culture is also visible at the level of regulation. One example is the use of principles, like the pathogenicity principle, to reduce uncertainty in decision-making about the risks of genetically modified products. For production, rules of Good Manufacturing Practice, Good Industrial Large Scale Process and Good Development Practices, play a similar role: they are becoming routines for the actors within biotechnology, and are authorized at the political level.

While the gap between the compartments is narrowing, certainly compared with the yawning chasm in nuclear technology, there are still myths involved, e.g., the genetically modified organism as a "monster" that has to be prevented at all costs. We are not implying that these myths should be done away with: Viable cultural orders must contain myths. The question is whether the myths are productive myths.

Practices, routines, and cultural legitimizations become a coherent whole through myths. In occupational cultures we see this in the way occupations define themselves as better, more courageous, and devoted to a high purpose, in order to manage and maintain their daily confrontation with danger. For chemicals, there is a myth of purity, especially with regard to "introducing" chemicals. The maze of norms and regulations accordingly has to create an order that maintains the distinction between pure and impure.[14]

At the moment, the myths of the runaway organism and of the "monster" prevail, implying that counter-myths about harmlessness and beneficence will be ineffective. In the face of concern about genetically modified tomatoes or cheese, the counter-myth is: "See, the tomatoes, or the cheese, are just like ordinary tomatoes or cheese (only better, or cheaper, or both)." The monster is declared to be harmless, even friendly. Whether this is actually the case or not, the point is that cultural patterns, like myths, do not change simply through declarations. However, they do change, but by other means, as the example of the computer shows.

6   In Conclusion

Evolving socio-technical orders show structural and cultural features which persist over time and across technologies. We have highlighted three socio-cultural patterns, using modern biotechnology as the main example but including episodes from nuclear technology, chemistry, and chemical technology. These socio-cultural patterns shape technological developments, as they themselves evolve, partly because of the specific plots of the technologies that are being played out. Socio-cultural patterns orient actors, enabling them to reduce the uncertainties introduced by new technologies. We have emphasized the antagonistic component in these patterns and have argued that these are more than circumstantial, or only a reflection of the general conflictual nature of society. They are part and parcel of societal reduction of uncertainty. By means of relatively stabilized but antagonistic patterns, coordination is created in the evolving socio-technical order. We call this process antagonistic coordination.

The next question then is how productive such antagonistic coordination is and can be. It is one way in which society learns to handle new technology (and itself), and it is important to recognize it as such because it is related to the possibility of technological development becoming reflexive, and being amenable to "management," though not from a center of power.

To discuss our findings and the question of learning, we shall make a little detour and ask whether antagonistic patterns shape all technologies, and are necessary for coming to terms with new technology. Micro-electronics and information and communication technologies appear at first sight, to be an instance where antagonistic coordination does not occur.

While there is no proponent-opponent dichotomy (at most local conflicts about automation) and satisfactory technology evolves without conflicts around the articulation of costs and benefits and the definition of requirements, at a deeper level one can see the antagonistic patterns and the typically modern division of labor around the introduction of new technology. Information and communication technologies are strongly linked with the labels of progress and modernity. Originally, progress was articulated as a better society to live in; by the early 1980s, a transformation occurred with the spread of micro-computers (the PC revolution, advocated by proponents as a positive transformation of our daily lives) and the idea that technical means for communication anytime, anyplace, could be, and had to be, developed.

The promise of information and communication technologies is played out at all levels, from the projections of politicians and industrialists that it will restore economic prosperity, to our daily practices, where the technical means for communication have become commonplace and one has to be computer literate to survive, and the symbolic order, where the computer has turned from a mysterious device behind glass walls, cared for by a white-coated priesthood, into a mobile device that is a cherished personal symbol of modernity. Being connected to the World-Wide Web is considered necessary, without it being clear why one has to join — actually, what one connects with is the label: one can demonstrate that one belongs by referring to one's surfing (or crawling) the web.

It is true that other patterns — as articulated around chemical technology, nuclear technology and modern biotechnology — are not applicable to information and communication technology, and do not shape their further development. The risk repertoire, with its focus on long-term, large-scale and possibly irreversible impacts on health and environment, does not seem applicable (it does relate to the material components, but these are rarely debated publicly). Another repertoire has emerged which highlights precision and virtual reality: information and communication technologies allow myth-building about, as well as actual investment in, surgically precise weapons, shields against star wars, and the end of war as we know it, because the battles can now be fought within the computers.

Antagonistic coordination as we described it does not, at the moment, apply to information and communication technologies, although some of the patterns we highlighted are recognizable. One could argue that antagonistic coordination, with its dynamic of interaction between proponents and opponents, occurs when new technologies are distant from daily life and appear to threaten (actually or symbolically) values like health, safety, and ecological sustainability that we agree should be protected. In such a situation, there is socio-cultural space for control of technology. If the conditions are different, as in the case of information and communication technologies at present, the issue of societal control cannot be put on the agenda convincingly (Van den Daele 1989).

Risk has become the accepted criterion for discussion and implementation of control of new technologies, but for other features there is no articulated socio-cultural pattern. For information and communication technologies, one can think of the ambivalence of jobless growth, and other effects of delegation to intelligent machines. For modern biotechnology, a discussion is just getting off the ground whether new products should also be assessed in terms of their societal value. The fact that it is called "the fourth hurdle" indicates the extent to which it is seen by proponents at least, as part of the storyline of proponents and opponents.

So we would argue that new technologies, because of their inherent ambivalences and their expected pervasiveness, will be more, rather than less, involved in antagonistic patterns, and that learning about the technology and its impacts will proceed through antagonistic coordination.

Will the antagonistic interactions be productive? Productivity is assessed by the actors themselves, in terms of their own immediate interests and goal achievement, but also with an eye to whether or not impasses will occur and conflicts become manageable. Productivity can also be assessed in a more distantiated way, by asking what the conditions are for a productive division of labor in antagonistic coordination.

From our analysis we conclude that one condition for learning how to handle a new technology is that it gets adopted in daily practices. Underlying tensions and problems can even become invisible in the resulting socio-technical order because of the learning that has occurred. The other condition appears to be that the nature of the socio-cultural patterns and the possibility of antagonistic coordination have to be recognized, and to some extent "managed." Learning how to learn (Jelsma 1995) requires reflexivity.

Notes

1 The concept of "storyline" is also used by analysts like Hajer (1995) to understand the dynamics of controversies. A "storyline" can unite groups and coalitions, more or less like "founding myths" reinforce the coherence of tribes and organizations.
2 The innovation race is a fact of life for technology actors (see micro-electronics for clear examples) and one that cannot be controlled by individual actors. So what happens is that actors make strategic assessments about what competitors will do. The accompanying cultural repertoire is about technological gaps (between countries and/or firms), and about compulsively following-up almost every lead, because if it can be done, one must assume it will be done, if not by oneself, than by one's competitors. One even speaks of "generations" of technology that will, inevitably, follow each other (see Van Lente and Rip, this volume).
3 Such processes of indirect network formation guided by shared reference to a label is quite general: It has been documented qualitatively for "ecology" by Thomas Söderqvist, and is taken up systematically by Siebe Talma for "biotechnology."
4 Note that naming occurs, and can become enshrined in management tools like SWOT analysis: Strengths and Weaknesses (of the firm) and Opportunities and Threats (from the environment).
5 Modernist introductors and their allies often use a specific gloss, by ascribing popular mistrust to ignorance and suggesting better information as the remedy. This continues in the recent debates on biotechnology. At a recent on cloning organized by the Rathenau Institute (for TA) the Dutch Minister of Public Health, Els Borst, declared: "There is much unrest because of ignorance. The Scottish sheep Dolly has encouraged all kinds of fantastic stories. This gives society the wrong idea of what is actually going on. We therefore have to start with knowledge transfer" (NRC, March 27, 1998).
6 Karl Weick (1979) has introduced the notion of "organizing" as a social process which reduces uncertainty and makes concerted action possible, and applied it to groups and organizations. We add a cultural backdrop, which itself has been built up, over time, through the "organizing" of practices and daily life.
7 Each time surgeons wash their hands for some minutes before entering the operating theater, they re-affirm "germ" as the justification of a cluster of risk-reducing practices in our own culture. This particular cluster of practices is now fully integrated in our world view, and is so stable that advertisements can use it to help sell products, as when toilet disinfectants are recommended "because they destroy all the germs."

8   The concept of "momentum" derives from Th. P. Hughes (1983). The issue of irreversible entrenchment of a new technology, before the consequences are widely visible, has been discussed by Collingridge as an "entrenchment dilemma."

9   Risk is a multi-faceted concept, and in order to capture the complexity, one should distinguish danger/safety (which is at the level of action) and risk (possible danger and effects, so important for decision making). One implication is that industrialists and government agencies have different ways of handling risks (Rip 1986).

10 The term "boundary object," as introduced by Star and Griesemer, refers to scientific objects "which are both plastic enough to adapt to local needs and the constraints of the several parties employing them, yet robust enough to maintain a common identity across sites. They are weakly structured in common use, and become strongly structured in individual-site use. These objects may be abstract or concrete." An example is the concept species. "This is a concept which in fact described no specimen, which incorporated both concrete and theoretical data and which served as a means of communicating across both worlds [the world of collectors, trappers and other non-scientists, and the world of scientists of natural history]" (Star and Griesemer 1989: 393, 410). Treating the terms "biotechnology" and "genetic engineering" as boundary objects means looking at them as terms with no fixed content or meaning used by all sorts of actors. Actors can link with other actors using these terms without consensus about their precise meaning. A boundary object thus provides an important means for coordination, especially in the early stages of development of a technology.

11 The definition was used at the founding of the Dutch Biotechnological Society (1978-1979), and used in a report to the Ministry of Economic Affairs which explored the economic potential of biotechnology.

12 The Dutch Biotechnological Society adopted a logo featuring all four of the basic disciplines, linked to different areas of application by the term "biotechnology."

13 The database consists of participation of persons and institutes in activities (joint research, committees and working parties, public events). Activities are called biotechnology and/or genetic engineering if one of these terms is used in the activity itself. FNV, ASVA, and VWO are the names of the social interest groups. Names beginning with RU (RUG, RUL, RUL), UvA, EUR, LUW refer to (departments of) universities. Names beginning with O&W, and V&W indicate offices within government departments. The industrial firms involved are AKZO, Gist-brocades, Shell, and Unilever.

14 A further example (to expand the argument to less manifestly "technical" dangers) is how the risks of life in the big city have generated routines for handling encounters, when and how to be careful, as well as "urban myths" that portray what may happen, and the "street wisdom" necessary to avoid these fates (see Wachs 1988).

Sources

A. Berg, P. (1974): "Letter to the USA National Academy of Sciences," *Nature,* 253, 23: 12.
B. *Consumentengids* (1990): "Erfelijk Knutselwerk Verlaat het Laboratorium: Steeds Meer Produkten door Moderne Biotechnologie," April, 4: 230-233.
C. Meer, R.R. van der (1980): *Biotechnologie en Innovatie.* Den Haag: Ministry of Economic Affairs.

Chapter 11
Getting Case Studies Together: Conclusions on the Coordination of Sociotechnical Order

Cornelis Disco and *Barend van der Meulen*

1 Introduction

This book moves beyond the current dominant focus on contingency and local practices in constructivist studies of science and technology. The contributors argue that technologies are shaped by the force of already existing and emerging socio-technical order. They look at more than how some bit of technology is developed in a local setting and do more than just "follow the actors." However, refusing to be locked in to contingency and local specificity, they also reject an opposite fixation on some particular structuralist or political/economic persuasion. Their "societal" constructivism is also a pragmatic constructivism. The authors steer a middle course which avoids the flatlands of either voluntarism or structuralism, either micro or macro analysis, and which instead infuses constructivist analysis of science and technology with a sensitivity to the dynamics taking place among and above locations.

The key to understanding how technology and technology's actors are constrained by prior and emergent socio-technical order is to recognize that implemented technology is never the direct result of any deliberate local actor strategy, but a phenomenon emerging simultaneously at local and global levels. Patterned linkages between local and global levels are the key to how actors coordinate (and are coordinated in) the collective and dispersed production of new technologies. As we argue in Section 1, this entails both the production of a global order and its reversal onto local practices. This dual movement has two main forms: first, abstraction of elements of local practices into constructs and their subsequent dispersal among locations as "resilient mobiles;" second, the institutionalization of practices at the global level.

Our argument for agency being a dialectic of discrete levels nonetheless fails to address an important challenge of technology studies: the specificity of coordination of new technology — as opposed to societal novelty in general. What makes technological renewal different from the renewal of social order in general? What is the role of technological knowledge and of technical artifacts? Is there something like socio-technical coordination, different from other coordination mechanisms like markets, hierarchies and networks? To answer these questions we examine socio-technical coordination in Section 2 and assess how the case studies presented in this book can contribute to a general theory of the coordination of socio-technical action. Section 3 concludes with brief remarks on what this perspective can contribute to more effective steering of technology development.

2 Local Orders and Global Orders

What do the various contributions to this book tell us about how coordination emerges from the linkage of local and global levels in technology development? How do global levels emerge from local practices? How are local practices enabled and constrained by global orders? How does the global level itself become a site for reflexive practices and what are the consequences for practices at local levels? How can the metaphors of local and global be exploited for describing how not only spatial, but also temporal and discursive locations are coordinated? As outlined in the introduction, the basic point is that coordination among local actors depends on their being constrained in particular ways by a global order of social constructs and agency which emerges from — but also stands apart from — local socio-technical practices. The global order has, or develops, a partly autonomous dynamic which reverses back onto the locations and enables or constrains actors in meshing goals and activities — whether in synergistic cooperation or in antagonistic games.

One classical example is of course the market. Markets are not just there, but function because of local exchanges which are mediated by prices. Price formation depends on these local exchanges, but is definitely a characteristic of the market, not of local exchange (Coleman 1990). In modern societies markets are regulated by all kinds of institutions because of the reflexive agency of governments, producers, and others who want to improve the market and prevent market failures. The idea of prices as only a global-level result of numerous local exchanges is an economistic myth. The example of markets as a form of coordination makes it clear that at more global levels patterns emerge which, while they depend on local levels, are not just aggregations of local interactions. This reciprocal determination of local and global levels is the source of coordination.

11 Getting Case Studies Together

A Genesis of Global Orders

Coordination is of course a ubiquitous feature of societies. In human societies coordination is achieved by means of the linkages of local practices with diverse sorts of global orders. This is what makes social order possible. Markets, political regimes, five-year plans, legal systems, religious beliefs, the prevailing technological state of the art or the current state of scientific theorizing — all are examples of global orders which coordinate local social action and stabilize social order. These global orders are all historical products, subject to change, but relative to any given new technology they are modes of coordination already in place that provide a basic socio-technical and cultural framework for the specific work of making new science and technology. We can treat these global orders as given social structure which constrains and enables any and all efforts at getting new technologies together. These basic global orders, in other words, define the specific socio-historical context of technological novelty. They provide the basic sociological frames for explaining, say, the design and construction of mediaeval cathedrals or the Woolworth Building in New York.

In addition to these existing global orders, getting new technologies together also entails the creation of ad hoc global orders which serve to coordinate the relevant technology actors themselves as they go about defining and developing the new technology. These ad hoc orders are always built up in the context of the long-term global orders and, while possibly synergetic with them, may also be attempts to overcome their structural constraints: for example, to influence market demand or to define and legitimize new technologies which are initially beyond the commercial, political, ethical or religious pale.

In this book much less attention is given to stabilized modes of social coordination already in place, that is, to the historically given points of departure and general social contexts for new technologies, than to modes of coordination that are developed on the fly, that is, as adjuncts to getting new technologies together. In this sense our project differs from that of fundamental socio-historical critics of technology like Marx, Mumford, or Ellul. In this book, global cultural, cognitive and historical orders are pretty much taken for granted; our aim is to understand how the actors closest to new technologies produce new bits of global order and how these subsequently coordinate the definition and development of new technologies.

How do global orders emerge from heterogeneous local practices in the first place? The basic mechanisms are abstraction and aggregation. Abstraction is the process of extracting the general features from an indexical local practice and representing these in some abstract (symbolic) form. This is often an adjunct to local practices themselves inasmuch as actors will generally try to comprehend and record enough of what they are doing to be able to reproduce successful strategies at a later time and to cope competently with non-routine variations. Moreover, as Henderson (1991, 1995) has shown, such abstractions are necessary to coordinate action at local sites where there is typically a division of design labor and responsibility (see also Disco, Duncker and Disco, Van der Meulen in this volume). In any case, abstraction as the symbolic re-representation of local technological practices opens the possibility for salient features to become mobile across multiple locations. This means, to simplify somewhat, that good ideas can be picked up at other locations and re-contextualized into prevailing practices there.

Classical studies of technological diffusion (Ogburn 1945; Gilfillan 1935) simply assume that new innovations spread from the location of their inception to become standard practice in an entire population of relevant firms (Abernathy and Clark 1985). But what actually diffuses is not the local practice which performs the innovation, or even the innovative artifact itself, but some resilient and mobile representation of it: A drawing, a technical model, a patent, a verbal description, a report. Such transportation of salient features by means of abstract representations is described, for example, by Stemerding and Hilgartner in their chapter on classical taxonomy and the human genome project. They describe how successive systems of classification developed by keepers of cabinets as part of the local practice of ordering and managing their burgeoning collections, are transformed into texts, drawings and verbal reports and so become adoptable at numerous other locations (allowing comparisons and cumulation of knowledge across the many spatial locations of the contemporary botanical and zoological worlds). Neither the cabinets nor the local practices of classification actually move through space; what circulates are resilient mobiles (i.e. the texts, drawings and verbal reports by itinerant keepers and naturalists).[1] Here, the global order is based on decontextualized constructs abstracted from local practices at the various cabinets and circulating among them. Stemerding and Hilgartner argue that the present-day human genome project is similarly dependent on the circulation of standard methods for classification or sequencing. These textualized (or computerized) standard methods become the objective framework of a global order which allows the locations to produce meaningful cumulative work.

But global orders do not exist merely as equivalences and linkages among locations secured by the circulation of mobile abstractions.[2] They can also develop sites for reflexive agency based on the aggregation (and further development) of the resilient mobiles circulating among the locations. This type of activity can be a strategic historical opportunity for particular kinds of actors, (e.g., professional engineers) and can lead to the emergence of global actors. As Disco et al. (1992) have argued, global orders can develop as divisions of design labor based both on the hierarchical composition of artifacts and on the specialized labor of developing standards and technical models (Van der Meulen, this volume). This meta design is the historical role first of polytechnical schools and later of universities, engineering consultants and corporate research laboratories. It is instrumental in the establishment of global technological orders. In this volume, in more specific ways, global level agency appears, for example, in the form of the Dutch Poultry Research Institute "Spelderholt" (Van de Poel), as Dutch automation research institutes and fora (Van Oost), as accelerator program committees (Disco), and as the Dutch HDTV platform (Van Lente and Rip).

B Reversal

The genesis of a global order is a necessary but insufficient condition for the coordination of new technologies. In order for coordination to occur, reversal must take place — that is, the global order must assume the nature of a structure of possibilities and constraints for local agency. This does not necessarily mean that actors are aligned in some functional sense — as, for example, in an interorganizational division of labor or in supplier-customer chains. It may also mean that actors can take account of one another as competitors or rivals in structured games. Strategic rivalries in market competition or technological tugs-of-war with the state as arbiter are examples of the kinds of antagonistic coordination which articulated global constructs (like prices, cultural symbols, laws, technical standards, technical models) make possible.

The idea of reversal does not view local actors as sociological dopes. Actors are never simply compelled by the contents of global orders. Rather actors can exploit the global level as a set of resources (for example because it enables them to solve local problems or because it provides information for making rational decisions) or they can accept it as a set of constraints (because it ranges particular choices and behaviors along specific cost gradients). However, actors can be indifferent to resources (and opt to bear the associated costs of passing them up) or they can choose to disregard constraints (and opt to bear the associated costs of deviance). In short, the possibilities and limitations inherent in the global level are always relative to specific costs, and actors are in principle at liberty to evaluate these as they see fit and take the consequences. Sometimes, however, the costs are very high and the margins for choice small.

For example, Van de Poel asks what it is that aligns poultry farmers so uniformly behind the battery cage and what chances there are for opposition movements to break this spell. He shows that poultry farmers and their political and scientific allies invariably argue that the economics of egg production force them to use battery cages. They argue that these systems are the most cost-effective means of producing eggs and in view of unbridled competition on the national and international egg market, any farmer not producing at this level of efficiency will simply go out of business. So according to farmers' accounts, they are prisoners of the antagonistic coordination imposed by the egg-market. However, given the emergence of a growing market segment in specially marked scratching eggs — the volume and price elasticity of which remains to be seen — as well as the availability of alternative hen-housing systems, Van de Poel finds this an unsatisfactory argument. He explains poultry farmers' obstinacy not as an inescapable economic constraint, but as a personal choice which is encouraged by a broad commitment to efficiency. This idea is embodied in a wide range of ideological expressions, texts, artifacts, design heuristics, technical models, and ethological theories. Van de Poel explores how the idea of efficiency — a major modernist idol — has become a guiding principle in the poultry world, i.e. a basic rhetorical point of reference for judging whether particular practices and technologies are, to put it simply, good or bad. The producers of battery cages exploit this shared worship of efficiency to promote their wares. In their advertisements they use rhetorics of efficiency which are available as cultural resources at the global level. However, just as the market is slowly being transformed by the economics of the scratching egg, so is the cultural sphere being transformed by a re-evaluation of the primacy of efficiency — in this case in relation to the values of animal welfare. Now poultry farmers face a stubborn opposition, which argues that animal welfare is an important value too. This, as Van de Poel notes, is visible in the adoption of welfare arguments by defenders of the battery cage themselves. However, the changes in these two domains — the economic and the cultural — have not arrested the spread of the battery cage. Van de Poel suggests this is because there are additional modes of coordination impinging on poultry farmers which allow them to persevere in embracing efficiency and to resist possible new economic opportunities or novel values. Among such modes of coordination, Van de Poel mentions politics and legislation, as well as prevailing scientific paradigms in the field of poultry research.

Van de Poel calls this multiply structured field of poultry husbandry a design regime, intending to convey how different types of global levels together constrain the design of systems of chicken housing and, in this case, support the entrenchment of the battery cage. At the same time, his analysis underscores the precariousness of this seemingly robust and resilient regime. Were a few more props to be knocked out, a situation could develop in which the actual costs to poultry farmers of opting for battery cages would outweigh the perceived costs of changing to new systems for egg production. The idea of design regime suggests that when this situation arises, new structures of supply and demand, guiding principles, heuristics, cultural rhetorics, paradigms, legal norms etc. will be articulated as global pendants of emerging new forms of local practice.

The other contributions to this book also play out this theme. The actors and the contexts differ, but the idea of reversal as the insinuation of a global level of constraints and resources into the reflexive strategies and practices of actors remains basic. Some accounts emphasize how the global order of resilient mobiles enables functional divisions of labor across locations, or synergetic coordination, while others emphasize how the global level helps actors maintain conflict-ridden strategic games, i.e. antagonistic coordination. In the next sections we will articulate some examples of synergetic coordination, leaving a discussion of antagonistic coordination to our reflections on coordination in general.

We distinguish two modalities of reversal, corresponding to the two sorts of contents we have described in global order, that is, constructs and global agency. Constructs refer to all manner of mnemonic and coordinating devices like ethical prescriptions, scientific theories, or mechanical devices which may enable actors to align behaviors in ongoing streams of social action. Global agency refers to purposeful and reflexive action at the global level itself, either on the part of local actors investing part of their time and energy in global-level events (meetings, standards committees, lobbying, etc.) or on the part of dedicated global personnel and organizations. In practice these two modalities cannot easily be seperated; analytically they correspond roughly to the distinction between non-human and human actors.

1 Reversal of Global Constructs

Van de Poel's story already provides a number of examples of the reversal of global constructs. Van Oost's account of the automation of the Dutch Postcheque and Clearing Service (PCGD) also centers on this type of reversal. She provides a good example of how global cultural patterns impinge on the design of a new local division of labor. The basic challenge facing the PCGD administration was how to introduce computer processing of transfer orders while avoiding both stagnation in the work flow and resistance by employees. Moreover, given the existing state of the art, data input via punch cards appeared to be the only reliable short-term option. This required an intermediate phase of tedious key-punching (which it was felt would be superseded by automatic reading systems within the foreseeable future). Given the PCGD's constraints, the tricky question was: Who was to occupy which positions in the new division of labor — in particular, who was to get the interesting jobs with career perspectives and who was to do the drudgery? Van Oost argues that on the grounds of the existing gender distribution in middle management within the PCGD, many more women should have been selected for interesting career-track jobs in automation than in fact was the case. She shows that PCGD management was able to solve its problems: to avoid employee dissatisfaction, to maintain production and to find new workers to do the drudgery of key-punching, by mobilizing a global level of norms defining gendered conceptions of appropriate work. Dutch women were not expected to have challenging careers and so those already entrenched in the organization could be passed up for interesting jobs without causing overt rebellion. Most women, insofar as they were employed at all, expected to work only a few years and then quit on marriage — hence they were considered ideally suited to fill tedious and temporary slots in divisions of labor — which, as keypunch operators, they did for a surprising number of years.

Duncker and Disco provide a further example of the coordinative force of global level constructs. In their account of the Twente Micro-optics project, they show how emergent global level of "rules of transformation," "dictionaries," and "hybrid repertoires" is gradually helping to transform what started out as a strategic alliance of heterogeneous disciplinary groups into a mission-oriented division of labor. In the case of the Brookhaven g-2 project, this global level is already present as a linguistic resource native to high-energy physics and routinely enables different disciplines and specialties to work together in the context of local practices. Disco's separate account of this latter project shows how the global frames accumulated from past bricolage provide common goals and task specifications for successive phases of bricolage. In this way, global-level structures produced in the course of experimentation itself help to coordinate subsequent agency, both in space and time. In a similar vein, Van der Meulen shows how globally available technical models coordinate the efforts of heterogeneous actors in the domain of ship propeller design. Consensus on such technical models, (a key feature of Van de Poel's design regimes) allows propeller manufacturers, shipyards, insurance companies, researchers, ship owners etc. to agree on evaluations of performances in spite of very different competencies and interests. The evolution of these technical models simultaneously mirrors the shifting dynamics within the actor networks themselves and thus coordination of the interactions through time is ensured as well.

2 Reversal of Global Agency

Global orders are not only repositories of more or less compelling constructs which allow actors to coordinate interaction — like the houses and hotels in Monopoly. They also include emergent and partly autonomous agents and interacting organizations. Their agency, which as always can be reflexive and innovative, can reverse onto local agency in purposive and strategic ways. This global level of agency cannot be interpreted as simply a summation of punctualized micro levels as Callon (1991) seems to suggest.[3] The local-global distinction is not equivalent to a simple micro-macro distinction, of which at least some variants argue that macro-interactions are only micro-interactions writ large: that macro agency is only micro agency punctualized and represented at a more inclusive level. Macro level agency as described in our accounts is distinct in kind from agency at the micro level for three reasons. First, in getting new technologies together agency is distributed, not only over locations, but also between local and global levels. This means that there are divisions of labor between locations and global levels and that neither the goals nor the content of agency at local and global levels are the same. Second, local situations are generally much more malleable for actors because they are directly accessible and often encapsulated within organizations in which hierarchical authority can be mobilized. Acting at global levels, because relations among actors there are primarily mediated by contracts, trust, or mutual need, is much more uncertain and in any case requires a different approach. Third, local and global settings are coupled in specific ways — via mutual delegation, aggregation and reversal — giving rise to a much more complex dynamic than that entailed in the punctualization-representation relation between micro-agency and macro-agency.

An example may help here. Smit, Elzen, and Enserink tell the story of what seems to be a macro-actor in Callon's (1991) specific sense of the term, i.e. an actor network, the European Fighter Aircraft network, composed of many punctualized national military aircraft networks. In Callon's (and Latour's) conception, the macro network is a network of networks, i.e. the relationships and content of the macro network consist of represented micro networks. Our argument is that the macro network must be seen as a global network with emergent contents (intermediaries, resilient mobiles) and actors which are not simply a summation of all the micro/local networks taken together. Smit, Elzen, and Enserink show that the EFA network is indeed a global order in our sense because the major stake of the interaction is the specification of military requirements and guiding principles for a fighter which is not specific to French, or British or German needs, but which must be a sensible weapon for all participants — not only in terms of military strategy but in terms of the commercial survival of domestic defense contractors. So, while each country has already developed specific needs and requirements and has adapted its defense industry accordingly, what is at issue at the global EFA level is the specification of new needs and requirements which are not simply the sum of existing local needs and requirements and which will in fact compel national defense contractors to depart from beaten paths. The point in the present context is that settling on mutually acceptable requirements is facilitated by new institutions which do not belong to any of the local networks and which do not represent specific local interests, but which are instituted to serve the common interest in having a joint fighter.

At the global level of the EFA network we see a complex set of new actors, some of which, like the series of conferences of Ministers of Defense or those of Air Force Chiefs of Staff, are indeed macro–actors in the sense that they consist of direct representatives of the national networks. Nonetheless they did produce global–level texts like the "Outline European Staff Target" of 1983 whose list of military requirements presumably compromised some of the interests of all the participants. This seems to apply less to the aerospace consortia that looked to the EFA as a vehicle for piggybacking projects already under development. These macro–actors tended to submit proposals aimed at steering the EFA project toward their specific commercial interests, in pursuit whereof they nearly succeeded in ripping it apart as the French withdrawal confirmed. But in addition to strictly representational macro–actors, whose role is to articulate and defend local interests in the global interaction, the global network also contains new actors whose role is to facilitate and manage interaction at the global level, for example by helping to articulate requirements, acting as legal representative of the global network to third parties, and managing internal conflicts. In the EFA case, this role (or portions of it) was delegated by the ministries of defense of the member countries to an office called the NATO European Fighter Management Agency (NEFMA).

The dramatic crux of the EFA story as told by Smit, Elzen, and Enserink is that the emergent global network is in fact constantly being jeopardized by the divisive interventions of its constituent pre–existing networks. What could save the network–in–the–making? Two things might. Either the alignment among the interests of the various constituent networks could improve so that the centrifugal forces acting on the EFA are tempered, or global level dynamics could become more strongly institutionalized so that there is a countervailing power to divisive localism. The dramatic crux of the EFA story can now be rephrased as follows: Can the EFA network become a robust global level network with its own dynamic agency or will it simply remain a field of conflict for the represented interests of punctualized sub-networks? Can it produce emergent institutions that, by dint of reflexive steering and conflict management, can improve its survival chances or will it succumb to the heterogeneous interests of its punctualized constituent networks?

While Smit et al. describe a case in which emergent agency at the global level remains quite marginal compared to other dynamics, most of the chapters describe well-institutionalized and successful examples of global agency. For instance, in the cases Van Lente and Rip use to illustrate their thesis of expectations building prospective structures, a well-defined institutionalization of the global level is either present or emerging. To take just one of them, Moore's law on the rate of development of chip size could not work as a global level constraint on chip designers in the absence of active commentators, publicists, trade journals, and a robust design regime in the field of chip design. Indeed, the very concept of field presupposes the reflexive activity of persons and organizations at the global level of knowledge production, performance and evaluation standards, market and technology forecasting, etc. Only such dense interconnections at the global level could give something like Moore's law the compelling force it has. Another example is Van der Meulen's discussion of the Lips Corporation's efforts to reestablish trust in their Grim Vane Assemblies in the wake of the *Queen Elizabeth II* disaster and hence to establish the use of Grim Vane Wheels as an element of the global design regime in ship propulsion systems. This similarly makes reference to reflexive actors who are "native" to the global level, for example insurance companies, test facilities, and various organizational loci of the international design regime in ship propulsion systems. Van der Meulen argues that reestablishing faith in failed constructions, or introducing new ones — that is, durably modifying elements of professional common-sense at the global level of the field as a whole and developing resources to support the transformation — can only succeed on the basis of a long march through the institutions aimed at recruiting the assent of the major actors.

C Local and Global as Metaphors: Coordinating Heterogeneous Time and Discourses

The categories of local and global (as to some extent also those of micro and macro) evoke a spatial, almost geographical, sense of the distribution of technological agency. This simply reproduces the commonsense perception that technologies are gotten together in many geographically distributed firms and laboratories, connected and coordinated by global orders like markets, standards, professional cultures, journals, supplier chains etc. Because these organizational actors are at different physical locations, it is easy to reify the local-global distinction as simply a topology of physical space. The local organizational settings are readily seen as sites where creative agency produces innovations which are constrained and subsequently selected at a global level. Some STS approaches, for example the quasi-evolutionary school of technology development (Van den Belt and Rip 1987) reinforce this spatial topology more than others.

This organizational-spatial topology pervades many of the contributions to this book. However, some suggest that broader conceptions of the local-global dynamic are possible. In several chapters time and discourse are also portrayed as having "locality" and "globality." This suggests that local and global can be seen as multi-dimensional concepts, referring not only to the coordination of agency through space, but also to the coordination of agency through time and sometimes across discourses. In this book locations are in fact not confined to settings in physical space, but may refer to settings in time (see the chapters by Disco, Rip and Talma, Van Lente and Rip, and Van der Meulen) or to settings in universes of discourse (Duncker and Disco, Rip and Talma). Each of these dimensions involves its own particular kind of relationship between local and global levels and its own specific organization of global levels — including the modes of coordination which emerge there. The upshot is that the local-global dynamic is not confined to the coordination of agency across spatially dispersed organizational sites, but also involves the simultaneous coordination of agency across moments in time and across different discourses.

1 Time

A number of chapters show how actors at specific locations in time (which are of course also spatial-organizational and discursive locations as well) produce constructs which constitute a global order in time. This global order enables varieties of cumulation to occur across many temporal locations and hence coordinates agency across locations in time. In his conclusion Van der Meulen notes that any actor-network is only an artifact of the investigator. It is not only an arbitrary enclosure of some particular set of interactions in space, but — the salient point here — a particular moment in the unfolding of this set of interactions through time. What keeps all the moments in time together, what governs the logic of their unfolding so that one can speak of the history of this actor network and can, in principle at least, show how later moments have unfolded from earlier ones? Van der Meulen argues that the coherence of these cross-sections in time, the coordination of the successive states of the network is ensured by a global order organized in this case around specific technical models. Disco makes a similar point, but pitched at the level of a single scientific experiment. He shows how global-level constructs emerge as products of the localized practices of organizing and carrying out the experiment. Hence, the experiment evolves at two levels: the local level of tactical improvisation (i.e. bricolage) and an emergent global level of texts, artifacts, contracts, and data. Disco argues that this global level of objective constructs, ultimately shored up into two major frames, is the main resource for actors in coordinating the orderly unfolding of the experiment through complexes of time, space, and discourses.

De la Bruhèze takes this theme to the level of interorganizational politics. He shows that the Atomic Energy Commission's post-war policy of maintaining control over radioactive waste disposal at dispersed government facilities entailed the production of a global level of rules of procedure which he characterizes as "bureaucratic" and "technological." This global level of texts (rules for on-site radwaste procedure and supporting documentation) and bureaucratic organization was the Atomic Energy Commission's main weapon in maintaining control over the coordination of radwaste procedures over space and time. However, in an ironic turn, De la Bruhèze shows how the efforts required to maintain this order, including the suppression of groups and texts critical of Atomic Energy Commission policies, produced a latent "legacy of exclusions" (of texts, persons, and groups) which was eventually forged (by means of issue linkage) into a forceful opposition to the Atomic Energy Commission policy of salt-mine deposition — at least at Lyons. This legacy, involving mutual positioning of actors, can be seen as a global level of structural alignments emerging out of the impromptu tactics of the Atomic Energy Commission as it sought to maintain boundaries and its hegemony over radwaste technology. Here, a gradually emerging global order (an unintended spin-off of the maintenance of a prevailing mode of coordination) undermines that mode of coordination in the course of time — at least once actors appear on the scene with enough power to tie the legacy together into a forceful opposition.

2 Discourse

The chapter by Duncker and Disco traces how a shared global level of discourse can emerge from different local discourses. The authors argue that an interest in communicating across discursive boundaries, such as those between disciplines and research groups in interdisciplinary research programs, can stimulate groups to establish rules of transformation and dictionaries for understanding exotic utterances. These rules and dictionaries are ancillary to the actual local disciplinary practices and are relevant only for coordinating activities among different local groups. Each local group has its own specific set of (active or passive) transformation rules for each of the other groups with which it needs to interact. The sum total of these transformation rules (or "dictionary entries" as Duncker and Disco call them) constitutes a global level of discourse within the collaboration and is in fact the medium through which the collaboration can be discursively coordinated in spite of the different systems of symbolic representations being used. This is portrayed as a self-organizing system which emerges even in the absence of reflexive central management.

Rip and Talma tacitly incorporate De la Bruhèze's ironies as they analyze how societal conflicts around new risk-ridden technologies can produce abiding global orders of texts, expectations, repertoires and social roles which coordinate further conflicts across time and discourses. Rip and Talma show how the fierce antagonisms engendered in the Dutch debate on nuclear power generated a new global level constellation of expectations and repertoires. Because they provide a definition of the situation and a script for what to do in the face of chronic uncertainty, these global resources have come to pre-define actors' behavior around every new risky technology, for example, recombinant DNA technology. The global level produced in the nuclear debate stipulates that there will be proponents and critics and it provides each of them with appropriate scripts for enacting the technodrama of societal evaluation and choice. Although this coordinates situated agency through time, it does this in part by effecting ritualized translations — and hence stereotyped understandings — between the incommensurable and mutually incomprehensible discourses of proponents and opponents. So now, while the two parties are de facto no closer to a true understanding of the texts and utterances of their opponents, they are armed with prefab dictionaries which allow them to interpret the other's incomprehensible discourse as a sequence of recognizable moves in the game which they are playing around the (now calculated and ritualized) introduction of new and risky technologies. One might imagine the development of just such societal repertoires in the course of repeated confrontations between the Atomic Energy Commission and its self-generated opposition around the disposal of radwaste.[4]

D Local-Global is More than Agency-Structure or Micro-Macro

It is enticing to reduce the local-global distinction to more familiar oppositions like agency-structure or micro-macro. Indeed, aspects of both these other dichotomies figure in the local-global linkage as well. Insofar as a global order is perceived as a robust heritage of past generations which shapes present action, it will effectively do what we normally attribute to social structure. Similarly, global orders do incorporate local practices and are indeed in a minimal sense what Callon (1991) calls punctualizations of the latter. So it is not far-fetched to consider the global order as a macro-level of constructs and interactions which incorporates many local micro-level systems or networks of interaction.

However, there are reasons to prefer the local–global distinction over both agency–structure and micro–macro distinctions. In the first place, locations as well as the global level contain both agency and structure so that a simple equation of local with agency and global with structure will not do. If we consider agency as Weberian social action, and structure to be sedimented constraints of various sorts on that action, then it is clear that local settings and global orders as we have defined them contain both these elements. The local is not the domain of pure agency and the global is not the repository of pure structure. Not only are there artifacts, rules, beliefs etc. which function like structures in local settings, there is also, as many of the accounts emphasize, agency at the global level. Insofar as there is a linkage between agency and structure, it also obtains within the local and the global and not only between them.

In the second place, as opposed to the usual interpretation of the agency–structure linkage, the local–global dynamic not only entails reversal, that is, how some stabilized and overarching order shaping local action, it also includes the ad hoc production of (at least parts of) that global order as well. While it is often argued that structure is a product of past agency, it is seldom shown just how agency at T_1 is aggregated and stabilized to form the sedimented structures which subsequently constrain future agency at T_2. In other words, while some kind of agency–structure dialectic is admitted in theory, it is rarely demonstrated empirically. This may be because constraints on present action are considered more interesting than the etiology of those constraints. Of course we have to acknowledge structure as sedimented, normalized, and stabilised socio–technical order which imposes specific cost/benefit gradients and thus constrains agency. We recognize historically sedimented constellations. But we add a dynamic analysis: structure is produced ad hoc, agency is enacted by local and global actors.

Local–global dynamics also argue against an interpretation of micro and macro levels of action as simply isomorphic layers of action, whereby the more micro levels are incorporated via punctualization into the more macro levels (like the scalar recursiveness of Russian dolls). The interesting feature of the local–global distinction is precisely the transformation of local practices as they are re-contextualized into global settings. Local practices are abstracted, generalized, and aggregated to form global orders; there is more than mere representation and black-boxing of local complexity involved.

The global order contains constructs and practices which relate to local orders, but which have their own emergent innovative dynamics. Sectoral research institutes, government laboratories, rules of symbolic transformation, gender scripts — all of these emerge only indirectly from local contexts. Novelty emerges independently at the global level but also, of course, as locations key in to the emergent opportunities and constraints.

3 Coordination and Sociotechnical Orders

We have analyzed coordination of new technologies as a societal phenomenon and as a spin-off of local-global dynamics. We have not yet raised questions about the nature and quality of such coordination. A considerable literature in economics and policy-oriented social sciences has done just that. The main thrust has been to consider coordination as the solution of a modernistic challenge: How can society as a whole or specific social interactions be organized so as best to assure maximum payoffs at the collective level despite the self-interested, profit-maximizing behavior of individual rational actors? This has been cast as a search either for alternatives to coordination by liberal markets (Williamson 1975, 1981; Granovetter 1985; Powell 1990; Johanson and Mattson 1987; Bradach and Eccles 1989; Zucker 1991) or for ways to escape from classical game-theoretical conundra like the prisoner's and the chicken dilemmas (Tsebelis 1990; Axelrod 1984).

Although coordination is a central concept in such studies, there is little attempt to analyze how coordination is in fact realized. Most studies refer to ideal situations in which coordination would improve outcomes. It is a welfare-theoretic concept, indicating that there are situations in which unilateral choices by actors do not produce the optimal outcome (Scharpf 1993). Attempts to define coordination thus stress efficiency, organization and the avoidance of conflicts:

Coordination implies the bringing into a relationship of otherwise disparate activities or events. Tasks and efforts can be made compatible by coordinating them. Bottlenecks and disjunctures can be eliminated, so coordination is usually discussed under a sign of efficiency. By coordinating a set of items something can be achieved which otherwise would not be. It is the positive performative consequence of coordination that makes it such an attractive social practice and objective. Various agents and agencies can be 'ordered,' 'balanced,' 'brought into equilibrium,' and the like, by the act of coordination. Without coordination these agents and agencies might all have different and potentially conflicting objectives resulting in chaos and inefficiency (Thompson et al.1991: 3).

The basic assumption of rational actors maximizing utility has resulted in a fascination with optimal payoffs; this to the detriment of understanding how coordination is actually achieved and maintained. We need to know what is in fact coordinated, by what means, and to what ends.

To achieve this we must combine insights taken from game theory and the comparative studies of markets, hierarchies and networks with those embedded in studies of socio-technical coordination. The contributors to this book have taken up the challenge of analyzing the dialectics between action and socio-technical constructs at different levels of aggregation. Researchers, designers, and users act simultaneously in the context of both local and global orders. These orders are resources and references for their activities. At the same time, the activities augment these orders: they reproduce them or modify them. But the relationship to local and global orders is not identical. As noted above, from the perspective of local actors anyway, local orders are much more malleable, much more subject to voluntaristic interventions, than global orders. Global orders are, in part at least, just there and must often be dealt with by local actors as given states of affairs — or reformulated as a special challenge rather than being suffered as ongoing interaction. This means that global orders assume some of the compelling properties usually associated with social structure. On the other hand, as we have argued, global order also depends on local effort for its existence and upkeep — for example by delegating particular bits of local agency to the global level as in the form of standards committees, conferences, journals, etc. This interrelationship between levels of social order is the quintessence of the patterns we have identified in the previous section. And indeed, this is also the quintessence of solving the game-theoretical dilemmas of prisoners and chickens: a certain Pareto optimal equilibrium is realized because particular contingent strategies are enabled in a certain global order. But now, instead of defining coordination by its fruits (increased welfare) which are also seen as the reason why coordination is aimed at and achieved, coordination can now be looked at as itself the outcome of patterned relations between local levels and more global levels.

Van Lente and Rip show how actors can produce robust global orders over time out of promises and expectations. Promises and expectations are aggregated and generalized from local agency and become compelling frames of reference for the future concerted behavior of mutually dependent actors. Moore's "law," after all, was only an extrapolation from the collective behavior of number of chip producers over several years and has acquired binding force only because of the mutual attuning of expectations. It is true because everyone believes it to be true — or believes that everyone believes it to be true and is afraid of missing the boat — a vessel not even afloat yet. The Dutch membrane bandwagon is another instance of how texts, technical models, prototypes etc. which have been developed at local sites can be generalized and rhetorically expanded into incontrovertible prognoses and expectations which align actors in pursuit of the realization of those expectations. The more actors commit to the expectation, the more they in fact realize the promise. "Riding in the train you're pushing" is how Van Lente and Rip describe this Munchhausian *condition technologique* with respect to the coming of High Definition Television. Our point is that the train — with its paradoxical sort of existence — is in fact the global order of texts, promises, prototypes etc. available as a common focus for actors in realizing their shared objective-in-becoming.

Such conceptions of coordination expand the range of what can in fact be seen as coordination. As noted, the explicit literature on this topic tends to conceive coordination mostly as a welfare theoretic concept. It is a universal good, always worth striving for because it increases payoffs. However if coordination can willy-nilly emerge from chains of interactions, it is not necessarily limited to those situations in which positive gains are collectively realized, i.e. what we have called synergistic coordination. Then the idea of negative coordination or antagonistic coordination becomes conceivable. The role of a global level in producing and maintaining antagonistic coordination among local sites is foregrounded in several of the contributions to this book.

Smit, Elzen, and Enserink, see the global level as organized around resilient mobiles like military requirements and guiding principles. The international tug-of-war about the design of the European Fighter Aircraft remains coordinated only because all the parties are agreed that some unique set of functional requirements must be formulated. For all of the parties this is tantamount to a commitment to having a single European fighter, as France's withdrawal made abundantly clear. So the antagonistic interests are played out in the medium of a struggle over the appropriate military requirements. It is only the coordination of strategic competition via military requirements that keeps the network together at all — even if this is at least in part antagonistic togetherness.

In Van de Poel's account, global levels reverse to coordinate antagonistic local agency at a number of points. He describes how global cultural and economic resources coordinate the emerging antagonistic game between the various proponents of battery cages and their critics. It is a game, rather than a mere clash of interests, because the parties take account of one another's moves and adjust their own strategies accordingly — i.e. both are aware of the other party's considerable resources and power and must bargain rather than simply dictate terms. This game is possible because both parties must pay at least lip service to global norms of efficiency and animal welfare. In fact, part of the game involves a struggle about how these norms should be translated into situated local practices. The antagonistic game also depends on the use of visual symbols like pictures of suffering battery-cage chickens and the identification stamp for scratching eggs.

De la Bruhèze shows that the Atomic Energy Commission produced its own antagonists in the course of years of exclusionary practices. The legacy of antagonistic positioning and suppressed reports which hovered over the Atomic Energy Commission's stern coordination of radwaste practices at local laboratories and production sites from beginning to end was finally mobilized like an exterminating angel by the coalition of parties opposing radwaste deposition in the Lyons salt mine. Again, this is a case of structurally coordinated antagonism (rather than mere confrontation) because the parties are in a manner of speaking performing a grim dance, but to music which has never been played before. This theme is elaborated in the chapter by Rip and Talma, where the very steps of the dance itself are shown to be a global cultural resource for the societal processing of new and risky technologies.

A Socio-technical Coordination?

The literature on coordination generally distinguishes three basic forms of societal coordination: markets, hierarchies, and networks (Thompson et al. 1991). These forms are viewed as mediated, respectively, by prices, authority, and trust. The question that remains is whether there is something specific about the coordination processes analyzed in this volume that warrant introducing the idea of socio-technical coordination. Do the chapters in this book reveal a form of coordination that might be distinguished from markets, hierarchies and networks as a fourth form? No, and then a qualified yes.

Why might we answer "no?" Delineating yet another ideal-typical means of coordination would be at least inconsistent after having demonstrated the importance of empirically based, contextually contingent analyses of coordination and arguing against forcing socio-technical order into some n (or n+1) Procrustean beds of formal models of coordination. Instead, we would do better to laud recent technology studies for revealing the heterogeneity of making socio-technical order. As we focus on coordination this heterogeneity becomes even more apparent. Socio-technical processes are not only locally heterogeneous in terms of the kinds of elements (humans and non-humans, i.e. actants) involved, but also heterogeneous as to geographic, temporal, and discursive contexts. Coordination links heterogeneous actants as well as heterogeneous places, times, and discourses. In speaking of socio-technical coordination we would prefer not to think of it as a specific fourth form as such, but as denoting how plural modes of coordination embedded in global levels link social and technical actants located at different places, times and discourses.

However, we still want to emphasize that technology itself has an important coordinating force, both in ordering societies generally as well as in ordering the specific process of getting *new* technologies together. Technological knowledge and material artifacts coordinate agents across locations and levels. Hence, also "yes." The already well-defined role played by prices, authority and trust in the coordination models of markets, hierarchies and networks, brings the special coordinating force of artifacts and cognitive constructs, featured in several of the chapters, into sharp relief. It is clear that the understanding of coordination in and of socio-technical processes demands attention to specifically technological means of coordination in addition to the three classical modalities. The question here is how cognitive constructs and artefacts themselves are able to work as means of coordination, that is, how they can bridge differences and relate local practices via global levels.

Take cognitive constructs. There are a range of cognitions, from fundamental technological theories and associated symbolic repertoires (thermodynamics, applied mechanics) through pragmatic technical models, through a broad category of expeditious rules of thumb, to rhetorical claims for the effectiveness or promise of specific approaches and constructions. In a functionalistic mode of reasoning we could point to the positive outcomes that adopting and using such constructs have for local actors. The development of Site Tagged Sequencing as a classification and management tool for the Human Genome Project seems to be solving basic coordination problems of all participants in the project. That is, local findings can now be cast in a standardized form which is resiliently mobile (independently of centrally allocated standard samples) and which can be interpreted in uniform ways across all the local sites of the Human Genome Project. But such a functionalistic argument is of course inadequate. For one thing, it neglects the constraining side of coordination and the conflicts that might emerge on both local and global levels because of competing concepts and research practices. Site Tagged Sequencing is still a local proposition for a global methodology and as such must be rhetorically pushed within the community. Getting a particular concept adopted generally enough so that it can become a means of coordination requires struggle about technologies, professional politics, and even economics.

What is needed to wage such a struggle with a chance of success? In a recent study on how ideas affect organizational changes, Czarniawska and Joerges (1995) note that ideas that make a difference are often objectified in metaphors, labels, and platitudes. So, in struggles for dominance, complex

11 Getting Case Studies Together 343

concepts may be simplified as ideographs or shibboleths and become ideas-on-the-march. Classically, the spreading of such idea-objects is described as a process of diffusion. Actor-network theory, of which Czarniawska and Joerges are advocates, understands this as translation. Here, the malleability of the concepts is taken into account and the process is framed less mechanistically than in the usual models of diffusion. Czarniawska and Joerges (1995) stress the hybridized humans/technologies network as a material basis for more complex translation mechanisms:

> Ideas are communicated images, intersubjective creations, and therefore the "property" of a community rather than of a single person (although individuals tend to appropriate ideas and the narratives attribute them to heroes) (Czarniawska and Joerges 1995: 189).

So when we speak of global-level cognitive constructs coordinating local practices we must appreciate the accomplishment not only as the outcome of a professional-political process of pushing a particular model, theory, method, rule of thumb as the preferred modus operandi, but also as enabled by the prior existence of a hybridized humans/technologies network. The production of new means of coordination is itself dependent on the existence of prior coordination, i.e. of stabilized interactions among many local sites and a global level made possible by (strategically managed) resilient mobiles. Although Site Tagged Sequencing is being propagated as a tool for managing the Human Genome Project, its coordinating force is to a large extent dependent on its entrenchability within molecular biology in general and within the Human Genome Project in particular. Within a professional community like the Human Genome Project, whose key members are all trained molecular biologists, Site Tagged Sequencing ultimately derives its coordinating force as a method from its reducibility to molecular-biological theory and technical models of human DNA molecules. That is, genome researchers must be able to see at least the plausibility of the method on theoretical grounds. In addition, the translation process of Site Tagged Sequencing depends on extant coordination processes within the Human Genome project itself, processes maintained by journals, conferences, shared theories and networks of researchers. Similar mechanisms are visible in chapters by Duncker and Disco (the recourse to generic symbolic repertoires), Disco (the role of existing physics theory in generating an experimental proposal), Van der Meulen (technical models as the basis for new claims that restructure the design of ship propulsion systems), Smit, Elzen, and Enserink (guiding principles and military requirements embedded in underlying networks), Van Lente and Rip (the technical community within which Moore's law, or the promises of membrane technology, become compelling facts of life), De la Bruhèze (the framework of existent local laboratory practices in regard to radwaste treatment).

Compared to cognitive constructs, coordination by artifacts — technical constructs — can be more direct. The literature reveals at least three ways that artifacts can coordinate situated agency: by imposing scripts, by functioning as actants, or by physically connecting settings. These three roles are often mixed, but they are analytically distinguishable. Artifacts impose scripts when their structure and/or functioning compels particular kinds of human behavior. The notorious *locus classicus* is Langdon Winner's story of Robert Moses and the Long Island Viaducts (Winner 1977). Equally incisive are accounts by Latour (1992) and Akrich (1992) on, respectively, safety belts and photoelectric kits. Artifacts help to coordinate agency as actants when they interact with other agents in specific settings, generally in a predictable or automatic way. Latour's (1992) account of an automatic door closer as a mechanical stand-in for the traditional doorman, illustrates how such an artifact not only enscripts the behavior of its human users, but can also be seen as an independent source of agency to which humans (or other artifacts) may have to respond in creative ways. The coordinating force of artifactual agency becomes particularly salient, we would argue, in cases of failure or malfunctioning when new patterns of behavior have to be improvised by other actants involved in a particular setting. Finally, artifacts can coordinate situated agency by physically connecting different settings and in so doing transmitting constraints or opportunities among them. The large technological systems approach originally developed by Thomas Hughes (1983, 1986) is a well articulated conceptualization of how artifacts physically bind space and time. Large technological systems are literally held together by concatenations of artifacts; Hughes speaks of generators, copper wire, meters, light bulbs, electric trolleys, switches and so on in his analysis of early electric lighting systems. In other roles, i.e. as carriers of scripts and as actants, artifacts can also channel large system dynamics, and so coordinate technology development, for example, as foci of user (dis)satisfaction or as the physical core of reverse salients. In general, although Hughes' ambitious theory clearly shows how artifacts physically coordinate settings distributed across relatively large expanses of space and time, we think similar types of coordination also take place in more modest settings.

In this volume we also see persistent delineations of the role of artifacts in coordination, i.e. as elements of local-global dynamics. Different authors stress the specific ways in which artifacts can coordinate technology dynamics — as bearers of scripts, as actants, as physical links in large technological systems. At the same time there is a tendency to recognize that artifacts often coordinate along these different routes simultaneously.

Artifacts as bearers of compelling scripts are discussed both by Van Oost and by Disco, albeit in different ways. Van Oost shows how the computers

11 Getting Case Studies Together 345

introduced into the Postcheque and Clearing Service came already bearing scripts which appointed men, rather than women, as their appropriate operators. Although her account does not focus on the actual process of gendered enscripting of the machines, it is clear that global level ideologies of masculine virtuosity and control reinforced both the gendered design of computers as well as the acceptability of the script at local sites where they were used. Disco suggests that in HEP experiments, the completed detector itself contains a performative script which coordinates the actual process of data taking. Because the detector is designed to embody, to make humanly accessible, the theoretical aims of the experiment, its functioning (i.e. its self-monitoring and the actual data it produces) is the major object of study in the data-taking phase of the experiment. Its functioning and malfunctioning definitely inform and coordinate the activities of the physicists who themselves claim to be running the experiment.

The coordinative force of artifacts as actants is illustrated in accounts by Van der Meulen and Van de Poel. Van der Meulen argues that the acting-up of the *Queen Elizabeth II*'s Grim Vane Wheel during sea trials effectively shattered the brand new Grim Vane Wheel network. A formerly well-coordinated pattern of interaction among scientists, propeller manufacturers, ship owners, shipyards, and insurance companies simply fell apart. The coordination of the diverse actors in this network — i.e. divisions of labor, collective goals and tasks, delegations of tasks to local sites or global organizations, etc. — had clearly depended on specific behaviors of the Grim Vane Wheel. Specific performances (particularly sub-optimal ones) continually demanded re-coordination of aspects of the network as new patterns of action and interaction were initiated to improve performances. This took on dramatic proportions after the *Queen Elizabeth II* sea-trial fiasco when all the screws came loose at once and workable patterns of coordination had to be reestablished all over again on the basis not of a functioning artifact, but of promises based on improved technical models (of cavitation, vibration, fatigue etc.). In Van de Poel's account we see how a particular artifact, the specially marked scratching egg, becomes an important actant which re-coordinates existing market relationships between consumers of eggs and poultry farmers. Markets coordinate buyers and sellers (and supply and demand) via complex (and of course also reflexively manipulated) price mechanisms. Prices are in any case attached to specific commodities, which are also artifacts because they have been produced or at least modified by technological processes of one kind or another. Introducing a new artifact-commodity onto the market unleashes a collective process of establishing some price equilibrium, i.e. coordinating buyers and sellers, supply and demand. At the same time, the new product affects existing patterns of coordination for similar products, affecting prices and levels of supply and demand. Such market re-coordination presupposes locations at which the new commodity re-coordinates consumer behavior. It is precisely by restructuring the pattern of possible choices at local consumption junctions (Schwartz Cowan 1987) that a new commodity is able to influence global market coordination (influence prices and change patterns of supply and demand for related products) and ultimately to influence the viability of technologies which differentiate among alternative artifact-commodities. In Van de Poel's case the specially stamped scratching egg appeared as a new commodity on supermarket shelves and in that setting has become an actant which re-coordinates relations between consumers and products and, ultimately, between consumers and technologies of chicken husbandry. Instead of only brown or white eggs of different sizes, consumers can now also choose eggs whose production incorporates a different hierarchy of values in regard to efficiency and animal well-being. Because of its distinctive markings it can be specifically bought by consumers as a market-mediated statement of preference for "humane" eggs. In essence, the scratching egg is conveying forceful information to producers about the public acceptability of their production technologies and is thus slowly re-ordering established technical models, supplier relationships, scientific theories, and the lives of chickens. So, the scratching egg is an actant in the supermarket, and is by virtue thereof an actant in the global technological order of egg production as well.

How artifacts coordinate by physically linking sites in large (socio-) technical systems is hinted at in the chapters by Stemerding and Hilgartner, and Van Lente and Rip. If we can consider the Human Genome Project as in part a large technological system (for the distributed production of a single global sequence map of the human genome) then artifacts certainly play a role in tying its multiple production sites together into a single mapping system. The clearest example is the central clone library which enables the indexing of local findings to gradually accumulating global maps. This now appears to be being superseded by more textually based (and hence more cheaply mobile) means of aggregation, for example, by Site Tagged Sequencing. But what appears at first sight to be only tools for aggregating findings also turn out to be global modes of coordination. The tools presuppose specific procedures and the use of specific artifacts in order to guarantee the aggregation they promise. In this way their standardized use *nolens volens* shapes and structures interactions among research sites and coordinates practices and discourses in the Human Genome Project.

Van Lente and Rip describe the High Definition Television as little more than a bundle of performative promises which represents and also rhetorically coordinates the emergence of an entire large technological telecommunications system. The image of HDTV in the public mind is not the concatenation of cameras, computers, satellites, studios, cables, transmitters, amplifiers, decoders, etc. etc. which is necessary to implement HDTV as a system, but a simple screen of which they are told that it will rival movie theaters in the quality and brilliance of the images it portrays. So, the promise of the image breaks the surface of public consciousness as it punctualizes (and black-boxes) the iceberg of artifacts and subsystems which is necessary to produce that image. The point is that the screen is the political artifact-idea which ties all the other parts of the system together and which coordinates the public system-building activities of the proponents (and opponents) of the system. This external publicity, by promising specific performances, also sets the internal goals that have to be met and so also frames the emerging technical regime of HDTV.

The HDTV case skirts the fringes of yet another modality of artifacts as means of coordination which is visible in the literature but has not been discussed from the perspective of coordination.[5] We refer to the rhetorical role of artifacts in the antagonistic coordination of new technology. Artifacts can become rhetorical foci in antagonistic processes prior to their actually being realized or quite apart from their actual properties — although defining this latter may well be the major item on the agenda in technology debates. In this process, complex and technologically ambiguous artifacts, like nuclear power plants or genetically modified organisms, become transformed into labels which start functioning as shibboleths for different antagonistic social groups.

As Rip and Talma's account shows, "nuclear power plant" gradually ceases to refer to a concrete artifact with specific properties and risks (about which it might be possible to have a critical debate) but instead refers to an artifactual embodiment of transcendental values: either evil incarnate or universal beneficence. These labels both refer to some (as yet malleable) concrete artifact, but at the same time, and precisely by suppressing the actual technological complexity, serve to coordinate the societal introduction of risky technology as an antagonistic process — in which learning is roundabout, at best. Similar effects can be seen around another nuclear problem, the processing of radioactive wastes. When a sufficiently alienated and militant opposition has been produced by systematic exclusion, "radwaste," as De la Bruhèze argues, no longer refers to the differentiable and potentially neutralizable hazard its proponents argue that it might be. Instead it becomes not only an inherently poisonous product of, but also a symbol for, a treacherous nuclear industry which cannot be trusted and which must be opposed at every turn. In the eyes of antagonistic opponents, radwaste is something which ought never to have existed in the first place and whose premature permanent storage should be resisted by every available means. In this way the label becomes a rhetorical focus for the coordination of antagonistic interactions which have less and less to do with the actual properties and varieties of radwaste as an artifact.

To summarize this section, we have argued in contrast to prevailing theories which more or less take coordination for granted as a welfare-theoretic concept, that coordination itself needs explaining. It is not enough to argue that because coordination solves problems and maximizes benefits it will therefore come about. If we want to understand how new technologies are coordinated we have to combine the insights of theories of economic, organizational, and political coordination (with their sensitivity to global level effects) with insights gained in STS about technology and the independent coordinative role of knowledge and artifacts. A first step is to note that coordination does not always lead to more synergy but can in fact shape and conserve antagonisms and, in their way, make them productive as well. We also asked whether it made sense to think of socio-technical coordination as a fourth modality — in addition to prices, authority, and trust. This earned a qualified yes and on the basis of STS literature and the contributions to this book we explored various ways in which technology — as textual concepts and as artifacts — performed as part of global order and hence coordinated agency.

4 Steering Technology is Steering Sociotechnical Orders

In the first section we discussed the distribution of agency and constructs across local sites and global orders as the social locus of coordination. In the second section we discussed the nature and quality of coordination. An important remaining question is: what kind of coordination (and by implication technology) do you want? In a number of chapters it is obvious that actors frequently ask themselves this question and formulate answers. This raises the issue of steering technology development, i.e. influencing the collective and distributed process of getting new technologies together so that particular goals or values can be realized in the shape of new artifacts and systems. Part of the overall important lesson of this book is that you can't just steer technological developments from some external position — by remote control, as it were. You can only hope to steer technology by reshaping patterns of coordination that are in turn embedded in complex local-global dynamics.

Actors like governments can try to impose coordination via direct steering of local practices by passing laws, imposing regulations, or subsidizing desirable practices, i.e. by so-called technology forcing. This has the advantage — at least in principle — of explicitness and precision. Legal specification of parameters for poultry housing like minimal floor area (Van de Poel) are examples of such measures. A major problem with trying to regulate local socio-technical agency and configurations directly is that neither conformity to rules nor proper use of subsidies is ensured, but must be policed. There is a multitude of actors at different locations in space, time, and discourse. Deviance must be ferreted out and effective deterrents worked out and effectuated. Few actors, even governments, actually have enough resources and power to try to regulate local technological practices in this manner.

Instead of following the dictum "think globally and act locally," actors seeking to steer technologies — particularly non-governmental actors — might better be encouraged to "think locally and act globally." The aim then is to influence technology development by influencing global orders in such a way that local actors adapt their practices of their own accord. This of course requires insight into how local and global levels are integrated; thus thinking globally after all but now in a strategic mode. If resources for policing are the critical limit on direct steering of local practices, insight is the critical limit on steering via global orders. Still, this considerably increases the range of actors able to throw their hat into the steering ring with some hope of success. Activist groups, for example, have a fair record of being able to influence consumer preferences through intelligently planned campaigns of various kinds. This, coupled with the restructuring of product choice, can create profitable markets for alternative products — like the scratching egg described by Van de Poel. In the same account, the Dutch government showed global sagacity by certifying the scratching egg label and subsidizing research into alternatives for the battery cage at the poultry industry's central research institute, Spelderholt. These sorts of initiatives, in order to be successful, demand secure and fine-grained knowledge of the socio-technical terrain. An additional difficulty in managing such steering processes is that the activity of steering cannot principly be divorced from the ongoing interactions. Hence steering becomes entangled in all the other interactions and coordination processes, making it difficult to assess the efficacy of different steering strategies and, indeed, to distinguish success from failure. This makes learning to steer a difficult challenge indeed.

Part of the complexity is due to the fact that global steering can be pursued at many levels — and at many levels simultaneously. For example, the Dutch High Definition Television Platform described by Van Lente and Rip is an attempt to coordinate public acceptance, industry standards, and governmental regulation in an effort to align all relevant actors behind the coming of the new technology. Reconstructing the global television order requires mounting publicity campaigns in favor of the new television standard and lobbying for government support, as well as ensuring that producers of television hardware and producers of television programs coordinate their expectations and their actual technological practices. Rip and Talma show that actors attempt to steer the societal acceptance (or rejection) of new technologies by attempting to influence even the most general global levels of moral casuistry and techno-myths. Global actors do this by telling stories about new technologies and making new myths which bind local actors into new global (and often antagonistic) techno-normative orders. It is clear that making up imaginative narratives can become an important element in coordinating technology development, and in particular the societal acceptance (or rejection) of controversial technologies.

Still, as many of the other contributions suggest, the steering effectiveness of new techno-myths must be limited in view of the tenacity of other global orders: in particular those global orders in which design criteria and design solutions become stabilized and conventional, and take on the form of well-institutionalized design-regimes. One of the lessons of this book must then be that effective steering will necessarily have to impact on prevailing modes of coordination at these levels. Steering the development of new technologies therefore requires far more than epic vision or political power (though that may be necessary too). Whatever else, it inevitably requires specific knowledge of how local technological practices are coordinated via global orders, and then the strategic acumen to modify these orders productively.

Notes

[1] 'Resilient' is better than Bruno Latour's "immutable" (Latour 1990) because mobile constructs are always being modified as they are positioned in new locations. However, there is a definite limit to their malleability, which is why we call them resilient. See also the discussion on immutable mobiles and boundary objects in the chapter by Duncker and Disco.

[2] David Turnbull also develops an explicit dialectic of local and global levels in his analyses of the coordination over space and time of what he calls assemblages: gothic cathedrals, native pacific navigation, Inca empire, etc. However, intriguingly, he seems to conclude that in the end these "assemblages" consist only of concatenations of local practices (see Turnbull (1995); Watson-Verran and Turnbull (1995)).

[3] Recent work on coordination at the Centre de Sociologie de l'Innovation at the Ecole des Mines in Paris, especially that of Michel Callon, has a strong affinity with work at Twente. There are differences of emphasis, but also considerable agreement on the importance and nature of coordination of new technologies.

[4] Indeed, the confrontations around the transport and dumping of radwaste in Germany, most recently at Ahaus in March, 1998, have procceded according to this interpretation. The opposition is no longer interested in convincing the promotors of nuclear power of the error of their ways by arguments or even by political pressure.

Their strategy is now aimed simply at making transports of radwaste so expensive (because of the huge police force and safety precautions the state must muster to contain massive obstruction) that nuclear power will become *economically* non-viable.

5 See, for example, Wiebe Bijker's story about the various contested "frames" surrounding fluorescent lighting in the 1930s and how emerging criteria were forged in an effort to keep the peace among contending parties (Bijker 1995).

Bibliography

Abernathy, W.J. and K.B. Clark (1985): "Innovation: Mapping the Winds of Creative Destruction," *Research Policy,* 14: 3–22.
Achterhuis, H. (1988): *Het Rijk van de Schaarste.* Baarn: Ambo.
Acker, J. (1990): "Hierarchies, Jobs, Bodies: A Theory of Gendered Organizations," *Gender and Society,* 4, 2: 139–158.
Allen, D.E. (1976): *The Naturalist in Britain. A Social History.* Harmondsworth: Penguin Books.
Allison, G.T. and F.A. Morris (1975): "Armaments and Arms Control: Exploring the Determinants of Military Weapons," *Daedalus,* 104, 3: 99–129.
Akrich, M. (1992): "The De-Scription of Technical Objects," in: W.E. Bijker and J. Law (eds.): *Shaping Technology/Building Society. Studies in Sociotechnical Change.* Cambridge, MA: The MIT Press: 205–224.
Akrich, M. and B. Latour (1987): "A Summary of a Convenient Vocabulary for the Semiotics of Human and Nonhuman Assemblies," in: Bijker et al. (eds.): 259–264.
Angino, E.E., C.K. Bayne, and J. Halepaska (1972): *Preliminary Geological Investigations of Supplemental Radioactive Waste Repository Areas in the State of Kansas.* Lawrence, KS: Kansas University, Center for Research Inc.
Appel, T.A. (1987): *The Cuvier — Geoffroy Debate. French Biology in the Decades before Darwin.* New York: Oxford University Press.
Archer, M.S. (1982): "Morphogenesis versus Structuration: on combining structure and action," *British Journal of Sociology,* 33: 455–483.
Axelrod, R. (1984): *The Evolution of Cooperation.* New York: Basic Books.
Barnes, B. (1974): *Scientific Knowledge and Sociological Theory.* London: Routledge and Kegan Paul.
Becker, H.S. (1963): *Outsiders: Studies in the Sociology of Deviance.* New York: The Free Press.
Becker, H.S. (1982): *Art Worlds.* Berkeley, CA: University of California Press.
Bell, G.I. and T.G. Marr (eds.) (1990): *Computers and DNA.* Redwood City, CA: Addison-Wesley.
Belt, H. van den and A. Rip (1987): "The Nelson-Winter-Dosi Model and Synthetic Dye Chemistry," in: Bijker et al. (eds.): 135–158.
Bijker, W.E. (1995): *Of Bicycles, Bakelites and Bulbs. Toward a Theory of Sociotechnical Change: New Directions in the Sociology and History of Technology.* Cambridge MA: MIT Press.
Bijker, W.E., T.P Hughes, and T.J. Pinch (eds.) (1987): *The Social Construction of Technological Systems.* Cambridge, MA: MIT Press.

Bijker, W.E. (1987): "The Social Construction of Bakelite: Toward a Theory of Invention," in: Bijker et al. (eds.): 159–187.
Bijker, W.E. and J. Law (eds.) (1992): *Shaping Technology / Building Society: Studies in sociotechnical change.* Cambridge, MA: MIT Press.
Bloomfield, B.P. and T. Vurdubakis (1995): "Disrupted Boundaries: New Reproductive Technologies and the Language of Anxiety and Expectation," *Social Studies of Science,* 25: 533–551.
Bodewitz, H., G. de Vries, and P. Weeder (1988): "Towards a Cognitive Model of Technology-Oriented R&D Processes," *Research Policy,* 17, 4: 213–224.
Boffey, P.M. (1975): *The Brain Bank of America.* New York: McGraw-Hill.
Borg, M.T. and W.A. Smit (eds.) (1987): *Tactical Ballistic Missile Defence in Europe: Feasible, Affordable, Desirable?* Amsterdam: Free University Press.
Boudon, R. (1986): *Theories of Social Change.* Berkeley and Los Angeles: University of California Press.
Boudon, R. and F. Bourricaud (1989): *A Critical Dictionary of Sociology.* Chicago, Il: University of Chicago Press.
Bourdieu, P. (1977): *Outline of a Theory of Practice.* Cambridge: Cambridge University Press.
Boyer, P. (1985): *By the Bomb's Early Light. American Thought and Culture at the Dawn of the Atomic Age.* New York: Pantheon.
Bradach, J.L. and R.G. Eccles (1989): "Price, Authority and Trust. From Ideal Types to Plural Forms," *Annual Review of Sociology,* 15: 97–118.
Braverman, H. (1974): *Labor and Monopoly Capitalism.* New York: Monthly Review Press.
Bruhèze, A.A. de la (1992a): "Radiological Weapons and Radioactive Waste in the United States: Insiders and Outsiders Views, 1941-1955," *British Journal for the History of Science,* 25: 207–227.
Bruhèze, A. de la. (1992b): "Closing the Ranks: Definition and Closure of Radioactive Waste in the US Atomic Energy Commission, 1945-1960," in: W.E. Bijker and J. Law (eds.): 140–175.
Bruhèze, A.A. de la (1992c): *Political Construction of Technology. Nuclear waste disposal in the United States, 1945-1972.* PhD. Dissertation University of Twente, Enschede/Delft: Eburon.
Bucciarelli, L. (1988): "An Ethnographic Perspective on Engineering Design," *Design Studies,* 9: 159–168.
Bud, R. (1991): "Biotechnology in the Twentieth Century," *Social Studies of Science,* 21: 415–457.
Callon, M. (1986a): "Some Elements of a Sociology of Translation; Domestication of the Scallops and the Fishermen of St. Brieuc Bay," in: J. Law (ed.): *Power, Action and Belief: A New Sociology of Knowledge?* London: Routledge and Kegan Paul: 196–233.
Callon, M. (1986b): "The Sociology of an Actor-Network: The Case of the Electric Vehicle," in: M. Callon et al. (eds.): 19–34.
Callon, M. (1990): Techno-Economic Networks and Irreversibility. Paper presented at the workshop on sociological networks. Paris: 19–20 December.
Callon, M. (1991): "Techno-Economic Networks and Irreversibility," in: J. Law (ed.): *A Sociology of Monsters: Essays on Power, Technology and Domination.* London: Routledge: 132–161.
Callon, M. (1992): "The Dynamics of Techno-Economic Networks," in: R. Coombs, P. Saviotti, and V. Walsh (eds.): *Technological Change and Company Strategies.* London: Academic Press Ltd.: 72–102.

Callon, M. and B. Latour (1981): "Unscrewing the Big Leviathan: How Actors Macrostructure Reality and how Sociologists Help Them to Do So," in: K. Knorr-Cetina and A. Cicourel (eds.): *Advances in Social Theory and Methodology: Toward an Integration of Micro- and Macro-Sociologies*. London: Routledge and Kegan Paul: 277-303.

Callon, M., J. Law, and A. Rip (eds.) (1986): *Mapping the Dynamics of Science and Technology*. London: The MacMillan Press Ltd.

Caskey, C.T. (1992): "DNA-based Medicine: Prevention and Therapy," in: D.J. Kevles and L. Hood (eds.): *The Code of Codes: Scientific and Social Issues in the Human Genome Project*. Cambridge, MA: Harvard University Press: 112-135.

Clark, K.B. (1985): "The Interaction of Design Hierarchies and Market Concepts in Technological Evolution," *Research Policy*, 14: 235-251.

Clarke, A.E. (1991): "Social Worlds/Arenas Theory as Organizational Theory," in: D.R. Maines (ed.): *Social Organization and Social Process, Essays in Honor of Anselm Strauss*. New York: Aldine de Gruyter: 119-149.

Clarke, A.E. and J.H. Fujimura (eds.) (1992): *The Right Tool for the Job: At Work in the Twentieth-Century Life Sciences*. Princeton, NJ: Princeton University Press.

Coase, R. (1937): "The Nature of the Firm," *Economica*, 4: 386-405.

Cockburn, C. (1983): *Brothers: Male Dominance and Technological Change*. London: Pluto Press.

Cockburn, C. and S. Ormrod (1993): *Gender and Technology in the Making*. London: Sage.

Coleman, J.S. (1990): *Foundations of Social Theory*. Cambridge, MA: Belknap Press of Harvard University Press.

Coleman, W. (1964): *George Cuvier, Zoologist: A Study in the History of Evolution Theory*. Cambridge, MA: Harvard University Press.

Collins, H.M. (1985): *Changing Order. Replication and Induction in Scientific Practice*. London: Sage Publications.

Collingridge, D. (1980): *The Social Control of Technology*. London: Frances Pinter.

Colwell, R.A. (ed.), (D.G. Swartz and M.T. Macdonell, Associate Editors), (1989): *Biomolecular Data: A Resource in Transition*. New York: Oxford University Press.

Compton, A.H. (1956): *Atomic Quest — A Personal Narrative*. London: Oxford University Press.

Cook, T.D. (ed.) (1972): Underground Waste Management and Environmental Implications: Proceedings of the Symposium held 6-9 December, 1971, Houston, Texas. Tulsa, Oklahoma: The American Association of Petroleum Geologists (AAPG).

Cook-Degan, R. (1994): *The Gene Wars: Science, Politics, and the Human Genome*. New York: W. W. Norton.

Cooper, R. (1989): "Modernism, Post Modernism and Organizational Analysis 3: the Contribution of Jacques Derrida," *Organization Studies*, 10, 4: 479-502.

Corn, J.J. (1987): *Imagining Tomorrow. History, Technology and the American Future*. Cambridge, MA: MIT Press.

Corsi, P. (1982): "Models and Analogies for the Reform of Natural History. Features of the French Debate 1790-1800," in: W. Bernardi and A. La Vergata (eds.) *Lazzaro Spallanzani e la Biologia del Settecento. Teorie, Esperimenti, Istituzioni Scientifiche*. Florence: Olschki: 381-396.

Czarniawska, B. and Joerges, B. (1995): "Winds of Organizational Change; How Ideas Translate Objects and Actions," *Research in the Sociology of Organizations*, 13: 171-209.

Daele, W. van den (1989): "Kulturelle Bedingungen der Technikkontrolle durch Regulative Politik," in: P. Weingart (ed.): *Technik als sozialer Prozess*. Frankfurt am Main: Suhrkamp: 197–231.
Daudin, H. (1926a): *De Linné à Jussieu. Les méthodes de la Classification et l'Idée de Série du Botanique et du Zoologie (1740–1790)*. Paris: Alcan.
Daudin, H. (1926b): *Cuvier et Lamarck. Classes Zoologiques et l'Idée de Série Animale (1790–1830)*. Paris: Alcan.
Davies, D. and R. Harré (1990): "Positioning: The Discursive Production of Selves," *Journal for the Theory of Social Behavior*, 20: 43–63.
Dawson, F.G. (1976): *Nuclear Power. Development and Management of a Technology*. Seattle/London: University of Washington Press.
DelSesto, S.L. (1979): *Science, Politics and Controversy: Civilian Nuclear Power in the US, 1947–1974*. Boulder, CO: Westview Press.
DelSesto, S.L. (1987): "Wasn't the Future of Nuclear Engineering Wonderful?" in: J.J. Corn (ed.): *Imagining Tomorrow. History, Technology and the American Future*. Cambridge, MA: MIT Press: 58–76.
Diderot, D. and J. Le Rond d'Alembert (eds.) (1966): *Encyclopédie ou Dictionnaire Raisonné des Sciences, des Arts et des Métiers*. Facsimile of the First Edition of 1751–1780. Stuttgart, Bad Cannstadt: Frommann.
Disco, C. (1990): *Made in Delft; Professional Engineering in the Netherlands 1880–1940*. Amsterdam: Unpublished PhD. Dissertation, University of Amsterdam.
Disco, C., A. Rip, and B. van der Meulen (1992): "Technical Innovation and the Universities: Divisions of Labour in Cosmopolitan Technical Regimes," *Social Science Information*, 31: 465–507.
Divine, R.A. (1978): *Blowing on the Wind. The Nuclear Test Ban Debate, 1954–1960*. New York: Oxford University Press.
Dosi, G. (1982): "Technological paradigms and technological trajectories," *Research Policy*, 11: 142–167.
Downey, G.L. (1985): "Politics and Technology in Repository Siting. Military versus Commercial Nuclear Wastes at WIPP 1972–1985," *Technology in Society*, 1,7: 47–75.
Dumouchel, P. (1979): "L'Ambivalence de la Rareté," in: J.P. Dupuy and P. Dumouchel (eds.): *L'Enfer des Choses*. Paris: Seuil: 135–254.
Edwards, R. (1979): *Contested Terrain. The Transformation of the Workplace in the Twentieth Century*. New York: Basic Books.
Eggink, S. (1994): *Verpakking: Niet Weg te Denken. Een Onderzoek Naar Waarden en Technieken in de Case Verpakkingen en Milieu*. Master Thesis, Department of Philosophy of Science and Technology, University of Twente.
Eisenstein, E.L. (1979): *The Printing Press as an Agent of Change: Communications and Cultural Transformations in Early-Modern Europe*. Cambridge, MA: Cambridge University Press.
Ellul, J. (1964): *The Technological Society*. New York: Alfred A. Knopf.
Elzen, B., B. Enserink, and W.A. Smit (1990): "Weapon Innovation: Networks and Guiding Principles," *Science and Public Policy*, 17, 3: 171–193.
Enserink, B., W.A. Smit, and B. Elzen (1990): "Assessments and the B-1 bomber network," *Project Appraisal*, 5, 4: 235–254.
Enserink, B. (1993): *Influencing Military Technological Innovation*. Delft: Eburon.
Eriksson, G. (1980): "The Botanical Success of Linnaeus. The Aspect of Organization and Publicity," in: G. Broberg (ed.): *Linnaeus. Progress and Prospects in Linnaean Research*. Stockholm: Almqvist and Wiksell: 57–66.

Evangelista, M. (1988): Innovation and the Arms Race — How the United States and the Soviet Union develop New Military Technologies. London: Cornell University Press.

Eyerman, R. and A. Jamison (1991): Social Movements: A Cognitive Approach. Cambridge: Polity Press.

Fallows, S. (1979): "The Nuclear Waste Disposal Controversy," in: D. Nelkin (ed.): Controversy. Politics of Technical Decisions. Beverly Hills, CA: Sage: 87-111.

Farber, P.L. (1975): "Buffon and Daubenton: Divergent Traditions within the Histoire Naturelle," Isis, 66: 63-74.

Farber, P.L. (1982): The Emergence of Ornithology as a Scientific Discipline: 1760-1850. Dordrecht: Reidel.

Ferguson, E.S. (1992): Engineering and the Mind's Eye. Cambridge, MA: MIT Press.

Ford, D. (1984): The Cult of the Atom. The Secret Papers of the Atomic Energy Commission. New York: Simon and Schuster.

Fujimura, J.H. (1987): "Constructing 'do-able' Problems in Cancer Research: Articulating Alignment," Social Studies of Science, 17: 257-293.

Fujimura, J.H. (1992): "Crafting Sience: Standardized Practices, Boundary Objects, and "'Translation'," in: A. Pickering (ed.): 168-211.

Fujimura, J.H. (1995): "Ecologies of Action: Recombining Genes, Molecularizing Cancer, and Transforming Biology," in: S.L. Star (ed.): Ecologies of Knowledge, Work, and Politics in Science and Technology. New York: State University of New York Press: 302-346.

Galison, P. (1987): How Experiments End. Chicago: University of Chicago Press.

Galison, P. (1997): Image and Logic. A Material Culture of Microphysics. Chicago: University of Chicago Press.

Gandara, A. (1977): Electric Utility Decision-Making and the Nuclear Option. Santa Monica, CA: The Rand Corporation.

Giddens, A. (1981): "Agency, Institutions and Time-space Analysis," in: K. Knorr-Cetina and A. Cicourel (eds.): Advances in Social Theory and Methodology: Toward an Integration of Micro-and Macro-Sociologies. London: Routledge and Kegan Paul: 54-78.

Giddens, A. (1984): The Constitution of Society: Outline of the Theory of Structuration. Berkeley, CA: University of California Press.

Gilfillan, S.G. (1935): The Sociology of Invention. Cambridge, MA: MIT Press.

Granovetter, M. (1985): "Economic Action and Social Structures. The Problem of Embeddedness," American Journal of Sociology, 91: 481-510.

Groves, L.R. (1963): Now It Can Be Told. The Story of the Manhattan Project. London: André Deutsch Publishers.

Gunsteren, H.R. van (1972): The Quest for Control: A Critique of the Rational-Central-Rule Approach in Public Affairs, University of Leiden: Unpublished PhD. Dissertation.

Hacker, S. (1979): "Sex Stratification, Technology and Organizational Change: A Longitudinal Case Study of AT&T," Social Problems, 26: 539-557.

Hafner, D.L. and J. Roper (eds.) (1988): ATBMs and Western Security: Missile Defenses for Europe. Cambridge, MA: Ballinger Publishing Company.

Hajer, M. (1995): The Politics of Environmental Discourse. Ecological Modernisation and the Policy Process. Oxford: Clarendon Press.

Hales, C. (1987): Analysis of the Engineering Process in an Industrial Context, Unpublished PhD. Dissertation, Department of Engineering, University of Cambridge.

Haralambos, M. and M. Holborn (1990): *Sociology: Themes and Perspectives*. London: Collins.

Hård, M. (1993): "Beyond Harmony and Consensus: A Social Conflict Approach to Technology," *Science, Technology and Human Values*, 18, 4: 408-432.
Harding, S. (1986): *The Science Question in Feminism*. New York: Cornell University Press.
Harré, R. (1975): "Images of the World and Social Icons," in: K.D. Knorr, H. Strasser, and H.G. Zilan (eds.): *Determinants and Controls of Scientific Development*. Dordrecht: Reidel: 257-283.
Harré, R. (1990): "Some Narrative Conventions of Scientific Discourse," in: C. Nash (ed.): *Narrative in Culture*. London: Routledge: 81-101.
Harré, R. and L. van Langenhove (1992): "Varieties of Positioning," *Journal for the Theory of Social Behaviour*, 22, 4: 393-407.
Hatchett, R.L. and R.L. Keuter (1992): "Weapons Research and Development in the United States," in: W.A. Smit et al. (eds.): 85-94.
Hays, S.P. (1987): *Beauty, Health and Permanence: Environmental Policies in the US, 1955-1985*. Cambridge, MA: Cambridge University Press.
Heilbron, J. L. and D.J. Kevles (1988): "Finding a Policy for Mapping and Sequencing the Human Genome: Lessons from the History of Particle Physics," *Minerva*, 16, 3: 299-314.
Henderson, K. (1991): "Flexible Sketches and Inflexible Data Bases: Visual Communication, Conscription Devices, and Boundary Objects in Design Engineering," *Science, Technology and Human Values*, 16: 448-473.
Henderson, K. (1995): "The Political Career of a Prototype — Visual Representation in Design Engineering," *Social Problems*, 42, 2: 274-299.
Hertsgaard, M. (1983): *Nuclear Inc. The Men and Money behind Nuclear Energy*. New York: Pantheon Books.
Hilgartner, S. (1995a): "The Human Genome Project," in: S. Jasanoff, G. Markle, J. Petersen, and T. Pinch (eds.): *Handbook of Science and Technology Studies*. Newbury Park, CA: Sage: 302-15.
Hilgartner, S. (1995b): "Biomolecular Databases: New Communication Regimes for Biology?" *Science Communication*, 17, 2: 240-263.
Hilgartner, S. (forthcoming): "Data Access Policy in Genome Research," in: A. Thackray (ed.): *Private Science: Biotechnology and the Rise of the Molecular Sciences*. Philadelphia, PA: University of Pennsylvania Press:
Hilgartner, S., R.C. Bell, and R. O'Connor (1982): *Nukespeak, Nuclear Language, Visions and Mindset*. San Francisco, CA: Sierra Club Books.
Hilgartner, S. and S.I. Brandt-Rauf (1994): "Data Access, Ownership, and Control: toward Empirical Studies of Access Practices," *Knowledge: Creation, Diffusion, Utilization*, 15: 355-372.
Hilgartner, S. and S.I. Brandt-Rauf (forthcoming): "Controlling Data and Resources: Access Practices in Molecular Genetics," in: P.A. David and W.E. Steinmueller (eds.): *A Productive Tension: University-Industry Research Collaborations*. Stanford CA: Stanford University Press.
Hollander, W. (1981): *Abel Wolman. His Life and Philosophy, an Oral History*. Chapel Hill, NC: Universal Printing and Publishing Company.
Hood, L. (1992): "Biology and Medicine in the Twenty-First Century," in: L. Hood and D.J. Kevles (eds.): 126-163.
Hood, L. and D.J. Kevles (eds.) (1992): *The Code of Codes: Scientific and Social Issues in the Human Genome Project*. Cambridge, MA: Harvard University Press.
Hubka, V. (1982): *Principles of Engineering Design*. London: Butterworth Scientific.

Hughes, T.P. (1983): *Networks of Power: Electrification in Western Society 1880-1930*. Baltimore: Johns Hopkins University Press.

Hughes, T.P. (1986): "The Seamless Web: Technology, Science, Etcetera, Etcetera," *Social Studies of Science*, 16: 281-292.

Jelsma, J. (1991): "Technology and the Role of Ethics: A Question of Theory or a Question of Practice?" in: A.W. Munchenga, B. Voorzanger, and A. Soeteman (eds.): *Morality, Worldview and Law. The Idea of Universal Morality and Its Critics*. Assen: Van Gorcum: 297-309.

Jelsma, J. (1995): "Learning About Learning in the Development of Biotechnology," in: A. Rip et al. (eds.): 141-165.

Jelsma, J. and W.A. Smit (1986): "Risks of Recombinant DNA Research: from Uncertainty to Certainty," in: H.A. Becker and A.L. Porter (eds.): *Impact Assessment Today*. Utrecht: Uitgeverij Jan van Arkel: 715-740.

Joerges, B. (1988): "Large Technical Systems: Concepts and Issues," in: R. Mayntz and T.P. Hughes (eds.): *The Development of Large Technical Systems*. Frankfurt am Main and Boulder: Campus Verlag and Westview Press: 9-37.

Johanson, J. and L.G. Mattson (1987): "Interorganizational Relations in Industrial Systems: A Network Approach Compared with the Transaction-Cost Approach," *International Studies of Management and Organization*, 18, 1: 34-48.

Jordan, B. (1993): *Travelling Around the Human Genome: An In Situ Investigation*. France: John Libbey Eurotext.

Jordan, K. and M. Lynch (1992): "The Sociology of a Genetic Engineering Technique: Ritual and Rationality in the Performance of the "Plasmid Prep'," in: A.E. Clarke and J.H. Fujimura (eds.): 77-114.

Judson, H. (1992): "A History of the Science and Technology behind Gene Mapping and Sequencing," in: L. Hood and D.J. Kevles (eds.): 37-80.

Kingdon, J.W. (1984): *Agendas, Alternatives and Public Policies*. Boston and Toronto: Little, Brown and Company.

Knorr-Cetina, K. (1977): "Producing and Reproducing Knowledge: Descriptive or Constructive? Toward a Model of Research Production," *Social Science Information*, 16: 669-696.

Knorr-Cetina, K. (1979): "Tinkering towards Success," *Theory and Society*, 8, 3: 347-376.

Knorr-Cetina, K. (1981): *The Manufacture of Knowledge: An Essay on the Constructivist and Contextual Nature of Science*. Oxford: Pergamon Press.

Knorr-Cetina, K. (1992): "The Couch, the Cathedral, and the Laboratory: On the Relationship between Experiment and Laboratory in Science," in: A. Pickering (ed.): 113-138.

Knorr-Cetina, K. (1995): "How Superorganisms Change: Consensus Formation and the Social Ontology of High-Energy Physics Experiments," *Social Studies of Science*, 25: 119-147.

Knorr-Cetina, K. (1996): "The Care of the Self and Blind Variation: The Disunity of Two Leading Sciences," in: P. Galison and D. Stump (eds.). *The Disunity of Science. Boundaries, Contexts, and Power*. Stanford: Stanford University Press: 287-310.

Knorr-Cetina, K. (1998): *Epistemic Cultures*. Forthcoming.

Kohler, R.E. (1994): *Lords of the Fly: Drosophila Genetics and the Experimental Life*. Chicago: University of Chicago Press.

Kolodziej, E.A. (1987): *Making and Marketing Arms: The French Experience and its Implications for the International System*. Princeton, NJ: Princeton University Press.

Kopp, C. (1979): "The Origins of the American Scientific Debate over Fallout," *Social Studies of Science,* 9: 403-422.
Krimsky, S. (1982): *Genetic Alchemy. The Social History of the Recombinant DNA Controversy.* Cambridge, MA: MIT Press.
Kuhn, T. (1962): *The Structure of Scientific Revolutions.* Chicago: University of Chicago Press.
Langenhove, L. van and R. Harré (1990): Scientific Positioning, Unpublished Manuscript. Brussels, Free University.
Larson, J.L. (1971): *Reason and Experience: The Representation of Natural Order in the work of Carl von Linné.* Berkeley: University of California Press.
Latour, B. (1986): "Visualization and Cognition: Thinking with Eyes and Hands," *Knowledge and Society,* 6: 1-40.
Latour, B. (1987): *Science in Action. How to Follow Scientists and Engineers through Society.* Cambridge, MA: Harvard University Press.
Latour, B. (1990): "Drawing Things Together," in: M. Lynch and S. Woolgar (eds.): *Representation in Scientific Practice.* Cambridge, MA: MIT Press:19-68.
Latour, B. (1991. *Nous n'Avons Jamais été Modernes,* Paris: Editions La Découverte.
Latour, B. (1992): "Where Are the Missing Masses? The Sociology of a Few Mundane Artefacts," in: W.E. Bijker and J. Law (eds.): 225-258.
Latour, B. and S. Woolgar (1979): *Laboratory Life.* Beverly Hills, CA: Sage.
Law, J. (1986a): "Laboratories and Texts," in: M. Callon et al. (eds.): 35-50.
Law, J. (1986b): "On the Methods of Long-Distance Control: Vessels, Navigations and the Portuguese Route to India," in: J. Law (ed.): *Power, Action and Belief: A New Sociology of Knowledge?* London: Routledge and Kegan Paul: 234-263.
Law, J. (1992): "The Olympus 320 Engine: A Case Study in Design, Development, and Organizational Control," *Technology and Culture,* 33, 3: 409-440.
Law, J. and M. Callon (1992): "The life and death of an aircraft: a network analysis of technical change," in: W.E. Bijker and J. Law (eds.): 21-52.
Lear, J. (1972): "Radioactive Ashes in the Kansas Salt Cellar," *Science Saturday Review,* February, 19: 39-42.
Lente, H. van (1993): *Promising Technology. The Dynamics of Expectations in Technological Developments.* PhD Dissertation, University of Twente. Delft: Eburon.
Levi-Strauss, C. (1966): *The Savage Mind.* London: Weidenfeld and Nicholson.
Levi-Strauss, C. (1972): *The Savage Mind.* London: Weidenfeld and Nicholson.
Long, F.A. and J. Reppy (eds.) (1980): *The Genesis of New Weapons.* New York: Pergamon Press.
Long, F.A. and J. Reppy (1980): "Decision Making in Military R&D: An Introductory Overview," in: Long and Reppy (eds.): 3-20.
Lynch, M. (1985): *Art and Artifact in Laboratory Science. A Study of Shop Work and Shop Talk in a Research Laboratory.* London: Routledge and Kegan Paul.
MacIntyre, A. (1981): *After Virtue.* London: Duckworth.
MacKenzie, D. (1990): *Inventing Accuracy: A Historical Sociology of Nuclear Missile Guidance.* Cambridge, Mass.: MIT Press.
Mahony, K. and B. van Toen (1990): "Mathematical Formalism as a Means of Occupational Closure in Computing — Why "Hard" Computing Tends to Exclude Women," *Gender and Education,* 2, 3: 319-331.
Marin, B. and R. Mayntz (1991): "Introduction: Studying Policy Networks," in: B. Marin and R. Mayntz (eds.): *Policy Networks: Empirical Evidence and Theoretical Considerations.* Frankfurt am Main and Boulder: Campus Verlag and Westview Press: 7-20.

Mazuzan, G.T. and J.S. Walker (1984): *Controlling the Atom. The Beginnings of Nuclear Regulation, 1946-1962*. Berkeley, CA: University of California Press.
McCarthy, J.D. and M.N. Zald (1973): *The Trend of Social Movements in America: Professionalization and Resource Mobilization*. Morristown, N.J.: General Learning Press.
Metzger, P.H. (1972): *The Atomic Establishment*. New York: Simon and Schuster.
Mitchell, D.G. (ed.) (1979): *A New Dictionary of Sociology*. London: Routledge and Kegan Paul.
Molenaar, L. (1994): *Wij Kunnen het Niet Langer aan de Politici Overlaten. De Geschiedenis van het Verbond van Wetenschappelijke Onderzoekers 1946-1980*. PhD Dissertation, University of Amsterdam, Delft: Uitgeverij Elmar.
Mulkay, M. (1993): "Rhetorics of Hope and Fear in the Great Embryo Debate," *Social Studies of Science*, 23: 721-742.
Mumford, L. (1963): *Technics and Civilization*. New York: Harcourt, Brace and World.
Nelkin, D. (ed.) (1992): *Controversy. Politics of Technical Decisions*. Newbury Park, CA: Sage.
Nelson, R.R. and S.G. Winter (1982): *An Evolutionary Theory of Economic Change*. Cambridge MA: The Belknap Press of Harvard University Press.
Noble, D. (1984): *Forces of Production. A Social History of Industrial Automation*. New York: Albert A. Knopf.
Nothnagel, D. (1992): "Spatio-Temporal Exchange Patterns and the Production of Power and Sense in High Energy Physics," Paper Presented at 4S-EASST Joint Conference, 12-15 August.
Ogburn, W.F. (1945): *The Social Effects of Aviation*. Boston: Houghton Mifflin.
Oost, E.C.J. van (1994): *Nieuwe Functies, Nieuwe Verschillen. Genderprocessen in de Constructie van de Nieuwe Automatiseringsfuncties (1955-1970)*. PhD Dissertation, University of Twente, Delft: Eburon.
OTA (1990): *The Big Picture: HDTV and High-Resolution Systems*. Washington, DC: U.S. Government Printing Office. Report of the U.S. Congress, Office of Technology Assessment, OTA-BP-CIT-64.
Outram, D. (1984): *Cuvier. Vocation, Science and Authority in Post-Revolutionary France*. Manchester: Manchester University Press.
Outram, D. (1986): "Uncertain Legislator: George Cuvier's Laws of Nature in Their Intellectual Context," *Journal of History of Biology*, 19: 323-368.
Pacey, A. (1983): *The Culture of Technology*. Cambridge, Mass.: MIT Press.
Pahl, G. and W. Beitz (1977): *Konstruktionslehre*. Berlin: Springer Verlag.
Pels, D. (1985): "Rivaliteit en Relativisme," *Kennis en Methode*, 8, 2: 121-140.
Perry, R.L. (1967): *Innovation and Military Requirements: A Comparative Study*. RAND Memorandum RM-5182-PR. Santa Monica: RAND Corporation.
Pfau, L. (1984): *No Sacrifice Too Great. The Life of Lewis L. Strauss*. Charlottesville, VA: University Press of Virginia.
Pickering, A. (ed.) (1992): *Science as Practice and Culture*. Chicago: University of Chicago Press.
Pinch, T.J. and W.E. Bijker (1984): "The Social Construction of Facts and Artifacts: or How the Sociology of Science and the Sociology of Technology Might Benefit Each Other," *Social Studies of Science*, 14: 399-441.
Pirsig, R.M. (1974): *Zen and the Art of Motorcycle Maintenance: An Inquiry into Values*. New York: Bantam Books.
Plantenga, J. (1993): *Een Afwijkend Patroon. Honderd Jaar Vrouwenarbeid in Nederland en West-Duitsland*. Amsterdam: Socialistische Uitgeverij Amsterdam.

Poel, I. van de (1994): "De Wereld van de Legbatterij," *Kennis en Methode,* 18, 4: 315-340.
Poel, I. van de and C. Disco (1995): "Influencing Technology: Design Worlds and their Legitimacy," The Role of the Design in the Shaping of Technology, Proceedings COST-A4 Workshop, February, Lyon.
Powell, W.W. (1990): "Neither Market nor Hierarchy: Network Forms of Organization," *Research in Organizational Behavior,* 12: 295-336.
Price, D. L. de Solla (1963): *Little Science, Big Science.* New York: Columbia University Press.
Rice, P. (ed.) (1990): *HDTV: The Politics, Policies and Economies of Tomorrow's Television.* New York: Union Square Press.
Rip, A. (1986): "The Mutual Dependence of Risk Research and Political Context," *Science and Technology Studies,* 4 (3,4): 3-15.
Rip, A. (1990): "Societal Construction of Research and Technology," in: H. Krupp (ed.): *Technikpolitik Angesichts der Umweltkatastrophe.* Heidelberg: Physica Verlag.
Rip, A. (1991): "The Danger Culture of Industrial Society," in: R.E. Kasperson and P.J.M. Stallen (eds.): *Communicating Health and Safety Risks To the Public. International Perspectives.* Dordrecht: Reidel: 345-365.
Rip, A. (1992): "A Quasi-Evolutionary Model of Technology Development and a Cognitive Approach to Technology Policy," *RISESST,* 2: 69-103.
Rip, A., T.J. Misa, and J.W. Schot (eds.) (1995): *Managing Technology in Society.* London: Pinter Publishers.
Rip, A. and A.J. Nederhof (1986): "Between Dirigism and Laissez-Faire: Effects of Implementing the Science Policy Priority for Biotechnology in the Netherlands," *Research Policy,* 15: 253-268.
Rolph, E.S. (1979): *Nuclear Power and the Public Safety.* Lexington, MA: Lexington Books.
Ruina J. (1988): "Controlling Military Research and Development," *Global Problems and Common Security,* Proceedings of the Thirty-Eighth Pugwash Conference on Science and World Affairs, Dagomys, USSR: 400-405.
Sahal, D. (1985): "Technological Guideposts and Innovation Avenues," *Research Policy,* 14: 61-82.
Schaffler, A. (1990): "EFA Partners Discuss Establishing Single Base for Training all Pilots," *Aviation Week and Space Technology,* 12 March.
Scharpf, F.W. (ed.) (1993): *Games in Hierarchies and Networks, Analytical and Empirical Approaches to the Study of Governance Institutions.* Frankfurt am Main: Campus Verlag.
Schot, J.W. (1991): *Maatschappelijke Sturing van Technische Ontwikkeling. Constructief Technology Assessment als Hedendaags Luddisme,* Unpublished PhD. Dissertation. Enschede: Twente University, Faculty of Philosophy and Social Science Series.
Schwartz Cowan, R. (1987): "The Consumption Junction: A Proposal for Research Strategies in the Sociology of Technology," in: W. Bijker et al. (eds.): 261-286.
Scott, J. (1986): "Gender: a Useful Category of Historical Analysis," *American Historical Review,* 91: 1053-1075.
Sen, S. and S. Deger (1990): "The Reorientation of Military R&D for Civilian Purposes," Proceedings of the Fortieth Pugwash Conference on Science and World Affairs, Egham: 429-452.
Shills, D.L. (ed.) (1968): *International Encyclopedia of the Social Sciences.* New York: Macmillan.

Sloan, P.R. (1972): "John Locke, John Ray and the Problem of the Natural System," *Journal of the History of Biology*, 5: 1-53.
Sloan, P.R. (1976):"The Buffon-Linnaeus Controversy," *Isis*, 67: 356-375.
Sloan, P.R. and H.P. Siderius (1990): *Hoge Definitie Televiste. Een overzicht van de huidije stand van zaken met betrekking tot outwikkeling en introductie*. Den Haag and Amsterdam: SWOKA/VU.
Smelser, H.J. (ed.) (1988): *Handbook of Sociology*. Newbury Parl, CA.: Sage Publications.
Smit, W.A. (1989): "Defense Technology Assessment and the Control of Emerging Technologies," in: M. ter Borg and W.A. Smit (eds.): *Non-Provocative Defence as a Principle of Arms Reduction*. Amsterdam: Free University Press: 61-76.
Smit, W.A. (1991): "Steering the Process of Military Technological Innovation," *Defense Analysis*, 7, 4: 401-415.
Smit, W.A. (1993): "Intervening in Military Technological Development: a Comment on Donald MacKenzie's Inventing Accuracy," *Science and Public Policy*, 20, 6: 396-404.
Smit, W.A. (1995): "A Framework for a Sociology of Assessing and Intervening in Technology Development," *European Review*, 3, 1: 73-82.
Smit, W.A., J. Grin, and L. Voronkov (eds.) (1992): *Military Technological Innovation and Stability in a Changing World. Politically Assessing andIinfluencing Weapon Innovation and Military Research and Development*. Amsterdam: Free University Press.
Søderqvist, T. (1986): *The Ecologists; From Merry Naturalists to Saviours of the Nation*. Stockholm: Almqvist and Wiksell International.
Stafleu, F.A. (1971): *Linnaeus and the Linnaeans. The Spreading of Their Ideas in Systematic Botany 1735-1789*. Utrecht: Oosthoek.
Stanley, W.L. and J.L. Birkler (1986): *Improving Operational Suitability through Better Requirements and Testing*. A Project Air Force Report Prepared for the United States Air Force. Santa Monica: Rand Corporation.
Star, S.L. (1989): *Regions of the Mind: Brain Research and the Quest for Scientific Certainty*. Stanford: Stanford University Press.
Star, S.L. (ed.) (1995a.): *Ecologies of Knowledge, Work and Politics in Science and Technology*. Albany, NY: State University of New York Press.
Star, S.L. (1995b) "The Politics of Formal Representations: Wizards, Gurus and Organizational Complexity," in: Star, S.L. (ed.): 88-118.
Star, S.L. and J.R. Griesemer (1989): "Institutional Ecology, Translations and Boundary Objects: Amateurs and Professionals in Berkeley's Museum of Vertebrate Zoology, 1907-39," *Social Studies of Science*, 19: 387-420.
Staudenmaier, J.M. (1989): "The Politics of Successful Technologies," in: R.C. Post and S.H. Cutliffe (eds.): *In Context: History and the History of Technology. Essays in Honor of Melvin Kranzberg*. Bethlehem, PA: Lehigh University Press: 150-171.
Steenbergen, Th.J. (1960): *De Invloed van de Automatisering op het Bedrijf. Een Studie aan de Hand van Gegevens Omtrent de Automatisering in Amerikaanse Bedrijven*. PhD. Dissertation, Leiden: Stenfert Kroese.
Stemerding, D. (1991): *Plants, Animals, and Formulae. Natural History in the Light of Latour's Science in Action and Foucault's The Order of Things*. PhD. Dissertation University of Twente, Delft: Eburon.
Strauss, A. (1978): "A Social World Perspective," *Studies in Symbolic Interaction*, 1: 119-128.
Strauss, A. (1982): "Social Worlds and Legitimation Processes," *Studies in Symbolic Interaction*, 4: 171-190.

Strauss, A. (1984): "Social worlds and their Segmentation Processes," *Studies in Symbolic Interaction,* 5: 123-139.
Thee, M. (1988): "Science and Technology for War and Peace," *Bulletin of Peace Proposal,* 19, 3:261-292.
Thompson, G., J. Frances, F. Levacic, and J. Mitchell (1991): *Markets, Hierarchies and Networks — The Coordination of Social Life.* London: Sage Publications.
Titus, A.C. (1986): *Bombs in the Backyard. Atomic Testing and American Politics.* Las Vegas: University of Nevada Press.
Traweek, S. (1988): *Beamtimes and Lifetimes; The World of High Energy Physicists.* Cambridge, MA: Harvard University Press.
Truman, H.S. (1956): *Memoirs. Vol. 2: Years of Trial and Hope.* New York: Signet Books.
Tsebelis, G. (1990): *Nested Games, Rational Choice in Comparative Politics.* Berkeley: University of California Press.
Turkle, S. (1984): *The Second Self: Computers and the Human Spirit.* London: Granada.
Turnbull, D. (1995): "Rendering Turbulence Orderly," *Social Studies of Science,* 25: 9-33.
Turner, J.H. (1986): *The Structure of Sociological Theory.* Chicago: The Dorsey Press.
Twain, M. (1962): Roughing It. New York: American Library.
Vincenti, W. (1990a): *What Engineers Know and How They Know It.* Baltimore: Johns Hopkins University Press.
Vincenti, W. (1990b): "Engeneering Knowledge, Type Design, and Level of Hierarchy: Further Thoughts about What Engineers Know...," Paper for Conference on the Relation between Science and Technology, Eindhoven, 6-9 November.
Vincenti, W. (1991): "The Scope for Social Input in Engineering Outcomes: A Diagrammatic Aid to Analysis," *Social Studies of Science,* 21, 4: 761-767.
Waarden, F. van (1992): "Dimensions and Types of Policy Networks," *European Journal of Political Research,* 21, 29-52.
Wachs, E. (1988): *Crime-Victim Stories.* New York City's Urban Folklore. Bloomington: Indiana University Press
Wajcman, J. (1991): *Feminism Confronts Technology.* Cambridge: Polity Press.
Watson, J.D. (1990): "The Human Genome Project: Past, Present, Future," *Science,* 248: 44-49.
Watson-Verran, H. and D. Turnbull (1995): "Science and Other Indigenous Knowledge Systems," in: S. Jasanoff, G.E. Markle, J.C. Petersen and T. Pinch (eds.): *Handbook of Science and Technology Studies.* London: Sage: 115-139.
Weick, K. (1979): *The Social Psychology of Organizing.* Reading, MA: Addison Wesley.
Weinberg, A.M. (1961): "Impact of Large Scale Science on the United States," *Science,* 134, 21 July: 164.
Williamson, O. (1975): *Markets and Hierarchies.* New York: Free Press.
Williamson, O. (1981): "The Economics of Organizations: The Transaction Cost Approach," *American Journal of Sociology,* 87: 548-577.
Winner, L. (1977): *Autonomous Technology. Technics out of Control as a Theme in Political Thought.* Cambridge MA: MIT Press.
Wit, D. de (1994): *The Shaping of Automation. A Historical Analysis of the Interaction between Technology and Organization 1950-1985.* Hilversum: Verloren.
Wittgenstein, L. (1972): *Philosophical Investigations.* London: Blackwell.
York, H. (1971): *Race to Oblivion: A Participant's View of the Arms Race.* New York: Simon and Schuster.

Zucker, L.G. (1991): "Markets for Bureaucratic Authority and Control; Information Quality in Professions and Services," *Research in the Sociology of Organizations*, 8: 157-190.
Zuckerman, S. (1982): *Nuclear Illusion and Reality*. New York: The Viking Press.

Notes on Contributors

Adri A. de la Bruhèze studied political science at the University of Amsterdam (M.Sc. 1983) and received his Ph.D. in the field of history and sociology of technology at the University of Twente in 1992. His dissertation was on the development of radwaste technology in the U.S. Since 1994 he has been editorial secretary of the large collaborative research project "History of Technology in the Netherlands in the Twentieth century" (TIN-20). In his own research he has focussed on the history of food technology, the history of (local) bicycle use and traffic policy and the development of "city-technologies" in the Netherlands in the 20th century.

Cornelis Disco studied sociology at Yale University (B.A.1969) and holds a Ph.D. in Political and Socio-Cultural Sciences from the University of Amsterdam (1990). He taught at various universities in The Netherlands before becoming assistant professor at Twente in 1988. He is broadly interested in the dynamics of technologies in modern cultural and historical contexts. He has done work on the historical emergence of different segments of the Dutch engineering profession and on the building of experimental apparatus in high energy physics. He is currently completing a study of Dutch hydraulic-civil engineering in the twentieth century with an emphasis on the shifting cultural and technological conceptions of water. Some of this work will be published in the context of the TIN-20 project (see de la Bruhèze).

Elke Duncker studied Sociology and Political Science at the Technical University of Aachen (RWTH), Germany (M.A. 1981). After graduation she held a post at Aachen as assistant lecturer on sociological and cultural issues, group dynamics, and communication. She subsequently spent several years as a systems engineer at a computer company where she designed and implemented computer systems for industrial use. In 1992 she moved to the Department of Philosophy of Science and Technology at the University of Twente, Netherlands. She has taught introductory courses in sociology and sociology courses for technical students doing industrial internships. Her recent work is on the dynamics of multi-disciplinary/multi-professional cooperations. In 1995 she was a visiting scholar at the School for Library and Information Science at the University of Illinois.

Boelie Elzen has been with the department of Philosophy of Science and Technology at Twente University since 1996. During 1979-1989 he was also associated with the Department as a graduate student and researcher. He was trained as an electrical engineer at Twente (M.Sc.1979) where he also obtained a Ph.D. in technology studies (1988). His scholarly work has pursued the (theoretical) understanding of socio-technical change and related societal problems. In the 1980s he focused on problems related to the arms race. Since 1991 he has been working on traffic and transport related problems, including vehicle-emissions, road-congestion and energy security. He was coordinator of a study on 'personal mobility and culture' and is currently coordinator of an international project funded by the European Commission to develop an 'interactive technology policy' to deal with the problems of traffic and transport.

Bert Enserink is assistant professor of policy analysis at the Faculty of Systems Engineering, Policy Analysis and Management of Delft University of Technology. He holds a M.Sc. from Wageningen Agricultural University (rural planning consultant) and a Ph.D. from the University of Twente, where he completed a project on technology assessment and weapons development in the Department of Philosophy of Science and Technology. His main areas of interest are participatory policy analysis, quick-scan and rapid appraisal techniques and (constructive) technology assessment. Recent research topics: interactive appraisals and integral assessment of land-use and infrastructure projects.

Stephen Hilgartner is an assistant professor in the Department of Science & Technology Studies, Cornell University. His research focuses on social studies of contemporary biology, biotechnology, and medicine. He is conducting a long-term, prospective study of the social world of genome mapping and sequencing. His articles have appeared, among other places, in Social Studies of Science, Science, Technology & Human Values, Knowledge and the American Journal of Sociology.

Harro van Lente studied physics and philosophy. His research interests are technological promises and scenario-building. In his study of technology he focuses on rhetorical analysis, agenda-building and cultural studies. In 1993 he finished a Ph.D. thesis on the way expectations affect technological developments. He held post-doc positions at the University of Oviedo, and the University of Maastricht. Recently, he joined KPMG, an international audit and consultancy firm. He is research manager at KPMG Inspire Foundation, a new group that studies the future of organizations.

Barend van der Meulen (1964) is senior researcher at the Centre for Studies of Science, Technology and Society, University of Twente. He has worked on socio-economic impacts of science, the development of engineering sciences, the transition of research systems, and science and technology foresight. He uses recent advances in science and technology studies as well as policy sciences to further the understanding of science policy processes and of research-system dynamics and in order to advise on science and technology policy. Recent publications have appeared in Research Policy, Scientometrics, and Science and Public Policy.

Ellen van Oost is assistant professor in the field of Gender and Technology Studies. She has a degree in mathematical engineering (M.Sc.) and did a sociological-historical thesis on gender processes in the emerging occupational structure of computer science. She is currently active in a nationwide project on the History of Information Technology in the Netherlands. Her present research interests also concern contemporary and future influences of Information- and Communication Technologies (ICT's) on societal gender relations, e.g. teleworking and digital services.

Ibo van de Poel holds a B.Sc. in Mechanical Engineering and a Masters degree in the Philosophy of Science, Technology and Society. In 1992 he started working on a Ph.D. -project 'Influencing Design' at the Department of Philosophy of Science and Technology at the University of Twente. The article published in this book is one of the results of this research project. His dissertation *Changing Technologies; A Comparative Study of Eight Processes of Transformation of Technological Regimes*, was published in 1998. At present he teaches courses on technology and ethics at the Technical University of Delft. His main scholarly interests are in engineering design and in the social and ethical aspects of technology.

Arie Rip was educated as a chemist and philosopher and is a well-known figure in the Science, Technology and Society community. His own research has focused on science and technology dynamics and science and technology policy analysis. He was Professor of Science Dynamics at the University of Amsterdam (1984–1987) and is now Professor of Philosophy of Science and Technology at the University of Twente. He has served as Secretary of the European Association for the Study of Science and Technology (1981–1986) and as President of the Society for Social Studies of Science (1988–1989). He is active in science policy committees of various kinds, has advised on science policy in the U.K., Australia and South-Africa, and is engaged in projects on R&D Evaluation, Technology Assessment and Science and Technology Foresight. He was a member of the board of the Rathenau Institute, formerly NOTA, the Netherlands Organisation for Technology Assessment.

Wim A. Smit is Associate Professor of Science, Technology and Society (STS), at the University of Twente. After receiving his Ph.D. in Physics (1973), he went into STS studies. He has published on a range of issues: risk assessment, assessment of nuclear technology, nuclear proliferation, and assessment and dynamics of military technological developments. He has been a member of three advisory dommittees of the Netherlands Health Council: on External Safety (1980–1985); on Reassessing Nuclear Energy (1988–1989) and on Risk Evaluation (1991–1996). His current research focuses on technological regimes, scripts and socio-technical networks in the field of Information and Communication Technologies. A second interest is in the increasing integration of civil and military technology. He is also pursuing the development of methodologies for designing socio-technical scenarios.

Dirk Stemerding has a degree in biology (M.Sc.) and a Ph.D. in the field of history and sociology of biology. His current research interests are the social dynamics and implications of the introduction of new forms of genetic (DNA-)diagnosis in clinical genetics and other clinical practices. This is related to his wider interest in the entrenchment of new technologies in society. In the context of an EU program on Genetic Services in Europe, he investigated the introduction of DNA diagnosis in cancer genetics in Denmark, France, and The Netherlands. He is presently involved in a study of the introduction of so-called triple-screening in Dutch prenatal care, focussing on the (potential) role of technology assessment. He has recently started a project focusing on the various norms which are incorporated in different clinical practices and the ways in which these norms interfere in the emergent field of cancer genetics.

Siebe Talma is a graduate of the program on environmental studies at the Agricultural University of Wageningen (B.Sc.) and of sociology and science dynamics at the University of Amsterdam (B.Sc., M.Sc.). He has pursued research on the position of sociologists in environmental research in the Netherlands, on multidisciplinary research and on the development of networks in biotechnology in the Netherlands.

Index

A
abstraction 325–326
actant 38n
actor 75
actor network 28, 29
actor network theory 26–28, 36–37, 66, 343
actor, critical 77
administrative automation 179, 185–186
agency 204–205, 212–215, 219–224, 323–333
agency–structure 336–337, 267
agenda building 220–222
agenda of requirements 213
aggregation 31, 35, 326, 331
Akrich, M. 26, 27
alignment 17, 252, 256–258
animal economy 51–54
antagonistic coordination 207, 232, 259, 302, 319, 327–329, 340–341, 347
antagonistic interaction 302
antagonistic pattern 300–301, 307, 313
Atomic Energy Agency, AEC 231–256
automation 179–182, 185, 188–190, 197–200, 329–330
autonomy 169
Avion de Combat Experimentale 82

B
benchtop science 139–140
big science 139–140
biotechnology 313–315, 317
boundary maintenance 246, 252, 257–258
boundary object 62, 65, 139, 151, 312, 290–296
bricolage 114, 137–140
Brisson, Mathurin- Jacques 42, 43, 44, 46, 47, 48, 50, 52
Brookhaven National Laboratories 108, 119–123, 127–129, 132–135
bureaucratic trajectory 251
bureaucratization 231, 232, 256, 258

C
cabinet 43–44
CAD 25
Callon, M. 331
categorization 235
center of calculation 44, 65–67
CERN 107, 118–120
chips 206–207
classification 231, 235–236, 250, 326
classification, system of 40, 42, 50–52,
co–evolution 218, 222–223
cognitive constructs 342
cognitive coordination 167, 174n, 296
collaboration 107–111, 121–124, 137–138, 291–292
collaboration in high energy physics 111–112, 116
collaboration, international 87, 100
collaborative R&D 266–269, 294–296
collecting 42–43, 50, 55
collection 40–42, 44–45, 54–56
computer 181–185, 187, 191–192, 199–200
computer job 179, 184, 194–196
computer programmer 186, 197–199

computer, gender images of 200
construct, global 329–330
construction, summatological 112–116, 136
constructivism, social 3–5
constructivism, societal 6–10, 323
constructs 110–112
controversy 155, 300, 302
coordination 6–9, 40, 50–51, 60, 64–68, 77–79, 83, 143, 204, 208, 218, 221, 272, 284, 323–325, 338–342, 348
coordination, antagonistic 207, 232, 259, 302, 319, 327–329, 340–341, 347
coordination, by artifacts 344, 346–347
coordination, cognitive 167, 174n, 296
coordination, diffuse 212
coordination, dynamics of 33, 67–68
coordination, in high energy physics 136–139
coordination, in science 40–41, 68
coordination, inter-organizational 79
coordination, market 165–166
coordination, means of 40, 55, 60, 66–68, 343
coordination, mechanism of 165, 205
coordination, modes of 166–167, 204
coordination, of design 24, 34
coordination, of experiments 111, 136
coordination, political 158, 166, 170
coordination, politics of 62, 66
coordination, problems of 57–59,
coordination, rhetorical 151
coordination, socio-cognitive 78
coordination, socio-technical 314, 324
correlation 51–54
coupling 268–270, 273, 293, 295
critical actor 77, 99–103
cultural argument 151, 163–166
cultural pattern 300
cultural transformation 306–307
Cuvier, G. 51–53

D

danger culture 306, 316–317
data typist 190
Daubentor, L. 52–53
Davies, B. 221
dedicated network builders 77, 87–88, 100–103
defense industry 72, 74
design 1, 22, 148, 152, 167
design envelope 124
design hierarchy 124–125
design process 15–17, 22–25, 28–30
design regime 143, 167–172, 174n, 328–329
design traditions 17
detector 114, 123, 136–137
dichotomy, proponent-opponent 307, 319
dictionary 273, 279–284, 293–294, 330
dictionary, active 280
dictionary, passive 275
diffusion 326
discourse 335–336
division of labor 24, 329
DNA samples 56
dynamics of expectation 203, 217

E

economics 150, 161–162
efficiency 143, 148–152, 182, 184–185, 328
Eisenhower, president 72
Eisenstein 46–47
engineers 137
enrollment, coordination as 65
ethnomethod 111–112, 115
ethology 155
European Figher Aircraft, EFA 73, 81–87, 331–332
exclusion 232, 252, 257
exclusion legacy 250
expectation 203–205, 216–218, 220–225, 333, 339
experiment 111, 116–118, 139, 335
experiment, high energy physics 113–114, 136
Experimental Aircraft Program, EAP 82, 85

F

Farber 42–43
feminine virtues 184
femininity 180, 182
frame 109–110, 113–115, 123–124, 137–140, 182, 199–200

Index 373

frame, global 330
frame, summatological 113-114, 122
Fujimura, J. 65-67
function of network 75
functionalism 204

G
Galison, P. 125-126
game 340
game theory 339
gender 179-181
gender relations 190
gender structure 181, 187-189, 197
gender symbolism 181, 187
gender typing 199
gendered organization 180
generalization 34-35
generation 214
generic repertoire 266, 268-270, 293-294
genetic engineering 310, 312-314
global agency 329, 331
global construct 329-330
global frame 330
global level 7-8, 323-324
global order 324-326, 349
government 170
Griesemer, J. 65-67, 290-291
group, superconducting coil 128-134
guiding principle 78-79, 84, 102-103, 104n, 151-152, 163, 165-166, 169, 236

H
Harré, R. 221
Hess committee 239-243
heterogeneous actors 28
heuristics 148, 156-157, 171, 173n
High Definition Television, HDTV 212-218, 347
high energy physics 107-108, 118-123, 139, 274-275, 284-288
high energy physics experiment 113-114, 136
high energy physics, collaboration in 111-112, 116
human genome project 39-40, 54-59
hybrid repertoire 267-269, 284-288, 292-295, 330

I
IBM 191-195
ideology 151-152
immutable mobile 290-295
inclusion 77, 88, 99, 103, 232, 252
industrial society 306
innovation, weapon 72
inscription 46-47, 56-57, 61
inside maintenance 247-249, 252
inside world 251
insiders 249, 256, 301
institutionalization 154, 169
interdependency 74
intermediaries 28, 74-76
international collaboration 87, 100
inter-network dynamics 77
inter-network interactions 87
inter-organizational coordination 79
introduction of technology 300, 304-305, 319

K
KEK 131, 134-135
key punch operator 193-194
Knorr-Cetina, K. 108-109, 111-113, 138, 140n

L
labeling 300-301, 312
labels 300, 312-313, 315
large technological systems 344
Latour, B. 26, 65-66, 290-291, 344
learning 302, 320
legislation 156
Levi-Strauss, C. 137
lingua franca 268, 270, 273, 293
Linnaeus 42, 48, 49, 50
local 7-8, 291
local design 24, 33
local practices 323-324
local standards 235
local variety 236
local-global 9, 19, 336-337
localization 34-35

M
MacIntyre, A. 221
macro-actor 331-332
male gendering 196

Manhattan project 303-304
mapping 40-42, 55-58, 61-62, 66
market 150, 153, 324, 345-346
market coordination 165-166
market, institutionalization of 154, 169
masculine values 184
masculine virtuosity 182-183, 187, 200-201
masculinity 180, 187, 200-201
McNamara 79
mechanical data processing 191
membrane technology 208-212, 218-220
metaphor 300, 302
micro optics 269
micro-macro 336-337
military doctrine 79
military laboratories 72, 74
military needs 78-80, 88, 90, 102-103, 104n
military R&D 71, 102, 244-245
military requirements 78-80, 82-85, 88-90, 95, 98, 101-103, 104n
military security 80
military strategy 79
military technology 72-73
military-industrial complex 72-73, 303
molecular biology 57
molecular genetics 55, 59
Moore's Law 2, 206-207, 333
multidisciplinary R&D 265, 269, 290, 292
muon g-2 collaboration 116-121, 269, 274
mutual accommodation 221
mutual appropriation 283
mutual articulation 308
mutual delegation 331
mutual orientation 76
mutual positioning 218-220
mutual reinforcement 166

N
naming 300, 306
natural history 40-43
Neher Laboratory 188, 191-195
network 58
network, environment of 99
network, socio-technical 27, 37, 73-77, 82-83, 87, 90-95, 100-102

network-in-the-making 101-102
non-human actors 18
normalization 266
nuclear technology 308-309

O
operator 196-197
opponents 301, 304-305
organizational hierarchy 189
outside maintenance 250-253
outside world 251
outsiders 249, 301

P
paradigms 166-167
parameters 127, 135
policing 256
political coordination 158, 166, 170
positioning 221, 315
Postcheque and Clearing Service 181, 187-190
pre-existing networks 87-102, 332
printing 46-47
prisoner's dilemma 207
problem definition 232, 246
programmer 186, 197-199
promise 209-210, 213-216, 223, 299, 303, 307-310, 314, 317, 339
promise, technological 214-216, 233, 246, 249-250
promise-requirement cycle 217, 223, 308, 311
promising technology 206, 209-210, 217, 299, 308
proponent-opponent dichotomy 307, 319
proponents 301, 304-305
proposal 113-114, 122-124, 129
prospective structure 203, 206, 225, 333
protected space 222-223
punch cards 190-193

R
R&D trajectories 125-126
radioactive waste management 232-256
radioactive waste world 232-233, 243
Rafale 82, 85, 87
reality test 22
recombinant DNA technology 310

Index 375

re-coordination 165
regime, design 143, 167–172, 174n
regime, technical 8, 236, 245
regime, translation 35
regulation 304–305, 308
repertoire 308–309
repertoire, generic 266, 268–270, 293–294
repertoire, hybrid 267–269, 284–288, 292–295, 330
repertoire, semi specific 266, 270–273, 284, 293–294
repertoire, symbolic 265–267, 272–275, 281, 293–294
representation 19, 25, 28, 327, 331–332, 272, 291
representation, symbolic 268, 271, 278–284, 290
representation, system of 272
representing 42, 46, 50, 55–57
requirements 155–157, 215, 222
requirements, agenda of 213
requirements, military 78–80, 82–85, 88–90, 95, 98, 101–103, 104n
re-representation 291, 293
resilience 76–77, 83, 88, 94, 97–100
resilient mobiles 326
reversal 34–35, 137, 209, 327–331
rhetorical coordination 151
Ricoeur, P. 221
risk 307–309, 312–314, 317
risk repertoire 308, 310–312

S
scale model 22
script 27, 203, 218, 222, 224–225, 344–345
self-fulfilling prophecy 206–207
self-justifying technology 214–215
sequence-tagged sites 60–64, 342–343, 346
sequencing 55–58
shipbuilding 17
social constructivism 3–5
social studies of technology 4, 17, 26, 36–37, 137–138, 205
social world 66–67, 290–291, 304
societal constructivism 6–10, 323
socio-cognitive coordination 78

socio-cognitive space 169–171, 320
socio-cultural space 320
socio-technical change 179–181
socio-technical coordination 324, 341
socio-technical network 27, 37, 73–77, 82–83, 87, 90–95, 100–102
socio-technical order 2, 5, 299, 302, 318, 323, 337
space, protected 222–223
space, socio- cultural 320
space, socio-cognitive 169–171, 320
standardization 48, 291–292, 294
standardized package 66
standardizing 42, 47, 50, 55–57
standards, local 235
Star, S.L. 65–67, 290–291
steering 10, 348–350
story line 205–206, 301, 320
structural pattern 300
structure 204–205, 217, 222–224
structure, gender 181, 187–189, 197
structure, prospective 203, 206, 225, 333
structure-agency 205, 215, 221
structure-agency, co-evolution of 218, 222–223
summatological construction 113–116, 136
summatological frame 113–114, 122, 136–137
summatology 113–114, 118
superconducting coil 120–121
symbolic interactionism 204
symbolic repertoire 265–267, 272–275, 281, 293–294
symbolic representation 268, 271, 278–284, 290
symmetrical analysis 26
symmetrical transformation 271

T
taxonomy 40–41, 46–50, 326
technical field 24
technical legacy 249, 257
technical model 18–24, 26, 30–35, 330, 335, 345
technical regime 8, 236, 245
technical trajectory 251
technological determinism 3–5, 216

technological innovation 303
technological modernism 305
technological promise 214–216, 233, 246, 249–250
technological traces 248–249
technologization 231–232, 256–258
technology dynamics 203–205
technology, military 72–73
technology, promising 206, 209–210, 217, 299, 308
technology, self-justifying 214–215
technology, social studies of 4, 17, 26, 36–37, 137–138, 205
time 334–335
Tornado 84–85
transformation 270, 273–274, 291–296
transformation mechanism 268–269
transformation, active 268, 273
transformation, cultural 306–307
transformation, passive 268, 273, 284
transformation, rules of 330
transformation, symmetrical 271
translation 28–29

translation regime 35
Turkle, S. 137–138

U
uncertainty 303, 306
unintended effects 204

V
Van Langenhove, L. 221
Vicq-d'Azyr, F. 52–53
Vincenti, W. 17–18
virtuosity, masculine 182–183, 187, 200–201

W
weapon innovation 72
Wolman, A. 238
women 198
women managers 189–190

Y
York, H. 72

Z
zoology 51–53